SOLUTIONS MANUAL
FOR DEVORE'S

PROBABILITY AND STATISTICS

FOR ENGINEERING AND THE SCIENCES

FIFTH EDITION

JULIE ANN SEELY

DUXBURY
THOMSON LEARNING™

Australia • Canada • Mexico • Singapore • Spain • United Kingdom • United States

DUXBURY

THOMSON LEARNING™

For more information about this or any other Duxbury products, contact:
DUXBURY
511 Forest Lodge Road
Pacific Grove, CA 93950 USA
www.duxbury.com
1-800-423-0563 (Thomson Learning Academic Resource Center)

Printed in the United States of America.

5 4 3 2 1

ISBN 0-534-37284-8

CONTENTS

Chapter 1

Section 1.1

1. **a** Houston Chronicle, Des Moines Register, Chicago Tribune, Washington Post
b Capital One, Campbell Soup, Merrill Lynch, Pulitzer
c Bill Jasper, Kay Reinke, Helen Ford, David Menendez
d 1.78, 2.44, 3.5, 3.04

2. **a** 29.1 yd., 28.3 yd., 24.7 yd., 31.0 yd.
b 432, 196, 184, 321
c 2.1, 4.0, 3.2, 6.3
d .07 g, 1.58 g, 7.1 g, 27.2 g

3. **a** In a sample of 100 VCRs, what are the chances that more than 20 need service while under warrantee? What are the chances than none need service while still under warrantee?
b What proportion of all VCRs of this brand and model will need service within the warrantee period?

4. **a** Concrete: All living U.S. Citizens, all mutual funds marketed in the U.S., all books published in 1980.

Hypothetical: All grade point averages for University of California undergraduates during the next academic year. Page lengths for all books published during the next calendar year. Batting averages for all major league players during the next baseball season.

b Concrete: Probability: In a sample of 5 mutual funds, what is the chance that all 5 have rates of return which exceeded 10% last year?

Statistics: If previous year rates-of-return for 5 mutual funds were 9.6, 14.5, 8.3, 9.9 and 10.2, can we conclude that the average rate for all funds was below 10%?

Conceptual: Probability: In a sample of 10 books to be published next year, how likely is it that the average number of pages for the 10 is between 200 and 250?

Statistics: If the sample average number of pages for 10 books is 227, can we be highly confident that the average for all books is between 200 and 245?

5. **a** No, the relevant conceptual population is all scores of all students who participate in the SI in conjunction with this particular statistics course.
b The advantage to randomly choosing students to participate in the two groups is that we are more likely to get a sample representative of the population at large. If it were left to students to choose, there may be a division of abilities in the two groups which could unnecessarily affect the outcome of the experiment.
c If all students were put in the treatment group there would be no results with which to compare the treatments.

6 One could take a simple random sample of students from all students in the California State University system and ask each student in the sample to report the distance form their hometown to campus. Alternatively, the sample could be generated by taking a stratified random sample by taking a simple random sample from each of the 23 campuses and again asking each student in the sample to report the distance from their hometown to campus. Certain problems might arise with self reporting of distances, such as recording error or poor recall. This study is enumerative because there exists a finite, identifiable population of objects from which to sample.

7 One could generate a simple random sample of all single family homes in the city or a stratified random sample by taking a simple random sample from each of the 10 district neighborhoods. From each of the homes in the sample the necessary variables would be collected. This would be an enumerative study because there exists a finite, identifiable population of objects from which to sample.

8 **a** Number observations equal 2 x 2 x 2 = 8

b This could be called an analytic study because the data would be collected on an existing process. There is no sampling frame.

9 **a** There could be several explanations for the variability of the measurements. Among them could be measuring error, (due to mechanical or technical changes across measurements), recording error, differences in weather conditions at time of measurements, etc.

b This could be called an analytic study because there is no sampling frame.

Section 1.2

10. **a** Minitab generates the following stem-and-leaf display of this data:

```
 5|9
 6|33588
 7|00234677889
 8|127
 9|077           stem: ones
10|7             leaf: tenths
11|368
```

What constitutes large or small variation usually depends on the application at hand, but an often-used rule of thumb is: the variation tends to be large whenever the spread of the data (the difference between the largest and smallest observations) is large compared to a representative value. Here, 'large' means that the percentage is closer to 100% than it is to 0%. For this data, the spread is 11 - 5 = 6, which constitutes 6/8 = .75, or, 75%, of the typical data value of 8. Most researchers would call this a large amount of variation.

b The data display is not perfectly symmetric around some middle/representative value. There tends to be some positive skewness in this data.

c In Chapter 1, outliers are data points that appear to be *very* different from the pack. Looking at the stem-and-leaf display in part (a), there appear to be no outliers in this data. (Chapter 2 gives a more precise definition of what constitutes an outlier).

d From the stem-and-leaf display in part (a), there are 4 values greater than 10. Therefore, the proportion of data values that exceed 10 is 4/27 = .148, or, about 15%.

11.

A stem-and-leaf display of this data is:

```
 83|4
 84|3            stem: ones
 85|3            leaf: tenths
 86|77
 87|456789
 88|2333556679
 89|0233678899
 90|0113344456789
 91|0001112256688
 92|22236777
 93|023347
 94|2247
 95|6
 96|1
 97|
 98|8
 99|
100|3
```

Because the stem-and-leaf display is nearly symmetric around 90, a representative value of about 90 is easy to discern from the diagram. The most apparent features of the display are its approximate symmetry and the tendency for the data values to stack up around the representative value in a bell-shaped curve. Also, the spread of the data, 100.3-83.4 = 16.9 is a relatively small percentage (16.9/90 ≈ .18, or 18%) of the typical value of 90.

12 a

Stem	Leaf
5	9
6	33588
7	234677889
8	127
9	77
10	7
11	368

Stem=Ones
Leaf=Tenths

Possible representative values could be the most frequently occurring values (in the "7" row) or the median value of 7.7(the middle value after the data is ranked). The data appears to be fairly concentrated about the representative value.

b The data appears to be slightly skewed to the right, or positively skewed.

c There are no outliers.

d There are 4 out of the 27 observations, or 4/27 = 0.148 of the observations, that are above 10 Mpa.

13.

Crunchy		Creamy
	2	2
644	3	69
77220	4	145
6320	5	3666
222	6	258
55	7	
0	8	

Both sets of scores are reasonably spread out. There appear to be no outliers. The three highest scores are for the crunchy peanut butter, the three lowest for the creamy peanut butter.

14. a

beams		cylinders
9	5	8
88533	6	16
98877643200	7	012488
721	8	13359
770	9	278
7	10	
863	11	2
	12	6
	13	
	14	1

The data appears to be slightly skewed to the right, or positively skewed. The value of 14.1 appears to be an outlier. Three out of the twenty, 3/20 or .15 of the observations exceed 10 Mpa.

b The majority of observations are between 5 and 9 Mpa for both beams and cylinders, with the modal class in the 7 Mpa range. The observations for cylinders are more variable, or spread out, and the maximum value of the cylinder observations is higher.

c Dot Plot

```
 .  .   .  :..  : .: .  .  .     :         .        .        .
-+---------+---------+---------+---------+---------+-----cylinder
6.0       7.5       9.0       10.5      12.0      13.5
```

15

```
6l |034
6h |667899
7l |00122244
7h |                 Stem=Tens
8l |001111122344     Leaf=Ones
8h |5557899
9l |03
9h |58
```

This display brings out the gap in the data: There are no scores in the high 70's.

16 One method of denoting the pairs of stems having equal values is to denote the first stem by L, for 'low', and the second stem by H, for 'high'. Using this notation, the stem-and-leaf display would appear as follows:

```
3L|1
3H|56678
4L|000112222234
4H|5667888
5L|144
5H|58           stem: tenths
6L|2            leaf: hundredths
6H|6678
7L|
7H|5
```

The stem-and-leaf display on the previous page shows that .45 is a good representative value for the data. In addition, the display is not symmetric and appears to be positively skewed. The spread of the data is .75 - .31 = .44, which is.44/.45 = .978, or about 98% of the typical value of .45. This constitutes a reasonably large amount of variation in the data. The data value .75 is a possible outlier

16. **a**

Number Nonconforming	Frequency	RelativeFrequency(Freq/60)
0	7	0.117
1	12	0.200
2	13	0.217
3	14	0.233
4	6	0.100
5	3	0.050
6	3	0.050
7	1	0.017
8	1	0.017
doesn't add exactly to 1 because relative frequencies have been rounded		1.001

b The number of batches with at most 5 nonconforming items is 7+12+13+14+6+3 = 55, which is a proportion of 55/60 = .917. The proportion of batches with (strictly) fewer than 5 nonconforming items is 52/60 = .867. Notice that these proportions could also have been computed by using the relative frequencies: e.g., proportion of batches with 5 or fewer nonconforming items = 1- (.05+.017+.017) = .916; proportion of batches with fewer than 5 nonconforming items = 1 - (.05+.05+.017+.017) = .866.

c The following is a Minitab histogram of this data. The center of the histogram is somewhere around 2 or 3 and it shows that there is some positive skewness in the data. Using the rule of thumb in Exercise 1, the histogram also shows that there is a lot of spread/variation in this data.

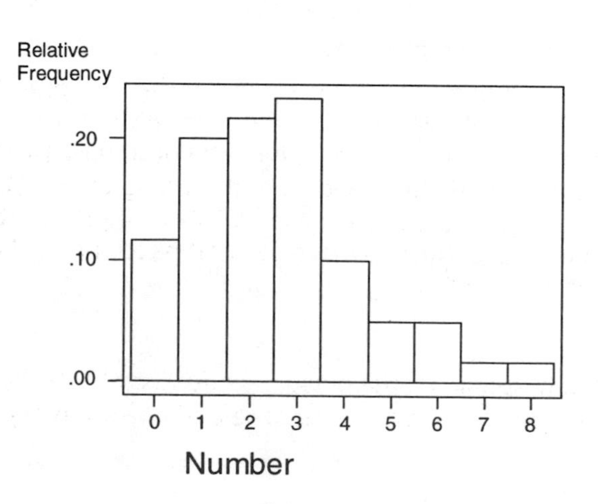

18. **a** The following histogram was constructed using Minitab:

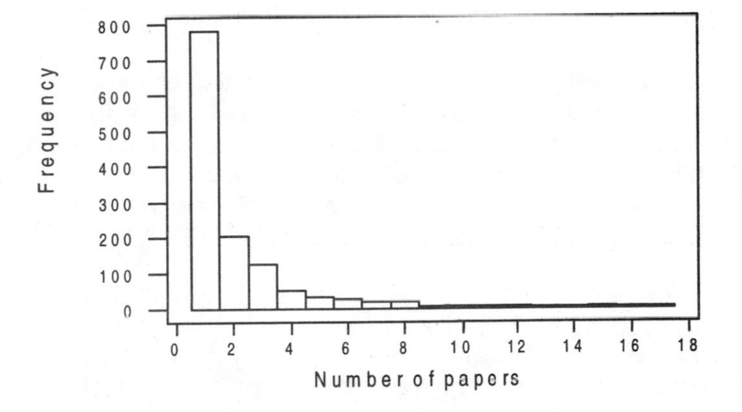

The most interesting feature of the histogram is the heavy positive skewness of the data.

Note: One way to have Minitab automatically construct a histogram from grouped data such as this is to use Minitab's ability to enter multiple copies of the same number by typing, for example, 784(1) to enter 784 copies of the number 1. The frequency data in this exercise was entered using the following Minitab commands:

```
MTB > set c1
DATA> 784(1) 204(2) 127(3) 50(4) 33(5) 28(6) 19(7) 19(8)
DATA> 6(9) 7(10) 6(11) 7(12) 4(13) 4(14) 5(15) 3(16) 3(17)
DATA> end
```

b From the frequency distribution (or from the histogram), the number of authors who published at least 5 papers is 33+28+19+…+5+3+3 = 144, so the proportion who published 5 or more papers is 144/1309 = .11, or 11%. Similarly, by adding frequencies and dividing by n = 1309, the proportion who published 10 or more papers is 39/1309 = .0298, or about 3%. The proportion who published more than 10 papers (i.e., 11 or more) is 32/1309 = .0245, or about 2.5%.

c No. Strictly speaking, the class described by ' ≥15 ' has no upper boundary, so it is impossible to draw a rectangle above it having finite area (i.e., frequency).

d The category 15-17 does have a finite width of 2, so the cumulated frequency of 11 can be plotted as a rectangle of height 6.5 over this interval. The basic rule is to make the area of the bar equal to the class frequency, so area = 11 = (width)(height) = 2(height) yields a height of 6.5.

19. **a** From this frequency distribution, the proportion of wafers that contained at least one particle is (100-1)/100 = .99, or 99%. Note that it is much easier to subtract 1 (which is the number of wafers that contain 0 particles) from 100 than it would be to add all the frequencies for 1, 2, 3,... particles. In a similar fashion, the proportion containing at least 5 particles is (100 - 1-2-3-12-11)/100 = 71/100 = .71, or, 71%.

b The proportion containing between 5 and 10 particles is (15+18+10+12+4+5)/100 = 64/100 = .64, or 64%. The proportion that contain strictly between 5 and 10 (meaning strictly *more* than 5 and strictly *less* than 10) is (18+10+12+4)/100 = 44/100 = .44, or 44%.

c The following histogram was constructed using Minitab. The data was entered using the same technique mentioned in the answer to exercise 8(a). The histogram is *almost* symmetric and unimodal; however, it has a few relative maxima (i.e., modes) and has a very slight positive skew.

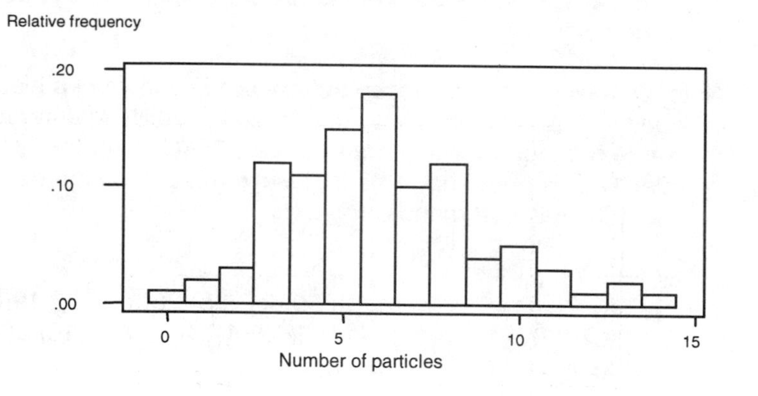

20. **a** The following stem-and-leaf display was constructed:

```
0|123334555599
1|00122234688    stem: thousands
2|1112344477     leaf: hundreds
3|0113338
4|37
5|23778
```

A typical data value is somewhere in the low 2000's. The display is almost unimodal (the stem at 5 would be considered a mode, the stem at 0 another) and has a positive skew.

b A histogram of this data, using classes of width 1000 centered at 0, 1000, 2000, 6000 is shown below. The proportion of subdivisions with total length less than 2000 is (12+11)/47 = .489, or 48.9%. Between 200 and 4000, the proportion is (7 + 2)/47 = .191, or 19.1%. The histogram shows the same general shape as depicted by the stem-and-leaf in part (a).

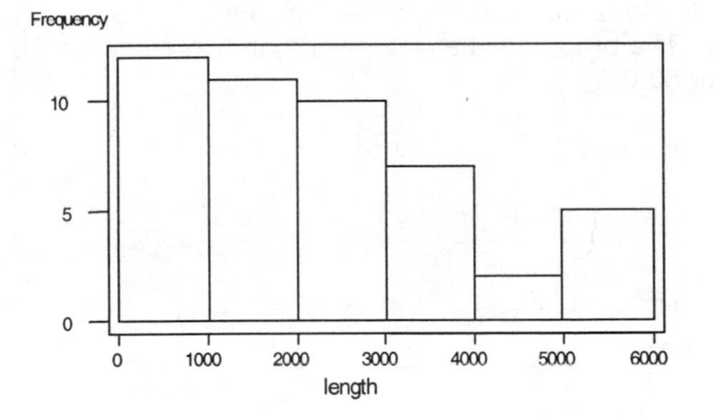

21. **a** A histogram of the y data appears below. From this histogram, the number of subdivisions having no cul-de-sacs (i.e., $y = 0$) is $17/47 = .362$, or 36.2%. The proportion having at least one cul-de-sac ($y \geq 1$) is $(47-17)/47 = 30/47 = .638$, or 63.8%. Note that subtracting the number of cul-de-sacs with $y = 0$ from the total, 47, is an easy way to find the number of subdivisions with $y \geq 1$.

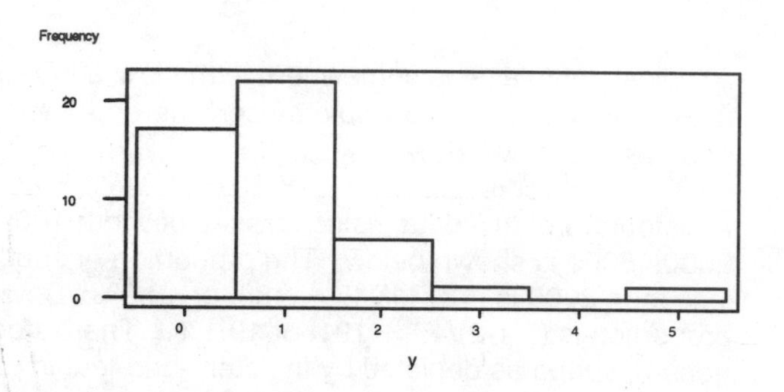

b A histogram of the z data appears below. From this histogram, the number of subdivisions with at most 5 intersections (i.e., $z \leq 5$) is $42/47 = .894$, or 89.4%. The proportion having fewer than 5 intersections ($z < 5$) is $39/47 = .830$, or 83.0%.

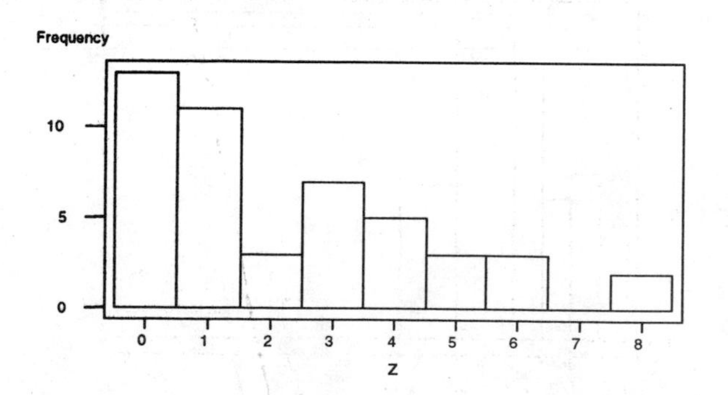

22. A very large percentage of the data values are greater than 0, which indicates that most, but not all, runners do slow down at the end of the race. The histogram is also positively skewed, which means that some runners slow down a *lot* compared to the others. A typical value for this data would be in the neighborhood of 200 seconds. The proportion of the runners who ran the last 5 km faster than they did the first 5 km is very small, about 1% or so.

23. **a**

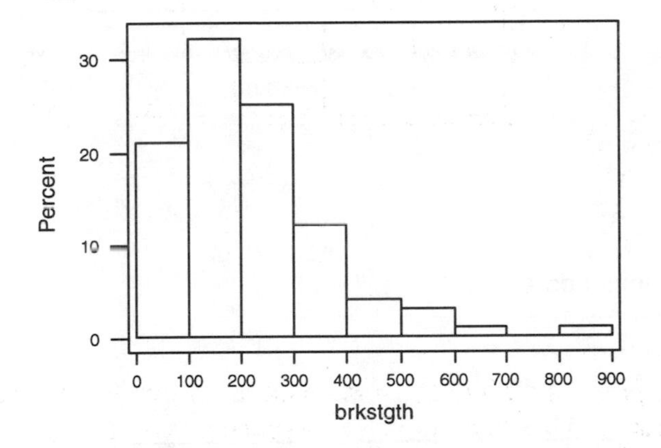

The histogram is skewed right, with a majority of observations between 0 and 300 cycles. The class holding the most observations is between 100 and 200 cycles.

b

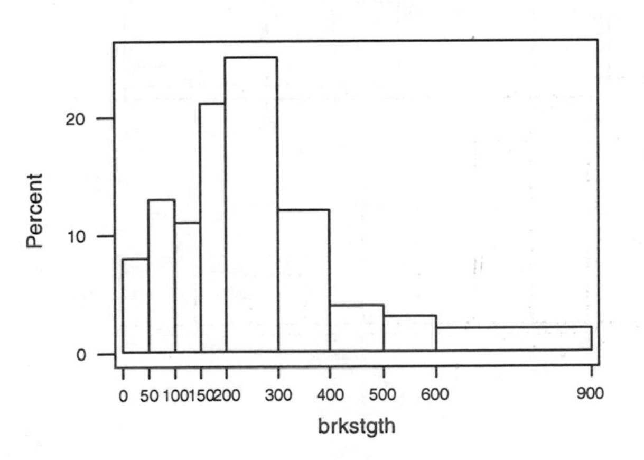

c [proportion $\geq$ 100] = 1 – [proportion < 100] = 1 - .21 = .79

24.

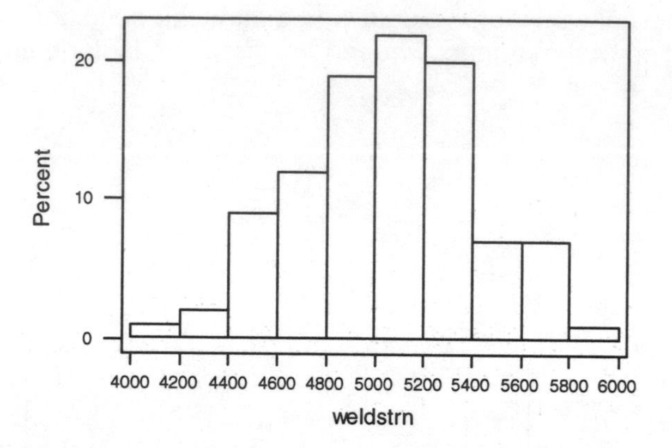

25. Histogram of original data:

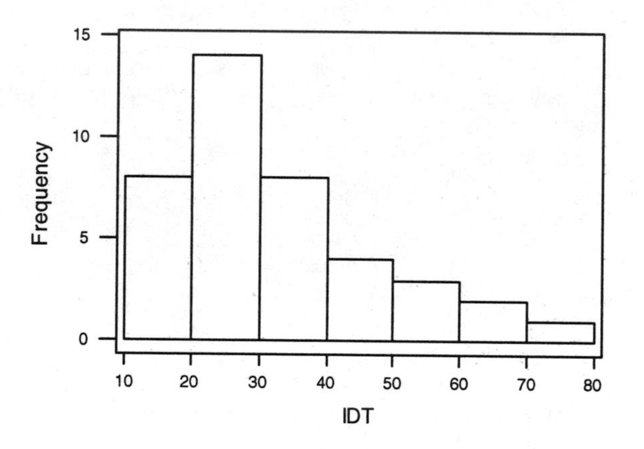

Histogram of transformed data:

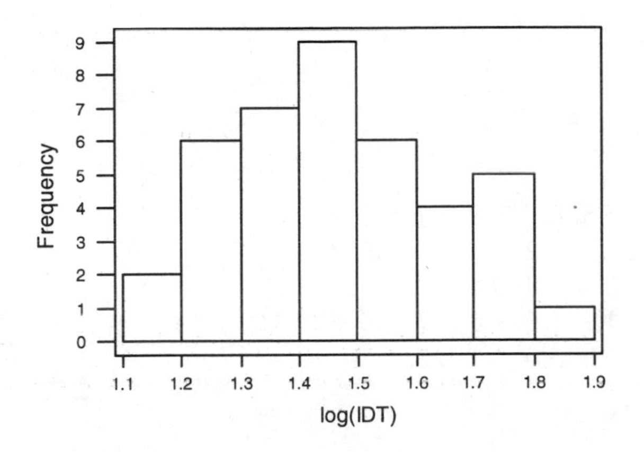

The transformation creates a much more symmetric, mound-shaped histogram.

26. **a**

Class Intervals	Frequency	Rel. Freq.
.15 -< .25	8	0.02192
.25 -< .35	14	0.03836
.35 -< .45	28	0.07671
.45 -< .50	24	0.06575
.50 -< .55	39	0.10685
.55 -< .60	51	0.13973
.60 -< .65	106	0.29041
.65 -< .70	84	0.23014
.70 -< .75	11	0.03014
	n=365	1.00001

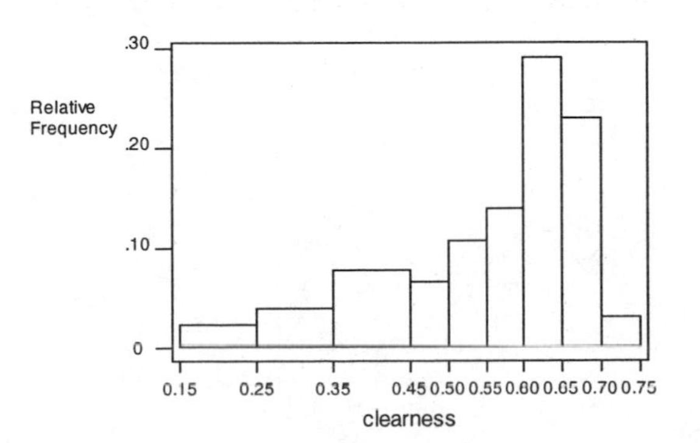

b The proportion of days with a clearness index smaller than .35 is $\frac{(8+4)}{365} = .06$, or 6%.

c The proportion of days with a clearness index of at least .65 is $\frac{(84+11)}{365} = .26$, or 26%.

27. **a**

```
2|7.1
3|0.0, 1.7, 3.8, 5.5, 6.7, 7.0, 9.1, 9.8                  stems = tens
4|0.0, 2.3, 4.6, 5.9, 7.2, 7.3, 8.0, 9.5
5|2.6, 5.8, 6.0, 6.3, 8.2, 9.1
6|0.6, 0.7, 0.9, 1.2, 1.5, 1.8, 2.3, 4.9, 5.0, 5.2, 5.8, 6.3, 8.2, 9.0, 9.3, 9.8
7|1.4, 1.7, 4.5, 5.3, 6.0, 7.1, 8.8
8|3.2, 7.1
9|1.3, 4.6
```

b There are observations that fall on a class boundary (e.g., 30.0) and each such observation should belong in exactly one interval.

c

class	frequency	relative frequency
20 -< 30	1	0.02
30 -< 40	8	0.16
40 -< 50	8	0.16
50 -< 60	6	0.12
60 -< 70	16	0.32
70 -< 80	7	0.14
80 -< 90	2	0.04
90 -< 100	2	0.04
	50	1.00

d .02+.16+.16 = .34; .32+.14+.04+.04 = .54

28.

There are seasonal trends with lows and highs 12 months apart.

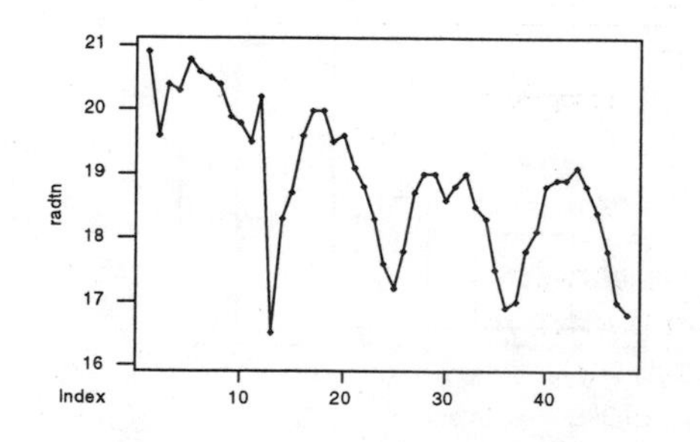

29.

Complaint	Frequency	Relative Frequency
B	7	0.1167
C	3	0.0500
F	9	0.1500
J	10	0.1667
M	4	0.0667
N	6	0.1000
O	21	0.3500
	60	1.0000

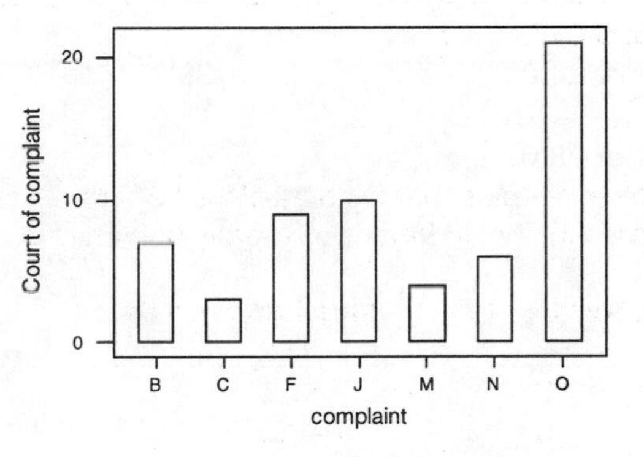

30.

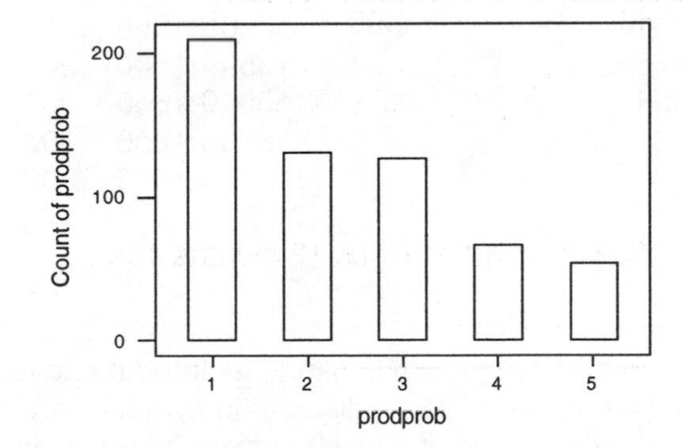

1. incorrect component
2. missing component
3. failed component
4. insufficient solder
5. excess solder

31.

Class	Frequency	Relative Frequency	Cumulative Relative Frequency
0.0 - under 4.0	2	2	0.050
4.0 - under 8.0	14	16	0.400
8.0 - under 12.0	11	27	0.675
12.0 - under 16.0	8	35	0.875
16.0 - under 20.0	4	39	0.975
20.0 - under 24.0	0	39	0.975
24.0 - under 28.0	1	40	1.000

32. **a** The frequency distribution is:

Class	Relative Frequency
0-< 150	.193
150-< 300	.183
300-< 450	.251
450-< 600	.148
600-< 750	.097
750-< 900	.066

Class	Relative Frequency
900-<1050	.019
1050-<1200	.029
1200-<1350	.005
1350-<1500	.004
1500-<1650	.001
1650-<1800	.002
1800-<1950	.002

The relative frequency distribution is almost unimodal and exhibits a large positive skew. The typical middle value is somewhere between 400 and 450, although the skewness makes it difficult to pinpoint more exactly than this.

b The proportion of the fire loads less than 600 is .193+.183+.251+.148 = .775. The proportion of loads that are at least 1200 is .005+.004+.001+.002+.002 = .014.

c The proportion of loads between 600 and 1200 is 1 - .775 - .014 = .211.
(continued)

d Since 500 and 1000 are not class boundaries, we have to approximate the proportion of loads between these two numbers. For the classes 600-<750 and 750-<900, no approximation is necessary; their accumulated proportions are simply .097 + .066. For the interval 500-<600, which is contained in the class 450-<600, we simply calculate what fraction the distance 600-500= 100 is compared to the entire class width of 600-450 = 150 and take that fraction of the corresponding class proportion.148. In this case, 100/150 = 2/3, so we estimate that about 2/3 of .148, or .0987, is the relative frequency to attribute the interval 500-<600. Similarly, the interval 900-<1000 should account for about (1000-900)/(1050-900) = 100/150 = 2/3 of the class frequency .019. Therefore, we estimate that the proportion of loads between 500 and 1000 is approximately .0987 + .097 + .066 + .0127 = .274.

33. **a** $\bar{x} = 192.57$, $\tilde{x} = 189$. The mean is larger than the median, but they are still fairly close together.

b Changing the one value, $\bar{x} = 189.71$, $\tilde{x} = 189$. The mean is lowered, the median stays the same.

c $\bar{x}_{tr} = 191.0$. $\frac{1}{14} = .07$ or 7% trimmed from each tail.

d For n = 13, Σx = (119.7692) x 13 = 1,557
For n = 14, Σx = 1,557 + 159 = 1,716

$$\bar{x} = \frac{1716}{14} = 122.5714 \text{ or } 122.6$$

34. **a** The sum of the n = 11 data points is 514.90, so $\bar{x}$ = 514.90/11 = 46.81.

b The sample size (n = 11) is odd, so there will be a middle value. Sorting from smallest to largest: 4.4 16.4 22.2 30.0 33.1 36.6 40.4 66.7 73.7 81.5 109.9. The sixth value, 36.6 is the middle, or median, value. The mean differs from the median because the largest sample observations are much further from the median than are the smallest values.

c Deleting the smallest (x = 4.4) and largest (x = 109.9) values, the sum of the remaining 9 observations is 400.6. The trimmed mean $\bar{x}_{tr}$ is 400.6/9 = 44.51. The trimming percentage is 100(1/11) ≈ 9.1%. $\bar{x}_{tr}$ lies between the mean and median.

35. **a** The sample mean is $\bar{x}$ = (100.4/8) = 12.55.

The sample size (n = 8) is even. Therefore, the sample median is the average of the (n/2) and (n/2) + 1 values. By sorting the 8 values in order, from smallest to largest: 8.0 8.9 11.0 12.0 13.0 14.5 15.0 18.0, the forth and fifth values are 12 and 13. The sample median is (12.0 + 13.0)/2 = 12.5.

The 12.5% trimmed mean requires that we first trim (.125)(n) or 1 value from the ends of the ordered data set. Then we average the remaining 6 values. The 12.5% trimmed mean $\bar{x}_{tr(12.5)}$ is 74.4/6 = 12.4.

All three measures of center are similar, indicating little skewness to the data set.

b The smallest value (8.0) could be increased to any number below 12.0 (a change of less than 4.0) without affecting the value of the sample median.

c The values obtained in part (a) can be used directly. For example, the sample mean of 12.55 psi could be re-expressed as

(12.55 psi) x $\left(\dfrac{1ksi}{2.2psi}\right) = 5.70ksi$.

36. **a** A stem-and leaf display of this data appears below:

Stem	Leaf
32	55
33	49
34	
35	6699
36	34469
37	03345
38	9
39	2347
40	23
41	
42	4

stem: ones
leaf: tenths

The display is reasonably symmetric, so the mean and median will be close.

b The sample mean is $\bar{x}$ = 9638/26 = 370.7. The sample median is $\tilde{x}$ = (369+370)/2 = 369.50.

(continued)

c The largest value (currently 424) could be increased by any amount. Doing so will not change the fact that the middle two observations are 369 and 170, and hence, the median will not change. However, the value x = 424 can not be changed to a number less than 370 (a change of 424-370 = 54) since that *will* lower the values(s) of the two middle observations.

d Expressed in minutes, the mean is (370.7 sec)/(60 sec) = 6.18 min; the median is 6.16 min.

37. $\bar{x} = 12.01$, $\tilde{x} = 11.35$, $\bar{x}_{tr(10)} = 11.46$. The median or the trimmed mean would be good choices because of the outlier 21.9.

38. a The reported values are (in increasing order) 110, 115, 120, 120, 125, 130, 130, 135, and 140. Thus the median of the reported values is 125.

b 127.6 is reported as 130, so the median is now 130, a very substantial change. When there is rounding or grouping, the median can be highly sensitive to small change.

39. a $\Sigma x_l = 16.475$ so $\bar{x} = \dfrac{16.475}{16} = 1.0297$

$$\tilde{x} = \frac{(1.007 + 1.011)}{2} = 1.009$$

b 1.394 can be decreased until it reaches 1.011(the largest of the 2 middle values) – i.e. by 1.394 – 1.011 = .383, If it is decreased by more than .383, the median will change.

40. $\tilde{x} = 60.8$

$\bar{x}_{tr(25)} = 59.3083$

$\bar{x}_{tr(10)} = 58.3475$

$\bar{x} = 58.54$

All four measures of center have about the same value.

41. a $7/10 = .70$

b $\bar{x} = .70$ = proportion of successes

c $\dfrac{s}{25} = .80$ so s = (0.80)(25) = 20

total of 20 successes

20 – 7 = 13 of the new cars would have to be successes

42. **a** $\bar{y} = \frac{\Sigma y_i}{n} = \frac{\Sigma(x_i + c)}{n} = \frac{\Sigma x_i}{n} + \frac{nc}{n} = \bar{x} + c$

$\tilde{y}$ = the median of $(x_1 + c, x_2 + c, ..., x_n + c)$ = median of $(x_1, x_2, ..., x_n) + c = \tilde{x} + c$

b $\bar{y} = \frac{\Sigma y_i}{n} = \frac{\Sigma(x_i \cdot c)}{n} = \frac{c\Sigma x_i}{n} = c\bar{x}$

$\tilde{y} = (cx_1, cx_2, ..., cx_n) = c \cdot median(x_1, x_2, ..., x_n) = c\tilde{x}$

43. median = $\frac{(57+79)}{2} = 68.0$, 20% trimmed mean = 66.2, 30% trimmed mean = 67.5.

Section 1.4

44. **a** range = 49.3 – 23.5 = 25.8

b

x_i	$(x_i - \bar{x})$	$(x_i - \bar{x})^2$	x_i^2
29.5	-1.53	2.3409	870.25
49.3	18.27	333.7929	2430.49
30.6	-0.43	0.1849	936.36
28.2	-2.83	8.0089	795.24
28.0	-3.03	9.1809	784.00
26.3	-4.73	22.3729	691.69
33.9	2.87	8.2369	1149.21
29.4	-1.63	2.6569	864.36
23.5	-7.53	56.7009	552.25
31.6	0.57	0.3249	998.56
$\Sigma x = 310.3$	$\Sigma(x_i - \bar{x}) = 0$	$\Sigma(x_i - \bar{x})^2 = 443.801$	$\Sigma(x_i^2) = 10{,}072.41$

$\bar{x} = 31.03$

$$s^2 = \frac{\sum_{i=1}^{n}(x_i - \bar{x})^2}{n-1} = \frac{443.801}{9} = 49.3112$$

c $s = \sqrt{s^2} = 7.0222$

d $$s^2 = \frac{\Sigma x^2 - (\Sigma x)^2/n}{n-1} = \frac{10{,}072.41 - (310.3)^2/10}{9} = 49.3112$$

45 **a** $\bar{x} = \frac{1}{n}\sum_i x_i = 577.9/5 = 115.58$. Deviations from the mean:

116.4 - 115.58 = .82, 115.9 - 115.58 = .32, 114.6 -115.58 = -.98, 115.2 - 115.58 = -.38, and 115.8-115.58 = .22.

b $s^2 = [(.82)^2 + (.32)^2 + (-.98)^2 + (-.38)^2 + (.22)^2]/(5-1) = 1.928/4 = .482$, so s = .694.

c $\sum_i x_i^2 = 66{,}795.61$, so $s^2 = \frac{1}{n-1}\left[\sum_i x_i^2 - \frac{1}{n}\left(\sum_i x_i\right)^2\right] =$

$[66{,}795.61 - (577.9)^2/5]/4 = 1.928/4 = .482$.

46 **a** $\bar{x} = \frac{1}{n}\sum_i x_i = 14438/5 = 2887.6$. The sorted data is: 2781 2856 2888 2900 3013, so the sample median is $\tilde{x} = 2888$.

b Subtracting a constant from each observation shifts the data, but does not change its sample variance (Exercise 16). For example, by subtracting 2700 from each observation we get the values 81, 200, 313, 156, and 188, which are smaller (fewer digits) and easier to work with. The sum of squares of this transformed data is 204210 and its sum is 938, so the computational formula for the variance gives $s^2 = [204210-(938)^2/5]/(5-1) = 7060.3$.

47. The sample mean, $\bar{x} = \frac{1}{n}\sum x_i = \frac{1}{10}(1{,}162) = \bar{x} = 116.2$.

The sample standard deviation,

$$s = \sqrt{\frac{\sum x_i^2 - \frac{\left(\sum x_i\right)^2}{n}}{n-1}} = \sqrt{\frac{140{,}992 - \frac{(1{,}162)^2}{10}}{9}} = 25.75$$

On average, we would expect a fracture strength of 116.2. In general, the size of a typical deviation from the sample mean (116.2) is about 25.75. Some observations may deviate from 116.2 by more than this and some by less..

48. Using the computational formula, $s^2 = \frac{1}{n-1}\left[\sum_i x_i^2 - \frac{1}{n}\left(\sum_i x_i\right)^2\right] =$

$[3{,}587{,}566-(9638)^2/26]/(26-1) = 593.3415$, so s = 24.36. In general, the size of a typical deviation from the sample mean (370.7) is about 24.4. Some observations may deviate from 370.7 by a little more than this, some by less.

49. a $\Sigma x = 2.75 + ... + 3.01 = 56.80$, $\Sigma x^2 = (2.75)^2 + ... + (3.01)^2 = 197.8040$

b $s^2 = \dfrac{197.8040 - (56.80)^2/17}{16} = \dfrac{8.0252}{16} = .5016,\ s = .708$

50. a n = 7, $\Sigma x = 374.1$, so $\overline{x} = \dfrac{374.1}{7} = 53.44$, whereas $\tilde{x} = 66.3$

The fact that $\overline{x} < \tilde{x}$ suggests that observations on the lower end of the sample are more stretched out than those on the upper end (a longer lower tail than upper tail).

b $\Sigma x_i^2 = 23{,}488.03$, so $s^2 = \dfrac{[23{,}448.03 - (374.1)^2/7]}{6}$

$= \dfrac{3495.057}{6} = 582.51$, and $s = \sqrt{582.51} = 24.14$

51. a $\Sigma x = 2563$ and $\Sigma x^2 = 368{,}501$, so

$s^2 = \dfrac{[368{,}501 - (2563)^2/19]}{18} = 1264.766$ and $s = 35.564$

b If y = time in minutes, then y = cx where $c = \frac{1}{60}$, so

$s_y^2 = c^2 s_x^2 = \dfrac{1264.766}{3600} = .351$ and $s_y = cs_x = \dfrac{35.564}{60} = .593$

52.

Let d denote the fifth deviation. Then $.3 + .9 + 1.0 + 1.3 + d = 0$ or $3.5 + d = 0$, so $d = -3.5$. One sample for which these are the deviations is $x_1 = 3.8,\ x_2 = 4.4,\ x_3 = 4.5,\ x_4 = 4.8,\ x_5 = 0$. (obtained by adding 3.5 to each deviation; adding any other number will produce a different sample with the desired property)

53. **a** lower half: 2.34 2.43 2.62 2.74 2.74 2.75 2.78 3.01 3.46
upper half: 3.46 3.56 3.65 3.85 3.88 3.93 4.21 4.33 4.52
Thus the lower fourth is 2.74 and the upper fourth is 3.88.

b $f_s = 3.88 - 2.74 = 1.14$

c f_s wouldn't change, since increasing the two largest values does not affect the upper fourth.

d By at most .40 (that is, to anything not exceeding 2.74), since then it will not change the lower fourth.

e Since n is now even, the lower half consists of the smallest 9 observations and the upper half consists of the largest 9. With the lower fourth = 2.74 and the upper fourth = 3.93, $f_s = 1.19$.

54. **a** The lower half of the data set: 4.4 16.4 22.2 30.0 33.1 36.6, whose median, and therefore, the lower quartile, is $\frac{(22.2+30.0)}{2}+26.1$.
The top half of the data set: 36.6 40.4 66.7 73.7 81.5 109.9, whose median, and therefore, the upper quartile, is $\frac{(66.7+73.7)}{2}=70.2$.
So, the IQR = (70.2 – 26.1) = 44.1

b A boxplot (created in Minitab) of this data appears below:

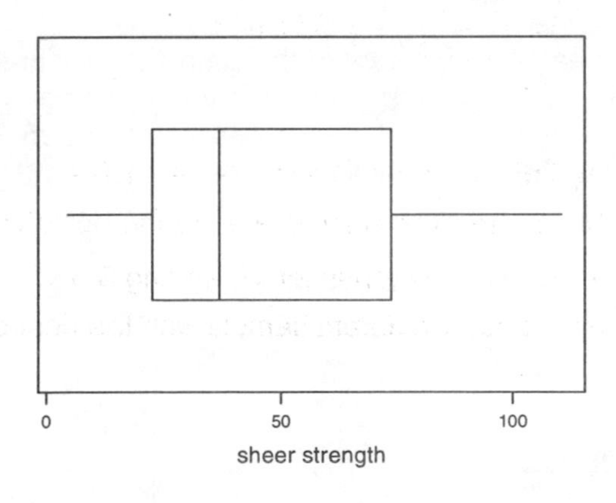

There is a slight positive skew to the data. The variation seems quite large. There are no outliers.
(continued)

c An observation would need to be further than 1.5(44.1) = 66.15 units below the lower quartile $[(26.1-66.15) = -40.05\ units]$ or above the upper quartile $[(70.2+66.15) = 136.35\ units]$ to be classified as a mild outlier. Notice that, in this case, an outlier on the lower side would not be possible since the sheer strength variable cannot have a negative value.

An extreme outlier would fall (3)44.1) = 132.3 or more units below the lower, or above the upper quartile. Since the minimum and maximum observations in the data are 4.4 and 109.9 respectively, we conclude that there are no outliers, of either type, in this data set.

d Not until the value x = 109.9 is lowered below 73.7 would there be any change in the value of the upper quartile. That is, the value x = 109.9 could not be decreased by more than (109.9 – 73.7) = 36.2 units.

55. a Lower half of the data set: 325 325 334 339 356 356 359 359 363 364 364 366 369, whose median, and therefore the lower quartile, is 359 (the 7th observation in the sorted list).
The top half of the data is 370 373 373 374 375 389 392 393 394 397 402 403 424, whose median, and therefore the upper quartile is 392.
So, the IQR = 392 - 359 = 33.

b 1.5(IQR) = 1.5(33) = 49.5 and 3(IQR) = 3(33) = 99. Observations that are further than 49.5 below the lower quartile (i.e., 359-49.5 = 309.5 or less) or more than 49.5 units above the upper quartile (greater than 392+49.5 = 441.5) are classified as 'mild' outliers. 'Extreme' outliers would fall 99 or more units below the lower, or above the upper, quartile. Since the minimum and maximum observations in the data are 325 and 424, we conclude that there are no mild outliers in this data (and therefore, no 'extreme' outliers either).

c A boxplot (created by Minitab) of this data appears below. There is a slight positive skew to the data, but it is not far from being symmetric. The variation, however, seems large (the spread 424-325 = 99 is a large percentage of the median/typical value)

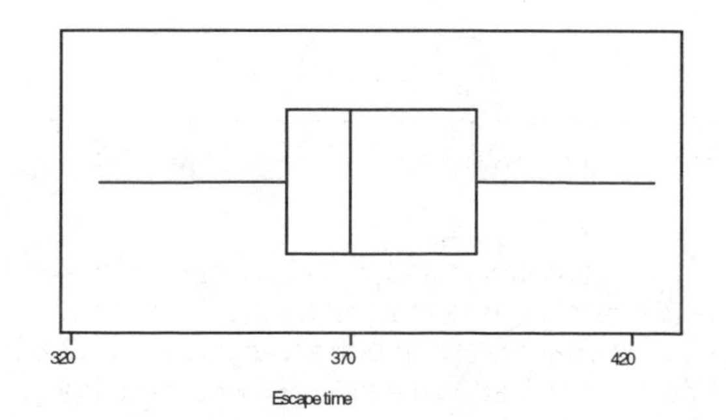

d Not until the value x = 424 is lowered below the upper quartile value of 392

e Not until the value x = 424 is lowered below the upper quartile value of 392 would there be any change in the value of the upper quartile. That is, the value x = 424 could not be decreased by more than 424-392 = 32 units.

56. A boxplot (created in Minitab) of thin data appears below.

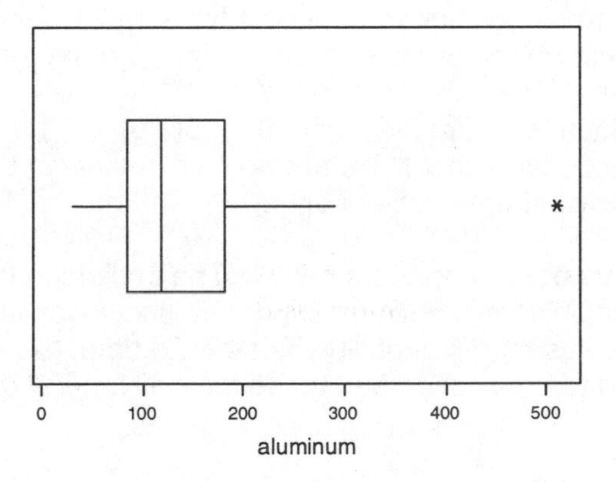

There is a slight positive skew to this data. There is one extreme outler (x=511). Even when removing the outlier, the variation is still moderately large.

57. **a** 1.5(IQR) = 1.5(216.8-196.0) = 31.2 and 3(IQR) = 3(216.8-196.0) = 62.4. Mild outliers: observations below 196-31.2 = 164.6 or above 216.8+31.2 = 248. Extreme outliers: observations below 196-62.4 = 133.6 or above 216.8+62.4 = 279.2. Of the observations given, 125.8 is an extreme outlier and 250.2 is a mild outlier.

b A boxplot of this data appears below. There is a bit of positive skew to the data but, except for the two outliers identified in part (a), the variation in the data is relatively small.

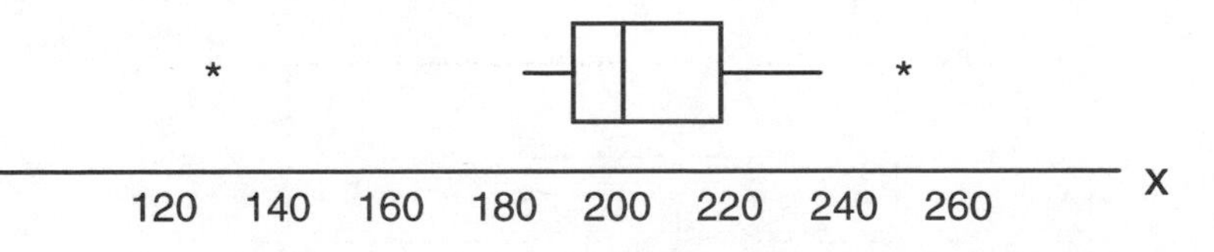

58. The most noticeable feature of the comparative boxplots is that machine 2's sample values have considerably more variation than does machine 1's sample values. However, a typical value, as measured by the median, seems to be about the same for the two machines. The only outlier that exists is from machine 1.

59. **a** ED: median = .4 (the 14th value in the *sorted* list of data). The lower quartile (median of the lower half of the data, including the median, since n is odd) is (.1+.1)/2 = .1. The upper quartile is (2.7+2.8)/2 = 2.75. Therefore, IQR = 2.75 - .1 = 2.65.

Non-ED: median = (1.5+1.7)/2 = 1.6. The lower quartile (median of the lower 25 observations) is .3; the upper quartile (median of the upper half of the data) is 7.9. Therefore, IQR = 7.9 - .3 = 7.6.

b ED: mild outliers are less than .1 - 1.5(2.65) = -3.875 or greater than 2.75 + 1.5(2.65) = 6.725. Extreme outliers are less than .1 - 3(2.65) = -7.85 or greater than 2.75 + 3(2.65) = 10.7. So, the two largest observations (11.7, 21.0) are extreme outliers and the next two largest values (8.9, 9.2) are mild outliers. There are no outliers at the lower end of the data.

Non-ED: mild outliers are less than .3 - 1.5(7.6) = -11.1 or greater than 7.9 + 1.5(7.6) = 19.3. Note that there are no mild outliers in the data, hence there can not be any extreme outliers either.

c A comparative boxplot appears below. The outliers in the ED data are clearly visible. There is noticeable positive skewness in both samples; the Non-Ed data has more variability then the Ed data; the typical values of the ED data tend to be smaller than those for the Non-ED data.

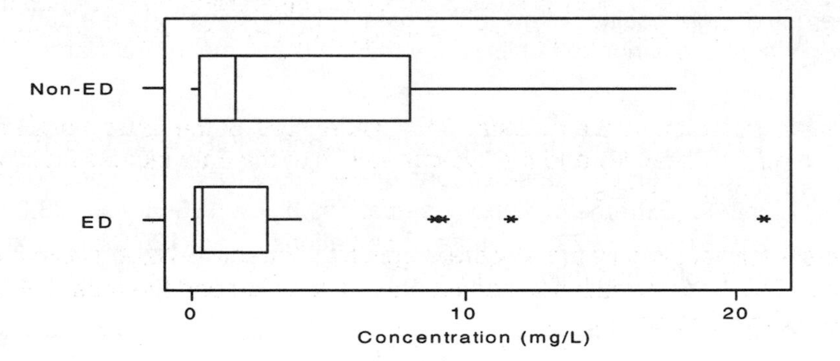

60. A comparative boxplot (created in Minitab) of this data appears below.

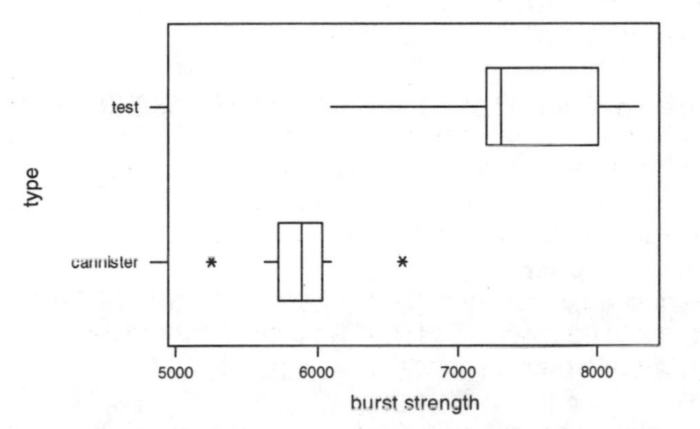

The burst strengths for the test nozzle closure welds are quite different from the burst strengths of the production canister nozzle welds.

The test welds have much higher burst strengths and the burst strengths are much more variable.

The production welds have more consistent burst strength and are consistently lower than the test welds. The production welds data does contain 2 outliers.

61. Outliers occur in the 6 a.m. data. The distributions at the other times are fairly symmetric. Variability and the 'typical' values in the data increase a little at the 12 noon and 2 p.m. times.

62. To somewhat simplify the algebra, begin by subtracting 76,000 from the original data. This transformation will affect each date value and the mean. It will not affect the standard deviation.

$x_1 = 683, \quad x_2 = 1{,}048, \quad \bar{y} = 831$

$n\bar{x} = (4)(831) = 3{,}324$ so, $x_1 + x_2 + x_3 + x_4 = 3{,}324$

and $x_2 + x_3 = 3{,}324 - x_1 - x_4 = 1{,}593$ and $x_3 = (1{,}593 - x_2)$

Next, $s^2 = (180)^2 = \left[\dfrac{\sum x_i^2 - \dfrac{(3324)^2}{4}}{3} \right]$

(continued)

So, $\sum x_i^2 = 2{,}859{,}444$, $x_1^2 + x_2^2 + x_3^2 + x_4^2 = 2{,}859{,}444$ and
$x_2^2 + x_3^2 = 2{,}859{,}444 - x_1^2 + x_4^2 = 1{,}294{,}651$

By substituting $x_3 = (1593 - x_2)$ we obtain the equation
$x_2^2 + (1{,}593 - x_2)^2 - 1{,}294{,}651 = 0$.
$x_x^2 - 1{,}593x_2 + 621{,}499 = 0$
Evaluating for x_2 we obtain $x_2 = 682.8635$ and $x_3 = 1{,}593 - 682.8635 = 910.1365$.
Thus, $x_2 = 76{,}683$ $x_3 = 76{,}910$.

63.

Flow rate	Median	Lower quartile	Upper quartile	IQR	1.5(IQR)	3(IQR)
125	3.1	2.7	3.8	1.1	1.65	.3
160	4.4	4.2	4.9	.7	1.05	.1
200	3.8	3.4	4.6	1.2	1.80	3.6

There are no outliers in the three data sets. However, as the comparative boxplot below shows, the three data sets differ with respect to their central values (the medians are different) and the data for flow rate 160 is somewhat less variable than the other data sets. Flow rates 125 and 200 also exhibit a small degree of positive skewness.

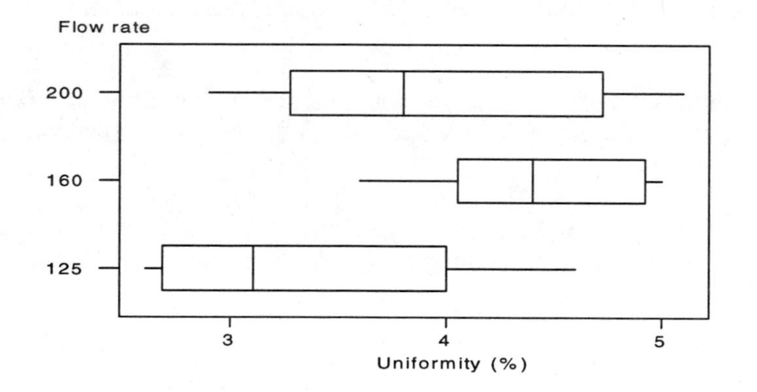

64

```
 6|34          stem=ones
 7|17          leaf=tenths
 8|4589
 9|01
10|012667789
11|122499
12|2
13|1
```

$\bar{x} = 9.9556, \tilde{x} = 10.6$

$s = 1.7594$

$n = 27$

$f_s = 2.3$

lower fourth = 8.85, upper fourth = 11.15

$8.85 - (1.5)(2.3) = 5.4$

$11.15 + (1.5)(2.3) = 14.6$

no outliers

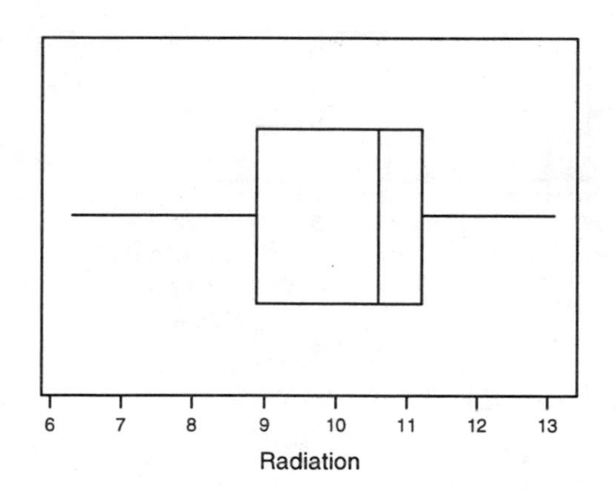

There are no outliers. The distribution is skewed to the left.

65 **a** HC data: $\sum_i x_i^2 = 2618.42$ and $\sum_i x_i = 96.8$,
so $s^2 = [2618.42 - (96.8)^2/4]/3 = 91.953$
and the sample standard deviation is $s = 9.59$.

CO data: $\sum_i x_i^2 = 145645$ and $\sum_i x_i = 735$, so $s^2 = [145645 - (735)^2/4]/3 =$ 3529.583 and the sample standard deviation is $s = 59.41$.

b The mean of the HC data is 96.8/4 = 24.2; the mean of the CO data is 735/4 = 183.75. Therefore, the coefficient of variation of the HC data is 9.59/24.2 = .3963, or 39.63%. The coefficient of variation of the CO data is 59.41/183.75 = .3233, or 32.33%. Thus, even though the CO data has a larger standard deviation than does the HC data, it actually exhibits *less* variability (in percentage terms) around its average than does the HC data.

66. **a** The histogram appears below. A representative value for this data would be x = 90. The histogram is reasonably symmetric, unimodal, and somewhat bell-shaped. The variation in the data is not small since the spread of the data (99-81 = 18) constitutes about 20% of the typical value of 90.

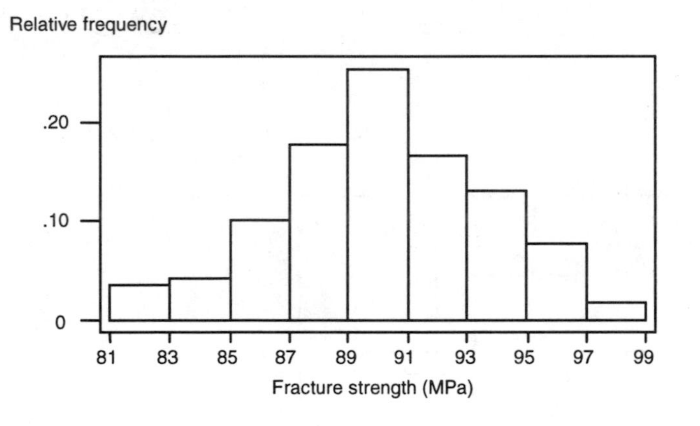

b The proportion of the observations that are at least 85 is 1 - (6+7)/169 = .9231. The proportion less than 95 is 1 - (22+13+3)/169 = .7751.

c x = 90 is the midpoint of the class 89-<91, which contains 43 observations (a relative frequency of 43/169 = .2544. Therefore about half of this frequency, .1272, should be added to the relative frequencies for the classes to the left of x = 90. That is, the approximate proportion of observations that are less than 90 is .0355 + .0414 + .1006 + .1775 + .1272 = .4822.

67.

$$\sum x_i = 163.2$$

$$100\left(\frac{1}{15}\right)\% trimmed mean = \frac{163.2 - 8.5 - 15.6}{13} = 10.70$$

$$100\left(\frac{2}{15}\right)\% trimmed mean = \frac{163.2 - 8.5 - 8.8 - 15.6 - 13.7}{11} = 10.60$$

$$\therefore \frac{1}{2}(100)\left(\frac{1}{15}\right) + \frac{1}{2}(100)\left(\frac{2}{15}\right) = 100\left(\frac{1}{10}\right) = 10\% trimmed mean$$

$$= \frac{1}{2}(10.70) + \frac{1}{2}(10.60) = 10.65$$

68. **a**

$$\frac{d}{dc\left\{\sum (x_i - c)^2\right\}} = \frac{\sum d}{dc(x_i - c)^2} = -2\sum (x_i - c) = 0 \Rightarrow \sum (x_i - c) = 0$$

$$\Rightarrow \sum x_i - \sum c = 0 \Rightarrow \sum x_i - nc = 0 \Rightarrow nc = \sum x_i \Rightarrow c = \frac{\sum x_i}{n} = \bar{x}.$$

b

$$\sum (x_i - \bar{x})^2 is smaller than \sum (x_i - \mu)^2.$$

69. **a**

$$\bar{y} = \frac{\sum y_i}{n} = \frac{\sum (ax_i + b)}{n} = \frac{a\sum x_i + b}{n} = a\bar{x} + b.$$

$$s_y^2 = \frac{\sum (y_i - \bar{y})^2}{n-1} = \frac{\sum (ax_i + b - (a\bar{x} + b))^2}{n-1} = \frac{\sum (ax_i - a\bar{x})^2}{n-1}$$

$$= \frac{a^2\sum (x_i - \bar{x})^2}{n-1} = a^2 s_x^2.$$

(continued)

b

$$x = {}^\circ C,\ y = {}^\circ F$$

$$\bar{y} = \frac{9}{5}(87.3) + 32 = 189.14$$

$$s_y = \sqrt{s_y^2} = \sqrt{\left(\frac{9}{5}\right)^2 (1.04)^2} = \sqrt{3.5044} = 1.872$$

70.

median = 46, lower fourth = 44, upper fourth = 52

$$f_s = 52 - 44 = 8$$
$$44 - (1.5)(8) = 32$$
$$52 + (1.5)(8) = 64$$
$$44 - (3)(8) = 20$$
$$52 + (3)(8) = 76$$
$$\text{min} = 33, \text{max} = 60$$

a There are no outliers, either mild or extreme.

b

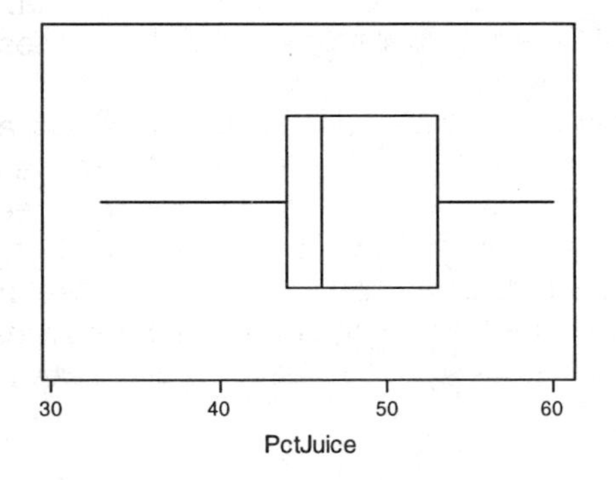

There are no outliers. The distribution is not symmetric. The measure of spread is not terribly large.

71.

```
0.7|8              stem=tenths
0.8|11556          leaf=hundredths
0.9|2233335566
1.0|0566
```

$\bar{x} = .9255, s = .0809, \tilde{x} = .93$

$lowerfourth = .855, upperfourth = .96$

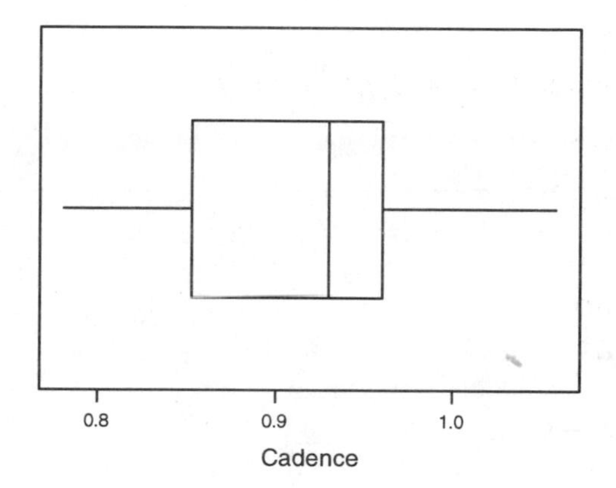

The data appears to be a bit skewed toward smaller values (negatively skewed). There are no outliers. The mean and the median are close in value.

72. **a** Mode = .93. It occurs four times in the data set.

b The Modal Category is the one in which the most observations occur.

73. **a** The median is the same (371) in each plot and all three data sets are very symmetric. In addition, all three have the same minimum value (350) and same maximum value (392). Moreover, all three data sets have the same lower (364) and upper quartiles (378). So, all three boxplots will be *identical.*

b A comparative dotplot is shown below. These graphs show that there are differences in the variability of the three data sets. They also show differences in the way the values are distributed in the three data sets.

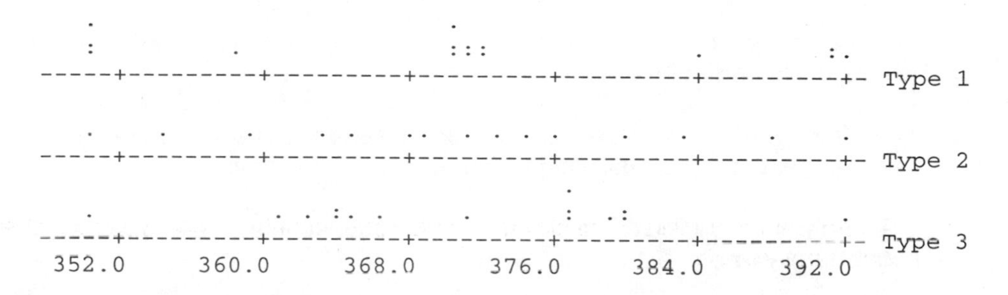

c The boxplot in (a) is not capable of detecting the differences among the data sets. The primary reason is that boxplots give up some detail in describing data because they use only 5 summary numbers for comparing data sets.

Note: The definition of lower and upper quartile used in this text is slightly different than the one used by some other authors (and software packages). Technically speaking, the median of the lower half of the data is not really the first quartile, although it is generally *very close.* Instead, the medians of the lower and upper halves of the data are often called the **lower** and **upper hinges.** Our boxplots use the lower and upper hinges to define the spread of the middle 50% of the data, but other authors sometimes use the *actual* quartiles for this purpose. The difference is usually very slight, usually unnoticeable, but not always. For example in the data sets of this exercise, a comparative boxplot based on the actual quartiles (as computed by Minitab) is shown below. The graph shows substantially the same type of information as those described in (a) except the graphs based on quartiles are able to detect the slight differences in variation between the three data sets.

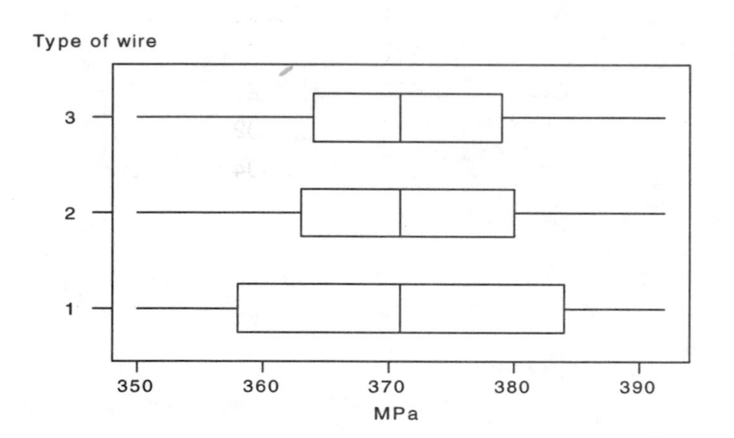

74. The measures that are sensitive to outliers are: the mean and the midrange. The mean is sensitive because all values are used in computing it. The midrange is sensitive because it uses only the most extreme values in its computation.

The median, the trimmed mean, and the midhinge are not sensitive to outliers.

The median is the most resistant to outliers because it uses only the middle value (or values) in its computation.

The trimmed mean is somewhat resistant to outliers. The larger the trimming percentage, the more resistant the trimmed mean becomes.

The midhinge, which uses the quartiles, is reasonably resistant to outliers because both quartiles are resistant to outliers.

75. **a**

Stem	Leaves
0	2355566777888
1	0000135555
2	00257
3	0033
4	0057
5	044
6	
7	05
8	8
9	0
10	3
HI	22.0 24.5

stem: ones
leaf: tenths

b

Interval	Frequency	Rel. Freq.	Density
0 -< 2	23	.500	.250
2 -< 4	9	.196	.098
4 -< 6	7	.152	.076
6 -< 10	4	.087	.022
10 -< 20	1	.022	.002
20 -< 30	2	.043	.004

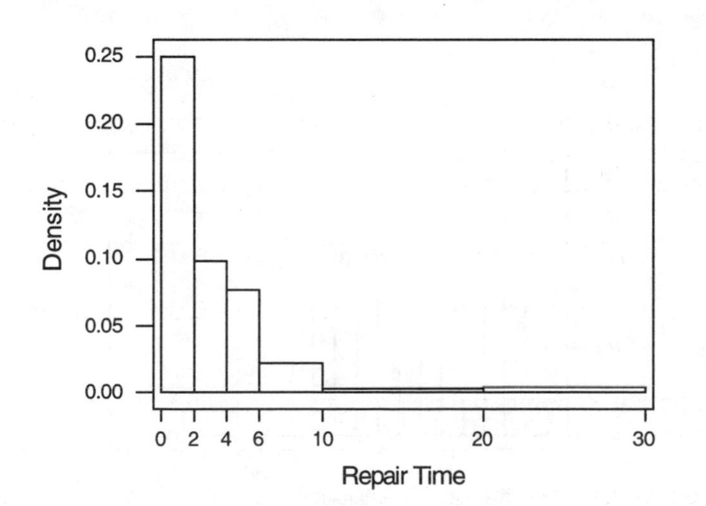

76. **a** Since the constant $\overline{x}$ is subtracted from each x value to obtain each y value, and addition or subtraction of a constant doesn't affect variability,

$$s_y^2 = s_x^2 \qquad\qquad s_y = s_x$$

b Let c = 1/s, where s is the sample standard deviation of the x's and also (by a) of the y's.

Then $s_z = cs_y = (1/s)s = 1$, and $s_z^2 = 1$. That is, the "standardized" quantities $z_1, \dots, z_n$ have a sample variance and standard deviation of 1.

77. **a**

$$\sum_{i=1}^{n+1} x_i = \sum_{i=1}^{n} x_i + x_{n+1} = n\overline{x}_n + x_{n+1}, so \overline{x}_{n+1} = \frac{[n\overline{x}_n + x_{n+1}]}{(n+1)}$$

b

$$ns_{n+1}^2 = \sum_{i=1}^{n+1}(x_i - \overline{x}_{n+1})^2 = \sum_{i=1}^{n+1} x_i^2 - (n+1)\overline{x}_{n+1}^2$$

$$= \sum_{i=1}^{n} x_i^2 - n\overline{x}_n^2 + x_{n+1}^2 + n\overline{x}_n^2 - (n+1)\overline{x}_{n+1}^2$$

$$= (n-1)s_n^2 + \left\{x_{n+1}^2 + n\overline{x}_n^2 - (n+1)\overline{x}_{n+1}^2\right\}$$

When the expression for $\overline{x}_{n+1}$ from **a** is substituted, the expression in braces

simplifies to the following, as desired: $\dfrac{n(x_{n+1} - \overline{x}_n)^2}{(n+1)}$

78. **a**

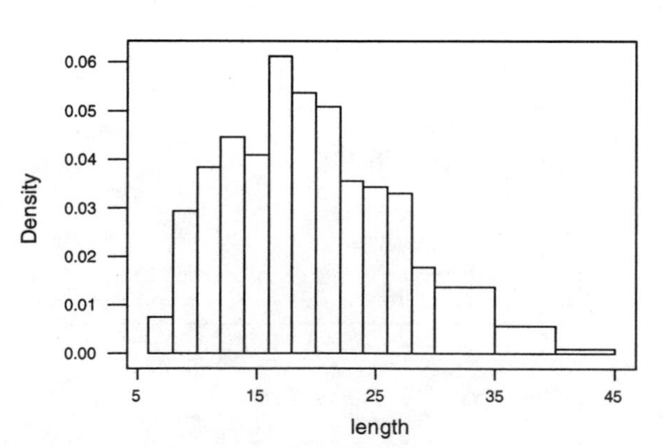

b Proportion less than $20 = \left(\frac{216}{391}\right) = .552$

Proportion at least $30 = \left(\frac{40}{391}\right) = .102$

c First compute (.90)(391 + 1) = 352.8. Thus, the 90th percentile should be about the 352nd ordered value. The 351st ordered value lies in the interval 28 - < 30. The 352nd ordered value lies in the interval 30 - < 35. There are 27 values in the interval 30 - < 35. We do not know how these values are distributed, however, the smallest value (i.e., the 352nd value in the data set) cannot be smaller than 30. So, the 90th percentile is roughly 30.

d First compute (.50)(391 + 1) = 196. Thus the median (50th percentile) should be the 196 ordered value. The 174th ordered value lies in the interval 16 - < 18. The next 42 observation lie in the interval 18 - < 20. So, ordered observation 175 to 216 lie in the intervals 18 - < 20. The 196th observation is about in the middle of these. Thus, we would say, the median is roughly 19.

79. Assuming that the histogram is unimodal, then there is evidence of positive skewness in the data since the median lies to the left of the mean (for a symmetric distribution, the mean and median would coincide). For more evidence of skewness, compare the distances of the 5th and 95th percentiles from the median: median - 5th percentile = 500 - 400 = 100 while 95th percentile -median = 720 - 500 = 220. Thus, the largest 5% of the values (above the 95th percentile) are further from the median than are the lowest 5%. The same skewness is evident when comparing the 10th and 90th percentiles to the median: median - 10th percentile = 500 - 430 = 70 while 90th percentile -median = 640 - 500 = 140. Finally, note that the largest value (925) is much further from the median (925-500 = 425) than is the smallest value (500 - 220 = 280), again an indication of positive skewness.

80. **a**

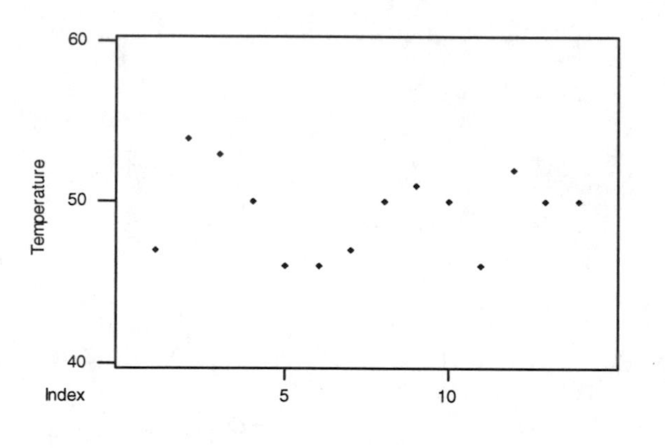

There is some evidence of a cyclical pattern.

b

$$\bar{x}_2 = .1x_2 + .9\bar{x}_1 = (.1)(54) + (.9)(47) = 47.7$$
$$\bar{x}_3 = .1x_3 + .9\bar{x}_2 = (.1)(53) + (.9)(47.7) = 48.23 \approx 48.2, etc.$$

(continued)

t	$\bar{x}_t$ for $\alpha = .1$	$\bar{x}_t$ for $\alpha = .5$
1	47.0	47.0
2	47.7	50.5
3	48.2	51.8
4	48.4	50.9
5	48.2	48.4
6	48.0	47.2
7	47.9	47.1
8	48.1	48.6
9	48.4	49.8
10	48.5	49.9
11	48.3	47.9
12	48.6	50.0
13	48.8	50.0
14	48.9	50.0

α= .1 gives a smoother series.

c

$$\begin{aligned}\bar{x}_t &= \alpha x_t + (1-\alpha)\bar{x}_{t-1} \\ &= \alpha x_t + (1-\alpha)[\alpha x_{t-1} + (1-\alpha)\bar{x}_{t-2}] \\ &= \alpha x_t + \alpha(1-\alpha)x_{t-1} + (1-\alpha)^2[\alpha x_{t-2} + (1-\alpha)\bar{x}_{t-3}] \\ &= \ldots = \alpha x_t + \alpha(1-\alpha)x_{t-1} + \alpha(1-\alpha)^2 x_{t-2} + \ldots + \alpha(1-\alpha)^{t-2}x_2 + (1-\alpha)^{t-1}\bar{x}_1\end{aligned}$$

Thus, $(x \text{ bar})_t$ depends on x_t and all previous values. As k increases, the coefficient on x_{t-k} decreases (further back in time implies less weight).

d Not very sensitive, since $(1-\alpha)^{t-1}$ will be very small.

81. **a** When there is perfect symmetry, the smallest observation y_1 and the largest observation y_n will be equidistant from the median, so

$$y_n - \bar{x} = \bar{x} - y_1$$

Similarly, the second smallest and second largest will be equidistant from the median, so

$$y_{n-1} - \bar{x} = \bar{x} - y_2$$

and so on. Thus, the first and second numbers in each pair will be equal, so that each point in the plot will fall exactly on the 45 degree line. When the data is positively skewed, y_n will be much further from the median than is y_1, so

$$y_n - \tilde{x}$$

will considerably exceed

$$\tilde{x} - y_1$$

and the point

$$(y_n - \tilde{x}, \tilde{x} - y_1)$$

will fall considerably below the 45 degree line. A similar comment aplies to other points in the plot.

b The first point in the plot is (2745.6 – 221.6, 221.6 0- 4.1) = (2524.0, 217.5). The others are: (1476.2, 213.9), (1434.4, 204.1), (756.4, 190.2), (481.8, 188.9), (267.5, 181.0), (208.4, 129.2), (112.5, 106.3), (81.2, 103.3), (53.1, 102.6), (53.1, 92.0), (33.4, 23.0), and (20.9, 20.9). The first number in each of the first seven pairs greatly exceed the second number, so each point falls well below the 45 degree line. A substantial positive skew (stretched upper tail) is indicated.

Chapter 2

Section 2.1

1.

a S = { 1324, 1342, 1423, 1432, 2314, 2341, 2413, 2431, 3124, 3142, 4123, 4132, 3214, 3241, 4213, 4231 }

b Event A contains the outcomes where 1 is first in the list:
A = { 1324, 1342, 1423, 1432 }

c Event B contains the outcomes where 2 is first or second:
B = { 2314, 2341, 2413, 2431, 3214, 3241, 4213, 4231 }

d The compound event $A \cup B$ contains the outcomes in A or B or both:
$A \cup B$ = {1324, 1342, 1423, 1432, { 2314, 2341, 2413, 2431, 3214, 3241, 4213, 4231 }

The compound event $A \cap B$ contains outcomes that are in BOTH A and B. Since there are none in common, this event has no outcomes:
$A \cap B$ ={ ∅ }.

The event A′ contains the outcomes that are not in A:
A′ = { 2314, 2341, 2413, 2431, 3124, 3142, 4123, 4132, 3214, 3241, 4213, 4231 }

2.

a Event A = { RRR, LLL, SSS }
b Event B = { RLS, RSL, LRS, LSR, SRL, SLR }
c Event C = { RRL, RRS, RLR, RSR, LRR, SRR }
d Event D = { RRL, RRS, RLR, RSR, LRR, SRR, LLR, LLS, LRL, LSL, RLL, SLL, SSR, SSL, SRS, SLS, RSS, LSS }
e Event D′ contains outcomes where all cars go the same direction, or they all go different directions:
D′ = { RRR, LLL, SSS, RLS, RSL, LRS, LSR, SRL, SLR }

Because Event D totally encloses Event C, the compound event $C \cup D = D$:
$C \cup D$ = { RRL, RRS, RLR, RSR, LRR, SRR, LLR, LLS, LRL, LSL, RLL, SLL, SSR, SSL, SRS, SLS, RSS, LSS }

Using similar reasoning, we see that the compound event $C \cap D = C$:
$C \cap D$ = { RRL, RRS, RLR, RSR, LRR, SRR }

3.

a Event A = { SSF, SFS, FSS }
b Event B = { SSS, SSF, SFS, FSS }

c For Event C, the system must have component 1 working (S in the first position), then at least one of the other two components must work (at least one S in the 2nd and 3rd positions:
Event C = { SSS, SSF, SFS }

d Event C′ = { SFF, FSS, FSF, FFS, FFF }
Event A∪C = { SSS, SSF, SFS, FSS }
Event A∩C = { SSF, SFS }
Event B∪C = { SSS, SSF, SFS, FSS }
Event B∩C = { SSS SSF, SFS }

4.

a

	Home Mortgage Number			
Outcome	1	2	3	4
1	F	F	F	F
2	F	F	F	V
3	F	F	V	F
4	F	F	V	V
5	F	V	F	F
6	F	V	F	V
7	F	V	V	F
8	F	V	V	V
9	V	F	F	F
10	V	F	F	V
11	V	F	V	F
12	V	F	V	V
13	V	V	F	F
14	V	V	F	V
15	V	V	V	F
16	V	V	V	V

b Outcome numbers 2, 3, 5 ,9

c Outcome numbers 1, 16

d Outcome numbers 1, 2, 3, 5, 9

e In words, the UNION described is the event that either all of the mortgages are variable, or that at most all of them are variable: outcomes 1,2,3,5,9,16. The INTERSECTION described is the event that all of the mortgages are fixed: outcome 1.

f The UNION described is the event that either exactly three are fixed, or that all four are the same: outcomes 1, 2, 3, 5, 9, 16. The INTERSECTION in words is the event that exactly three are fixed AND that all four are the same. This cannot happen. (There are no outcomes in common) : **b** ∩ **c** = ∅.

Chapter 2

5.

a

Outcome Number	Outcome
1	111
2	112
3	113
4	121
5	122
6	123
7	131
8	132
9	133
10	211
11	212
12	213
13	221
14	222
15	223
16	231
17	232
18	233
19	311
20	312
21	313
22	321
23	322
24	323
25	331
26	332
27	333

b Outcome Numbers 1, 14, 27

c Outcome Numbers 6, 8, 12, 16, 20, 22

d Outcome Numbers 1, 3, 7, 9, 19, 21, 25, 27

Chapter 2

6.

a

Outcome Number	Outcome
1	123
2	124
3	125
4	213
5	214
6	215
7	13
8	14
9	15
10	23
11	24
12	25
13	3
14	4
15	5

b Outcomes 13, 14, 15

c Outcomes 3, 6, 9, 2, 15

d Outcomes 10, 11, 12, 13, 14, 15

7.

a S = {BBBAAAA, BBABAAA, BBAABAA, BBAAABA, BBAAAAB, BABBAAA, BABABAA, BABAABA, BABAAAB, BAABBAA, BAABABA, BAABAAB, BAAABBA, BAAABAB, BAAAABB, ABBBAAA, ABBABAA, ABBAABA, ABBAAAB, ABABBAA, ABABABA, ABABAAB, ABAABBA, ABAABAB, ABAAABB, AABBBAA, AABBABA, AABBAAB, AABABBA, AABABAB, AABAABB, AAABBBA, AAABBAB, AAABABB, AAAABBB}

b {AAAABBB, AAABABB, AAABBAB, AABAABB, AABABAB}

Chapter 2

8.

a $A_1 \cup A_2 \cup A_3$

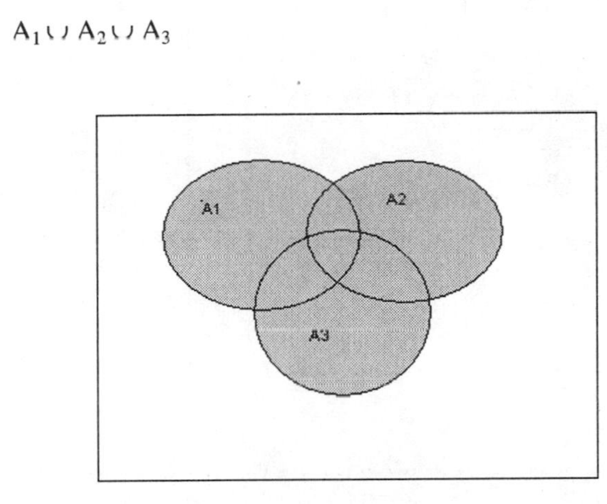

b $A_1 \cap A_2 \cap A_3$

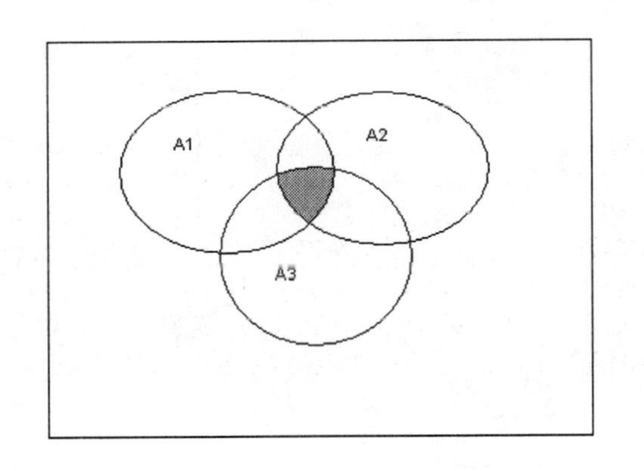

c $A_1 \cup A_2' \cup A_3'$

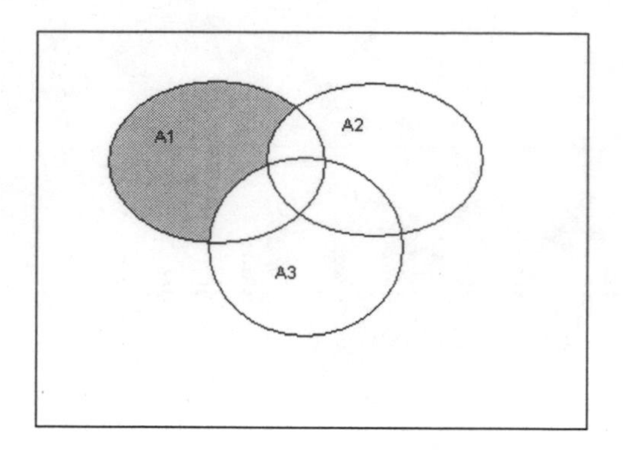

d $(A_1 \cap A_2' \cap A_3') \cup (A_1' \cap A_2 \cap A_3') \cup (A_1' \cap A_2' \cap A_3)$

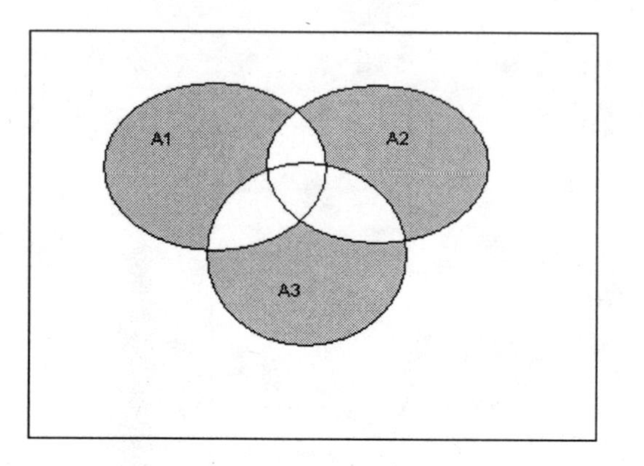

e $A_1 \cup (A_2 \cap A_3)$

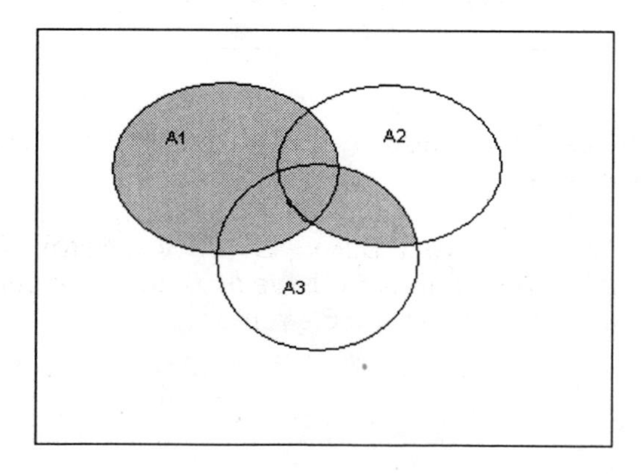

Chapter 2

9.

a In the diagram on the left, the shaded area is $(A \cup B)'$. On the right, the shaded area is A', the striped area is B', and the intersection $A' \cap B'$ occurs where there is BOTH shading and stripes. These two diagrams display the same area.

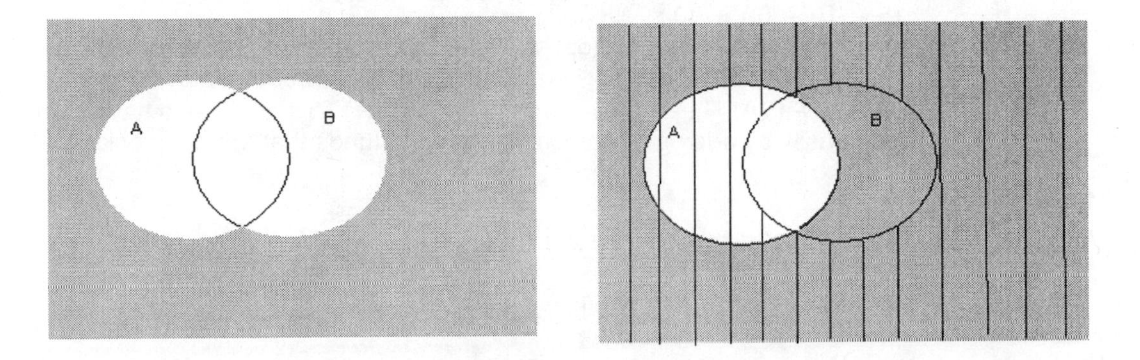

b In the diagram below, the shaded area represents $(A \cap B)'$. Using the diagram on the right above, the union of A' and B' is represented by the areas that have either shading or stripes or both. Both of the diagrams display the same area.

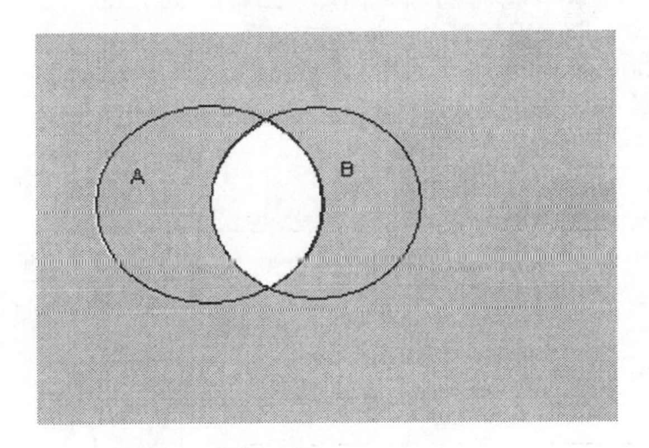

10.

a A = {Chev, Pont, Buick}, B = {Ford, Merc}, C = {Plym, Chrys} are three mutually exclusive events.

b No, let E = {Chev, Pont}, F = {Pont, Buick}, G = {Buick, Ford}. These events are not mutually exclusive (e.g. E and F have an outcome in common), yet there is no outcome common to all three events.

Chapter 2

Section 2.2

11.

a .07

b .15 + .10 + .05 = .30

c Let event A = selected customer owns stocks. Then the probability that a selected customer does not own a stock can be represented by $P(A') = 1 - P(A) = 1 - (.18 + .25) = 1 - .43 = .57$. This could also have been done easily by adding the probabilities of the funds that are not stocks.

12.

a $P(A \cup B) = .50 + .40 - .25 = .65$

b $P(A \cup B)' = 1 - .65 = .35$

c $A \cap B'$; $P(A \cap B') = P(A) - P(A \cap B) = .50 - .25 = .25$

13.

a awarded either #1 or #2 (or both):
$P(A_1 \cup A_2) = P(A_1) + P(A_2) - P(A_1 \cap A_2) = .22 + .25 - .11 = .36$

b awarded neither #1 or #2:
$P(A_1' \cap A_2') = P[(A_1 \cup A_2)'] = 1 - P(A_1 \cup A_2) = 1 - .36 = .64$

c awarded at least one of #1, #2, #3:
$$P(A_1 \cup A_2 \cup A_3) = P(A_1) + P(A_2) + P(A_3) - P(A_1 \cap A_2) - P(A_1 \cap A_3) - P(A_2 \cap A_3) + P(A_1 \cap A_2 \cap A_3)$$
$$= .22 + .25 + .28 - .11 - .05 - .07 + .01 = .53$$

d awarded none of the three projects:
$P(A_1' \cap A_2' \cap A_3') = 1 - P(\text{awarded at least one}) = 1 - .53 = .47.$

e awarded #3 but neither #1 nor #2:
$$P(A_1' \cap A_2' \cap A_3) = P(A_3) - P(A_1 \cap A_3) - P(A_2 \cap A_3) + P(A_1 \cap A_2 \cap A_3)$$
$$= .28 - .05 - .07 + .01$$
$$= .17$$

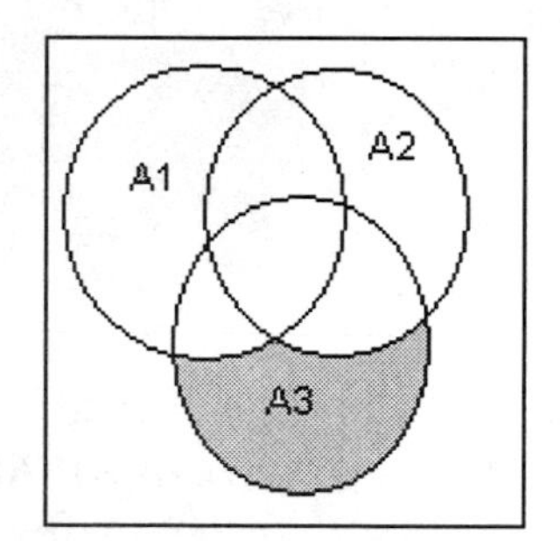

(continued)

f either (neither #1 nor #2) or #3:
$P[(A_1' \cap A_2') \cup A_3] = P(\text{shaded region})$
$= P(\text{awarded none}) + P(A_3)$
$= .47 + .28 = .75$

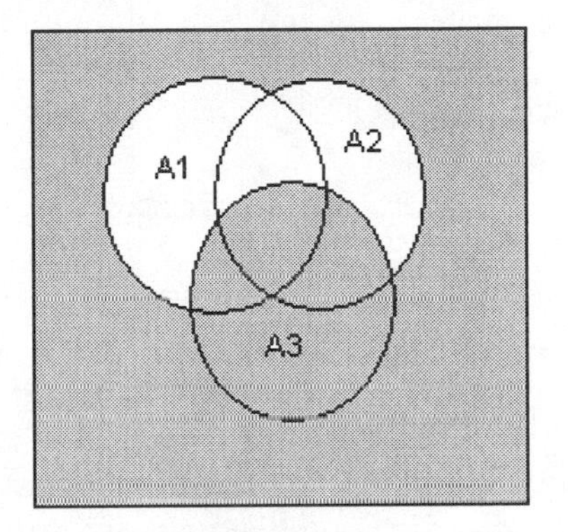

Alternatively, answers to **a** – **f** can be obtained from probabilities on the accompanying Venn diagram

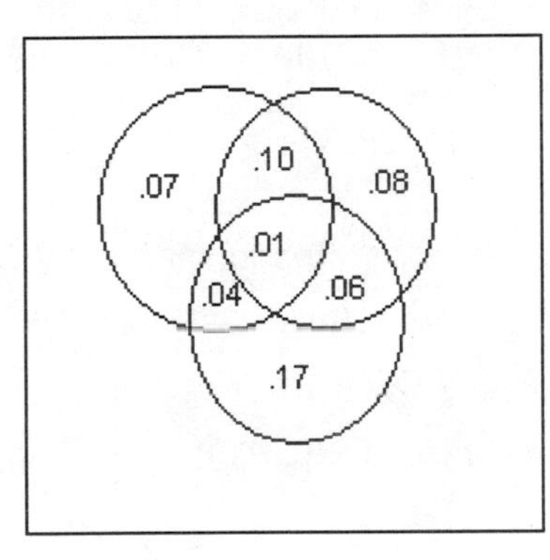

14.

a $P(A \cup B) = P(A) + P(B) - P(A \cap B)$,
so $P(A \cap B) = P(A) + P(B) - P(A \cup B)$
$= .8 + .7 - .9 = .6$

b

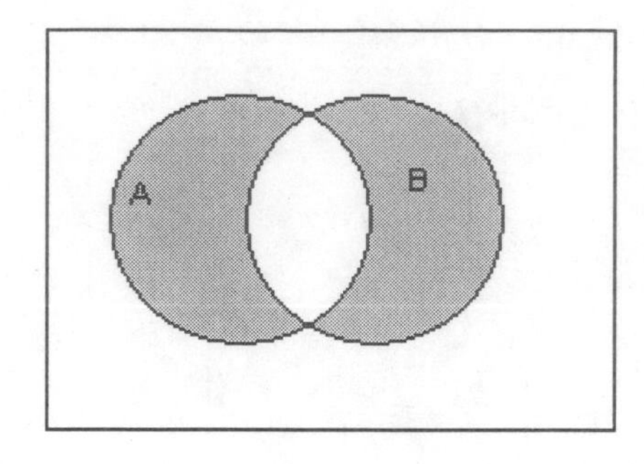

P(shaded region) = $P(A \cup B) - P(A \cap B) = .9 - .6 = .3$
Shaded region = event of interest = $(A \cap B') \cup (A' \cap B)$

15.

a Let event E be the event that at most one purchases an electric dryer. Then E′ is the event that at least two purchase electric dryers.
$P(E') = 1 - P(E) = 1 - .087 = .913$

b Let event A be the event that all five purchase gas. Let event B be the event that all five purchase electric. All other possible outcomes are those in which at least one of each type is purchased. Thus, the desired probability =
$1 - P(A) - P(B) = 1 - .0768 - .0102 = .913$

16.

a There are six simple events, corresponding to the outcomes CDP, CPD, DCP, DPC, PCD, and PDC. The probability assigned to each is $\frac{1}{6}$.

b P(C ranked first) = P({CPD, CDP}) = $\frac{1}{6}+\frac{1}{6}=\frac{2}{6}=.333$

c P(C ranked first and D last) = P({CPD}) = $\frac{1}{6}$

17.

a The probabilities do not add to 1 because there are other items besides Non-Fiction and Fiction books to be checked out from the library.

b $P(A') = 1 - P(A) = 1 - .35 = .65$

c $P(A \cup B) = P(A) + P(B) = .35 + .50 = .85$
(since A and B are mutually exclusive events)

d $P(A' \cap B') = P[(A \cup B)']$ (De Morgan's law)
$= 1 - P(A \cup B)$
$= 1 - .85 = .15$

18. This situation requires the complement concept. The only way for the desired event NOT to happen is if a 75 W bulb is selected first. Let event A be that a 75 W bulb is selected first, and P(A) = $\frac{6}{15}$. Then the desired event is event A′.

So $P(A') = 1 - P(A) = 1-\frac{6}{15}=\frac{9}{15}=.60$

Chapter 2

19. Let event A be that the selected joint was found defective by inspector A. P(A) = $\frac{724}{10,000}$. Let event B be analogous for inspector B. P(B) = $\frac{751}{10,000}$. Compound event $A \cup B$ is the event that the selected joint was found defective by at least one of the two inspectors. $P(A \cup B) = \frac{1159}{10,000}$.

a The desired event is $(A \cup B)'$, so we use the complement rule:
$P(A \cup B)' = 1 - P(A \cup B) = 1 - \frac{1159}{10,000} = \frac{8841}{10,000} = .8841$

b The desired event is $B \cap A'$. $P(B \cap A') = P(B) - P(A \cap B)$.
$P(A \cap B) = P(A) + P(B) - P(A \cup B)$,
$= .0724 + .0751 - .1159 = .0316$
So $P(B \cap A') = P(B) - P(A \cap B)$
$= .0751 - .0316 = .0435$

20. Let H1 and H2 represent the two health plans. Let D1 and D2 represent the two dental plans.

a The simple events are {H1,D1}, {H1,D2}, {H2,D1}, {H2,D2}.

b P({H1,D1}) = .25

c P({D2}) = P({H1,D2}, {H2,D2}) = .16 + .37 = .53

21.

a P({M,H}) = .10

b P(low auto) = P[{(L,N}, (L,L), (L,M), (L,H)}] = .04 + .06 + .05 + .03 = .18
Following a similar pattern, P(low homeowner's) = .06 + .10 + .03 = .19

c P(same deductible for both) = P[{ LL, MM, HH }] = .06 + .20 + .15 = .41

d P(deductibles are different) = 1 – P(same deductibles) = 1 - .41 = .59

e P(at least one low deductible) = P[{ LN, LL, LM, LH, ML, HL }]
= .04 + .06 + .05 + .03 + .10 + .03 = .31

f P(neither low) = 1 – P(at least one low) = 1 - .31 = .69

22.

a $P(A_1 \cap A_2) = P(A_1) + P(A_2) - P(A_1 \cup A_2) = .4 + .5 - .6 = .3$

b $P(A_1 \cap A_2') = P(A_1) - P(A_1 \cap A_2) = .4 - .3 = .1$

c $P(\text{exactly one}) = P(A_1 \cup A_2) - P(A_1 \cap A_2) = .6 - .3 = .3$

23. Possible outcomes are (1,2) (1,3) (1,4) (1,5) (2,3) (2,4) (2,5) (3,4) (3,5) and (4,5).

a P(both are first printings) = P[{ (1,2)}] = $\frac{1}{10}$=.10

b P(both are second printings) = P[{ (3,4) (3,5) (4,5) }] = $\frac{3}{10}$=.30

c P(at least one first printing) = 1 – P(no first printings)
= 1 – P(both are second printings)
= 1 – .30 = .70

d P(different printings) = 1 – P(same printings)
= 1 – [P(both firsts) + P(both seconds)]
= 1 - .40 = .60

24.

Since A is contained in B, then B can be written as the union of A and $(B \cap A')$, two mutually exclusive events. (See diagram).

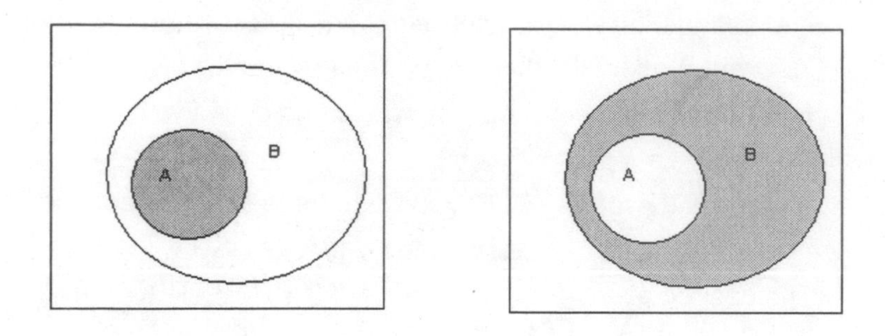

From Axiom 3, $P[A \cup (B \cap A')] = P(A) + P(B \cap A')$. Substituting P(B), $P(B) = P(A) + P(B \cap A')$ or $P(B) - P(A) = P(B \cap A')$. From Axiom 1, $P(B \cap A') \geq 0$, so $P(B) \geq P(A)$ or $P(A) \leq P(B)$.

For general events A and B, $P(A \cap B) \leq P(A)$, and $P(A \cup B) \geq P(A)$.

25. $P(A \cap B) = P(A) + P(B) - P(A \cup B) = .65$

$P(A \cap C) = .55$, $P(B \cap C) = .60$

$$P(A \cap B \cap C) = P(A \cup B \cup C) - P(A) - P(B) - P(C) + P(A \cap B) + P(A \cap C) + P(B \cap C)$$
$$= .98 - .7 - .8 - .75 + .65 + .55 + .60$$
$$= .53$$

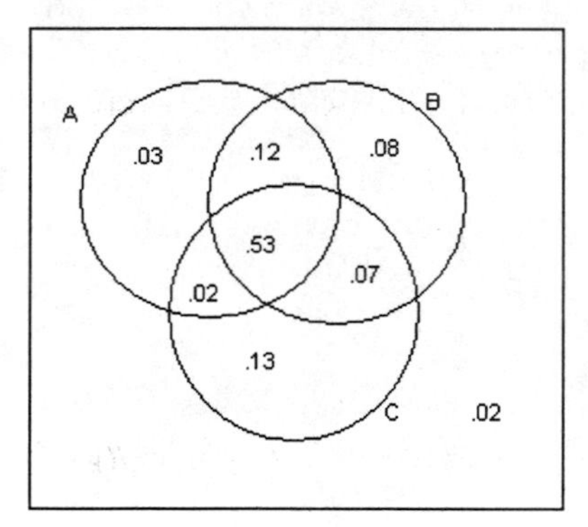

a $P(A \cup B \cup C) = .98$, as given.

b $P(\text{none selected}) = 1 - P(A \cup B \cup C) = 1 - .98 = .02$

c P(only automatic transmission selected) = .03 from the Venn Diagram

d P(exactly one of the three) = .03 + .08 + .13 = .24

26.

a $P(A_1') = 1 - P(A_1) = 1 - .12 = .88$

b $P(A_1 \cap A_2) = P(A_1) + P(A_2) - P(A_1 \cup A_2) = .12 + .07 - .13 = .06$

c $P(A_1 \cap A_2 \cap A_3') = P(A_1 \cap A_2) - P(A_1 \cap A_2 \cap A_3) = .06 - .01 = .05$

d P(at most two errors) $= 1 - P(\text{all three types})$
$$= 1 - P(A_1 \cap A_2 \cap A_3)$$
$$= 1 - .01 = .99$$

Chapter 2

27. Outcomes: (A,B) (A,C_1) (A,C_2) (A,F) (B,A) (B,C_1) (B,C_2) (B,F)
(C_1,A) (C_1,B) (C_1,C_2) (C_1,F) (C_2,A) (C_2,B) (C_2,C_1) (C_2,F)
(F,A) (F,B) (F,C_1) (F,C_2)

a P[(A,B) or (B,A)] = $\frac{2}{20} = \frac{1}{10} = .1$

b P(at least one C) = $\frac{14}{20} = \frac{7}{10} = .7$

c P(at least 15 years) = 1 – P(at most 14 years)
= 1 – P[(3,6) or (6,3) or (3,7) or (7,3) or (3,10) or (10,3) or (6,7) or (7,6)]
= $1 - \frac{8}{20} = 1 - .4 = .6$

28. There are 27 equally likely outcomes.

a P(all the same) = P[(1,1,1) or (2,2,2) or (3,3,3)] = $\frac{3}{27} = \frac{1}{9}$

b P(at most 2 are assigned to the same station) = 1 – P(all 3 are the same)
= $1 - \frac{3}{27} = \frac{24}{27} = \frac{8}{9}$

c P(all different) = [{(1,2,3) (1,3,2) (2,1,3) (2,3,1) (3,1,2) (3,2,1)}]
= $\frac{6}{27} = \frac{2}{9}$

Section 2.3

29.

a (5)(4) = 20 (5 choices for president, 4 remain for vice president)

b (5)(4)(3) = 60

c $\binom{5}{2} = \frac{5!}{2!3!} = 10$ (No ordering is implied in the choice)

30.

a (10)(9)(8) = 720

b $\binom{10}{3} = \frac{10!}{3!7!} = 120$

c P(all 3 are new) = (# of ways of visiting all new)/(# of ways of visiting)

$$= \frac{\binom{4}{3}}{\binom{10}{3}} = \frac{4}{120} = \frac{1}{30}$$

31.

a $(n_1)(n_2)$ = (9)(27) = 243

b $(n_1)(n_2)(n_3)$ = (9)(27)(15) = 3645, so such a policy could be carried out for 3645 successive nights, or approximately 10 years, without repeating exactly the same program.

32.

a 5×4×3×4 = 240

b 1×1×3×4 = 12

c 4×3×3×3 = 108

d # with at least on Sony = total # - # with no Sony = 240 – 108 = 132

e P(at least one Sony) = $\frac{132}{240} = .55$

P(exactly one Sony) = P(only Sony is receiver)
+ P(only Sony is CD player)
+ P(only Sony is deck)

$$= \frac{1\times3\times3\times3}{240} + \frac{4\times1\times3\times3}{240} + \frac{4\times3\times3\times1}{240} = \frac{27+36+36}{240}$$

$$= \frac{99}{240} = .413$$

33.

a $\binom{20}{5} = \frac{20!}{5!15!} = 15{,}504$

b $\binom{8}{4} \bullet \binom{12}{1} = 840$

c P(exactly 4 have cracks) = $\frac{\binom{8}{4}\binom{12}{1}}{\binom{20}{5}} = \frac{840}{15{,}504} = .0542$

d P(at least 4) = P(exactly 4) + P(exactly 5)

$$= \frac{\binom{8}{4}\binom{12}{1}}{\binom{20}{5}} + \frac{\binom{8}{5}\binom{12}{0}}{\binom{20}{5}} = .0542 + .0036 = .0578$$

34.

a $\binom{20}{6} = 38{,}760.$ P(all from day shift) = $\frac{\binom{20}{6}\binom{25}{0}}{\binom{45}{6}} = \frac{38{,}760}{8{,}145{,}060} = .0048$

b P(all from same shift) = $\frac{\binom{20}{6}\binom{25}{0}}{\binom{45}{6}} + \frac{\binom{15}{6}\binom{30}{0}}{\binom{45}{6}} + \frac{\binom{10}{6}\binom{35}{0}}{\binom{45}{6}}$

= .0048 + .0006 + .0000 = .0054

c P(at least two shifts represented) = 1 – P(all from same shift)
= 1 - .0054 = .9946

d Let A_1 = day shift unrepresented, A_2 = swing shift unrepresented, and A_3 = graveyard shift unrepresented. Then we wish $P(A_1 \cup A_2 \cup A_3)$.

$P(A_1)$ = P(day unrepresented) = P(all from swing and graveyard)

$$P(A_1) = \frac{\binom{25}{6}}{\binom{45}{6}}, \quad P(A_2) = \frac{\binom{30}{6}}{\binom{45}{6}}, \quad P(A_3) = \frac{\binom{35}{6}}{\binom{45}{6}},$$

$$P(A_1 \cap A_2) = \text{P(all from graveyard)} = \frac{\binom{10}{6}}{\binom{45}{6}}$$

$$P(A_1 \cap A_3) = \frac{\binom{15}{6}}{\binom{45}{6}}, \quad P(A_2 \cap A_3) = \frac{\binom{20}{6}}{\binom{45}{6}}, \quad P(A_1 \cap A_2 \cap A_3) = 0,$$

$$\text{So } P(A_1 \cup A_2 \cup A_3) = \frac{\binom{25}{6}}{\binom{45}{6}} + \frac{\binom{30}{6}}{\binom{45}{6}} + \frac{\binom{35}{6}}{\binom{45}{6}} - \frac{\binom{10}{6}}{\binom{45}{6}} - \frac{\binom{15}{6}}{\binom{45}{6}} - \frac{\binom{20}{6}}{\binom{45}{6}}$$

$$= .2939 - .0054 = .2885$$

35. There are 10 possible outcomes -- $\binom{5}{2}$ ways to select the positions for B's votes: BBAAA, BABAA, BAABA, BAAAB, ABBAA, ABABA, ABAAB, AABBA, AABAB, and AAABB. Only the last two have A ahead of B throughout the vote count. Since the outcomes are equally likely, the desired probability is $\frac{2}{10} = .20$.

36.

a $n_1 = 3$, $n_2 = 4$, $n_3 = 5$, so $n_1 \times n_2 \times n_3 = 60$ runs

b $n_1 = 1$, (just one temperature), $n_2 = 2$, $n_3 = 5$ implies that there are 10 such runs.

37. There are $\binom{60}{5}$ ways to select the 5 runs. Each catalyst is used in 12 different runs, so the number of ways of selecting one run from each of these 5 groups is 12^5. Thus the desired probability is $\frac{12^5}{\binom{60}{5}} = .0456$.

38.

a P(selecting 2 - 75 watt bulbs) = $\dfrac{\binom{6}{2}\binom{9}{1}}{\binom{15}{3}} = \dfrac{15 \cdot 9}{455} = .2967$

b P(all three are the same) = $\dfrac{\binom{4}{3}+\binom{5}{3}+\binom{6}{3}}{\binom{15}{3}} = \dfrac{4+10+20}{455} = .0747$

c $\binom{4}{1}\binom{5}{1}\binom{6}{1} = \dfrac{120}{455} = .2637$

d To examine exactly one, a 75 watt bulb must be chosen first. (6 ways to accomplish this). To examine exactly two, we must choose another wattage first, then a 75 watt. (9×6 ways). Following the pattern, for exactly three, $9 \times 8 \times 6$ ways; for four, $9 \times 8 \times 7 \times 6$; for five, $9 \times 8 \times 7 \times 6 \times 6$.

P(examine at least 6 bulbs) = 1 – P(examine 5 or less)
= 1 – P(examine exactly 1 or 2 or 3 or 4 or 5)
= 1 – [P(one) + P(two) + … + P(five)]

$$= 1 - \left[\frac{6}{15} + \frac{9\times 6}{15\times 14} + \frac{9\times 8\times 6}{15\times 14\times 13} + \frac{9\times 8\times 7\times 6}{15\times 14\times 13\times 12} + \frac{9\times 8\times 7\times 6\times 6}{15\times 14\times 13\times 12\times 11}\right]$$

= 1 – [.4 + .2571 + .1582 + .0923 + .0503]
= 1 - .9579 = .0421

39.

a We want to choose all of the 5 cordless, and 5 of the 10 others, to be among the first 10 serviced, so the desired probability is $\dfrac{\binom{5}{5}\binom{10}{5}}{\binom{15}{10}} = \dfrac{252}{3003} = .0839$

(continued)

b Isolating one group, say the cordless phones, we want the other two groups represented in the last 5 serviced. So we choose 5 of the 10 others, except that we don't want to include the outcomes where the last five are all the same.

So we have $\dfrac{\binom{10}{5}-2}{\binom{15}{5}}$. But we have three groups of phones, so the desired probability is $\dfrac{3\cdot\left[\binom{10}{5}-2\right]}{\binom{15}{5}} = \dfrac{3(250)}{3003} = .2498$.

c We want to choose 2 of the 5 cordless, 2 of the 5 cellular, and 2 of the corded phones: $\dfrac{\binom{5}{2}\binom{5}{2}\binom{5}{2}}{\binom{15}{6}} = \dfrac{1000}{5005} = .1998$

40.

a If the A's are distinguishable from one another, and similarly for the B's, C's and D's, then there are 12! Possible chain molecules. Six of these are:

$A_1A_2A_3B_2C_3C_1D_3C_2D_1D_2B_3B_1$, $A_1A_3A_2B_2C_3C_1D_3C_2D_1D_2B_3B_1$
$A_2A_1A_3B_2C_3C_1D_3C_2D_1D_2B_3B_1$, $A_2A_3A_1B_2C_3C_1D_3C_2D_1D_2B_3B_1$
$A_3A_1A_2B_2C_3C_1D_3C_2D_1D_2B_3B_1$, $A_3A_2A_1B_2C_3C_1D_3C_2D_1D_2B_3B_1$

These 6 (=3!) differ only with respect to ordering of the 3 A's. In general, groups of 6 chain molecules can be created such that within each group only the ordering of the A's is different. When the A subscripts are suppressed, each group of 6 "collapses" into a single molecule (B's, C's and D's are still distinguishable). At this point there are $\frac{12!}{3!}$ molecules. Now suppressing subscripts on the B's, C's and D's in turn gives ultimately $\frac{12!}{(3!)^4} = 369{,}600$ chain molecules.

b Think of the group of 3 A's as a single entity, and similarly for the B's, C's, and D's. Then there are 4! Ways to order these entities, and thus 4! Molecules in which the A's are contiguous, the B's, C's, and D's are also. Thus, P(all together) = $\frac{4!}{369.600} = .00006494$.

Chapter 2

41.

a P(at least one F among 1st 3) = 1 – P(no F's among 1st 3)

$$= 1 - \frac{4\times3\times2}{8\times7\times6} = 1 - \frac{24}{336} = 1 - .0714 = .9286$$

An alternative method to calculate P(no F's among 1st 3) would be to choose none of the females and 3 of the 4 males, as follows:

$$\frac{\binom{4}{0}\binom{4}{3}}{\binom{8}{3}} = \frac{4}{56} = .0714$$, obviously producing the same result.

b P(all F's among 1st 5) = $\frac{\binom{4}{4}\binom{4}{1}}{\binom{8}{5}} = \frac{4}{56} = .0714$

c P(orderings are different) = 1 – P(orderings are the same for both semesters) = 1 – (# orderings such that the orders are the same each semester)/(total # of possible orderings for 2 semesters)

$$= 1 - \frac{8\times7\times6\times5\times4\times3\times2\times1}{(8\times7\times6\times5\times4\times3\times2\times1\times)\times(8\times7\times6\times5\times4\times3\times2\times1)} = .99997520$$

42. Seats:

1	2	3	4	5	6

P(J&P in 1&2) $= \frac{2\times1\times4\times3\times2\times1}{6\times5\times4\times3\times2\times1} = \frac{1}{15} = .0667$

P(J&P next to each other) = P(J&P in 1&2) + … + P(J&P in 5&6)

$$= 5\times\frac{1}{15} = \frac{1}{3} = .333$$

P(at least one H next to his W) = 1 – P(no H next to his W)

We count the # of ways of no H next to his W as follows:

if orderings without a H-W pair in seats #1 and 3 and no H next to his W = 6* × 4 × 1* × 2# × 1 × 1 = 48

*= pair, #=can't put the mate of seat #2 here or else a H-W pair would be in #5 and 6.

# of orderings without a H-W pair in seats #1 and 3, and no H next to his W = 6 × 4 × 2# × 2 × 2 × 1 = 192

#= can't be mate of person in seat #1 or #2.

So, # of seating arrangements with no H next to W = 48 + 192 = 240

(continued)

And P(no H next to his W) = = $\frac{240}{6\times5\times4\times3\times2\times1} = \frac{1}{3}$, so

P(at least one H next to his W = 1 - $\frac{1}{3} = \frac{2}{3}$

43. # of 10 high straights = 4x4x4x4x4 (4 – 10's, 4 – 9's , etc)

P(10 high straight) = $\frac{4^5}{\binom{52}{5}} = \frac{1024}{2{,}598{,}960} = .000394$

P(straight) = $10\times\frac{4^5}{\binom{52}{5}} = .003940$ (Multiply by 10 because there are 10 different card values that could be high: Ace, King, etc.)

There are only 40 straight flushes (10 in each suit), so

P(straight flush) = $\frac{40}{\binom{52}{5}} = .00001539$

44. $\binom{n}{k} = \frac{n!}{k!(n-k)!} = \frac{n!}{(n-k)!k!} = \binom{n}{n-k}$

The number of subsets of size k = the number of subsets of size n-k, because to each subset of size k there corresponds exactly one subset of size n-k (the n-k objects not in the subset of size k).

Chapter 2

Section 2.4

45.

a $P(A) = .15 + .10 + .10 + .10 = .45$

$P(B) = .10 + .15 = .25$

$P(A \cap B) = .10$

b $$P(A|B) = \frac{P(A \cap B)}{P(B)} = \frac{.10}{.25} = .40$$

Knowing that the car is black, the probability that it has an automatic transmission is .40.

$$P(B|A) = \frac{P(A \cap B)}{P(A)} = \frac{.10}{.45} = .2222$$

Knowing that the car has an automatic transmission, the probability that it is black is .2222.

c $$P(A|C) = \frac{P(A \cap C)}{P(C)} = \frac{.15}{.30} = .50$$

The probability that the car has automatic transmission, knowing that the car is white is .50.

$$P(A|C') = \frac{P(A \cap C')}{P(C')} = \frac{.15}{.70} = .2143$$

Knowing that the car is not white, the probability that it has an automatic transmission is .2143.

46. Let event A be that the individual is more than 6 feet tall. Let event B be that the individual is a professional basketball player. Then $P(A|B)$ = the probability of the individual being more than 6 feet tall, knowing that the individual is a professional basketball player, and $P(B|A)$ = the probability of the individual being a professional basketball player, knowing that the individual is more than 6 feet tall. $P(A|B)$ will be larger. Most professional BB players are tall, so the probability of an individual in that reduced sample space being more than 6 feet tall is very large. The number of individuals that are pro BB players is small in relation to the # of males more than 6 feet tall.

47.

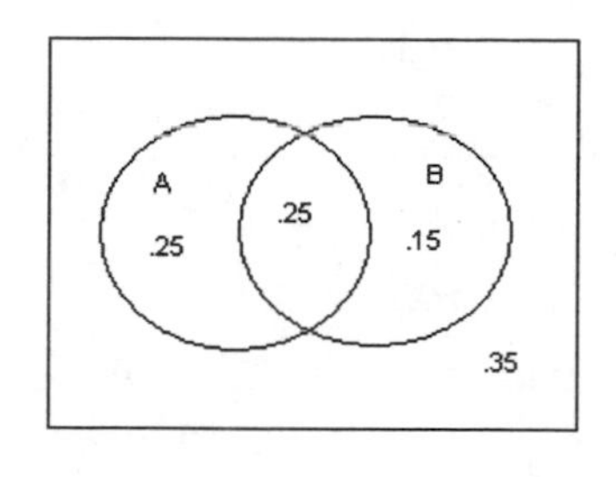

a $\quad$ P(B | A) = $\dfrac{P(A \cap B)}{P(A)} = \dfrac{.25}{.50} = .50$

b $\quad$ P(B′ | A) = $\dfrac{P(A \cap B')}{P(A)} = \dfrac{.25}{.50} = .50$

c $\quad$ P(A | B) = $\dfrac{P(A \cap B)}{P(B)} = \dfrac{.25}{.40} = .6125$

d $\quad$ P(A′ | B) = $\dfrac{P(A' \cap B)}{P(B)} = \dfrac{.15}{.40} = .3875$

e $\quad$ P(A | A∪B) = $\dfrac{P[A \cap (A \cup B)]}{P(A \cup B)} = \dfrac{.50}{.65} = .7692$

48.

a $\quad$ $P(A_2 | A_1) = \dfrac{P(A_1 \cap A_2)}{P(A_1)} = \dfrac{.06}{.12} = .50$

b $\quad$ $P(A_1 \cap A_2 \cap A_3 | A_1) = \dfrac{.01}{.12} = .0833$

c $\quad$ We want P[(exactly one) | (at least one)].

P(at least one) $= P(A_1 \cup A_2 \cup A_3)$

$= .12 + .07 + .05 - .06 - .03 - .02 + .01 = .14$

Also notice that the intersection of the two events is just the 1st event, since "exactly one" is totally contained in "at least one."

So P[(exactly one) | (at least one)]= $\dfrac{.04 + .01}{.14} = .3571$

49. The first desired probability is P(both bulbs are 75 watt | at least one is 75 watt).

P(at least one is 75 watt) = 1 – P(none are 75 watt)

$$= 1 - \frac{\binom{9}{2}}{\binom{15}{2}} = 1 - \frac{36}{105} = \frac{69}{105}.$$

Notice that P[(both are 75 watt)∩(at least one is 75 watt)]

$$= P(\text{both are 75 watt}) = \frac{\binom{6}{2}}{\binom{15}{2}} = \frac{15}{105}.$$

So P(both bulbs are 75 watt | at least one is 75 watt) = $\dfrac{\frac{15}{105}}{\frac{69}{105}} = \dfrac{15}{69} = .2174$

Second, we want P(same rating | at least one NOT 75 watt).

P(at least one NOT 75 watt) = 1 – P(both are 75 watt)

$$= 1 - \frac{15}{105} = \frac{90}{105}.$$

Now, P[(same rating)∩(at least one not 75 watt)] = P(both 40 watt or both 60 watt).

P(both 40 watt or both 60 watt) = $\dfrac{\binom{4}{2} + \binom{5}{2}}{\binom{15}{2}} = \dfrac{16}{105}$

Now, the desired conditional probability is $\dfrac{\frac{16}{105}}{\frac{90}{105}} = \dfrac{16}{90} = .1778$

Chapter 2

50.

a $P(M \cap LS \cap PR) = .05$, directly from the table of probabilities

b $P(M \cap Pr) = P(M,Pr,LS) + P(M,Pr,SS) = .05+.07=.12$

c P(SS) = sum of 9 probabilities in SS table = 56, P(LS) = 1 = .56 = .44

d $P(M) = .08+.07+.12+.10+.05+.07 = .49$

$P(Pr) = .02+.07+.07+.02+.05+.02 = .25$

e $$P(MISS \cap Pl) = \frac{P(M \cap SS \cap Pl)}{P(SS \cap Pl)} = \frac{.08}{.04+.08+.03} = .533$$

f $$P(SSIM \cap Pl) = \frac{P(SS \cap M \cap Pl)}{P(M \cap Pl)} = \frac{.08}{.08+.10} = .444$$

P(LSIM Pl) = 1 - P(SSIM Pl) = 1 - .444 = .556

51.

a $P(\text{R from } 1^{st} \cap \text{R from } 2^{nd}) = P(\text{R from } 2^{nd} \mid \text{R from } 1^{st}) \bullet P(\text{R from } 1^{st})$

$$= \frac{8}{11} \bullet \frac{6}{10} = .436$$

b P(same numbers) = P(both selected balls are the same color)

$$= \text{P(both red) + P(both green)} = .436 + \frac{4}{11} \bullet \frac{4}{10} = .581$$

52. Let A_1 be the event that #1 fails and A_2 be the event that #2 fails. We assume that $P(A_1) = P(A_2) = q$ and that $P(A_1 \mid A_2) = P(A_2 \mid A_1) = r$. Then one approach is as follows:

$P(A_1 \cap A_2) = P(A_2 \mid A_1) \bullet P(A_1) = rq = .01$

$P(A_1 \cup A_2) = P(A_1 \cap A_2) + P(A_1' \cap A_2) + P(A_1 \cap A_2') = rq + 2(1-r)q = .07$

These two equations give 2q - .01 = .07, from which q = .04 and r = .25. Alternatively, with $t = P(A_1' \cap A_2) = P(A_1 \cap A_2')$, t + .01 + t = .07, implying t = .03 and thus q = .04 without reference to conditional probability.

53. $$P(B \mid A) = \frac{P(A \cap B)}{P(A)} = \frac{P(B)}{P(A)} \text{ (since B is contained in A, } A \cap B = B)$$

$$= \frac{.05}{.60} = .0833$$

54. $P(A_1) = .22$, $P(A_2) = .25$, $P(A_3) = .28$, $P(A_1 \cap A_2) = .11$, $P(A_1 \cap A_3) = .05$, $P(A_2 \cap A_3) = .07$, $P(A_1 \cap A_2 \cap A_3) = .01$

a $$P(A_2 | A_1) = \frac{P(A_1 \cap A_2)}{P(A_1)} = \frac{.11}{.22} = .50$$

b $$P(A_2 \cap A_3 | A_1) = \frac{P(A_1 \cap A_2 \cap A_3)}{P(A_1)} = \frac{.01}{.22} = .0455$$

c $$P(A_2 \cup A_3 \mid A_1) = \frac{P[A_1 \cap (A_2 \cup A_3)]}{P(A_1)} = \frac{P[(A_1 \cap A_2) \cup (A_1 \cap A_3)]}{P(A_1)}$$

$$= \frac{P(A_1 \cap A_2) + P(A_1 \cap A_3) - P(A_1 \cap A_2 \cap A_3)}{P(A_1)} = \frac{.15}{.22} = .682$$

d $$P(A_1 \cap A_2 \cap A_3 \mid A_1 \cup A_2 \cup A_3) = \frac{P(A_1 \cap A_2 \cap A_3)}{P(A_1 \cup A_2 \cup A_3)} = \frac{.01}{.53} = .0189$$

This is the probability of being awarded all three projects given that at least one project was awarded.

55.

a $$P(A \quad B) = P(B|A) \bullet P(A) = \frac{2\times1}{4\times3} \times \frac{2\times1}{6\times5} = .0111$$

b P(two other H's next to their wives I J and M together in the middle)

$$\frac{P[(H - W.or.W - H) and (J - M.or.M - J) and (H - W.or.W - H)]}{P(J - M.or.M - J.in.the.middle)}$$

$$\text{numerator} = \frac{4\times1\times2\times1\times2\times1}{6\times5\times4\times3\times2\times1} = \frac{16}{6!}$$

$$\text{denominator} = \frac{4\times3\times2\times1\times2\times1}{6\times5\times4\times3\times2\times1} = \frac{48}{6!}$$

$$\text{so the desired probability} = \frac{16}{48} = \frac{1}{3}.$$

c P(all H's next to W's I J & M together)

= P(all H's next to W's – including J&M)/P(J&M together)

$$= \frac{\frac{6\times1\times4\times1\times2\times1}{6!}}{\frac{5\times2\times1\times4\times3\times2\times1}{6!}} = \frac{48}{240} = .2$$

56. If P(B|A) > P(B), then P(B'|A) < P(B').

Proof by contradiction. Assume P(B'|A) ≥ P(B').

Then 1 – P(B|A) ≥ 1 – P(B).

– P(B|A) ≥ – P(B).

P(B|A) ≤ P(B).

This contradicts the initial condition, therefore P(B'|A) < P(B').

57. $$P(A \mid B)+P(A' \mid B)=\frac{P(A\cap B)}{P(B)}+\frac{P(A'\cap B)}{P(B)}=\frac{P(A\cap B)+P(A'\cap B)}{P(B)}=\frac{P(B)}{P(B)}=1$$

58. $$P(A\cup B \mid C)=\frac{P[(A\cup B)\cap C)}{P(C)}=\frac{P[(A\cap C)\cup(B\cap C)]}{P(C)}$$

$$=\frac{P(A\cap C)+P(B\cap C)-P(A\cap B\cap C)}{P(C)}$$

= P(A|C) + P(B|C) – P(A ∩ B | C)

59.

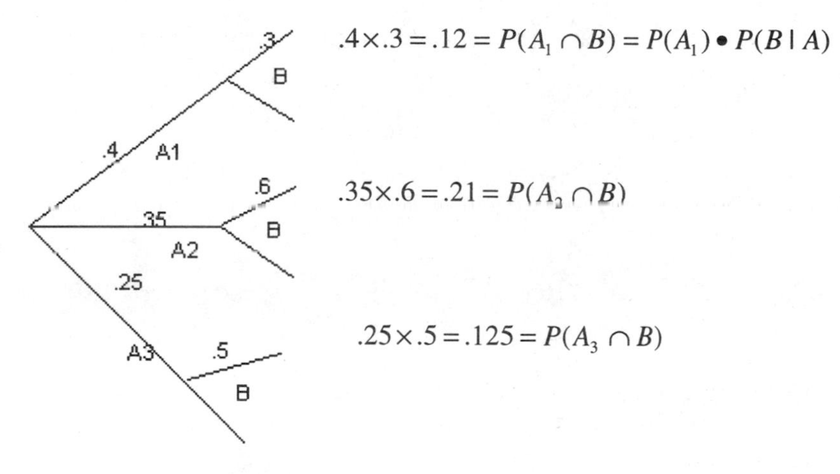

$.4\times.3=.12=P(A_1\cap B)=P(A_1)\bullet P(B\mid A)$

$.35\times.6=.21=P(A_2\cap B)$

$.25\times.5=.125=P(A_3\cap B)$

a $P(A_2 \cap B) = .21$

b $P(B) = P(A_1 \cap B) + P(A_2 \cap B) + P(A_3 \cap B) = .455$

c $P(A_1|B) = \dfrac{P(A_1\cap B)}{P(B)}=\dfrac{.12}{.455}=.264$

$P(A_2|B) = \dfrac{.21}{.455}=.462$, $P(A_3|B) = 1 - .264 - .462 = .274$

60.

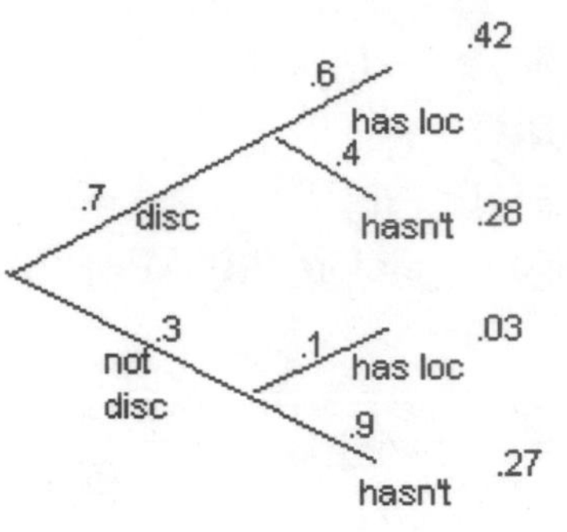

a P(not disc I has loc) = $\dfrac{P(not.disc \cap has.loc)}{P(has.loc)} = \dfrac{.03}{.03+.42} = .067$

b P(disc I no loc) = $\dfrac{P(disc \cap no.loc)}{P(no.loc)} = \dfrac{.28}{.55} = .509$

61. P(0 def in sample I 0 def in batch) = 1

P(0 def in sample I 1 def in batch) = $\dfrac{\binom{9}{2}}{\binom{10}{2}} = .800$

P(1 def in sample I 1 def in batch) = $\dfrac{\binom{9}{1}}{\binom{10}{2}} = .200$

P(0 def in sample I 2 def in batch) = $\dfrac{\binom{8}{2}}{\binom{10}{2}} = .622$

P(1 def in sample I 2 def in batch) = $\dfrac{\binom{2}{1}\binom{8}{1}}{\binom{10}{2}} = .356$

P(2 def in sample I 2 def in batch) = $\dfrac{1}{\binom{10}{2}} = .022$

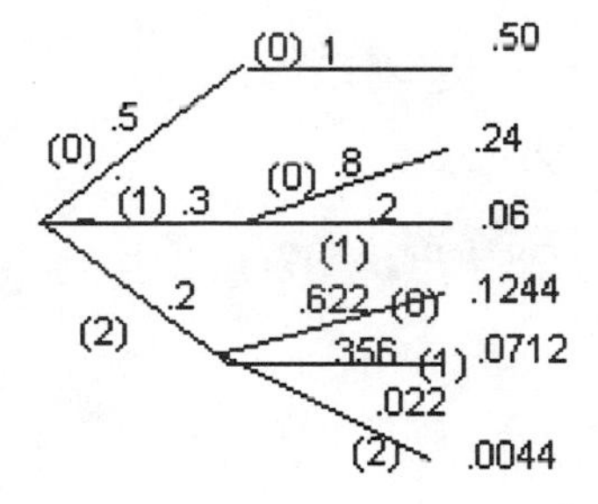

a P(0 def in batch | 0 def in sample) = $\dfrac{.5}{.5+.24+.1244} = .578$

P(1 def in batch | 0 def in sample) = $\dfrac{.24}{.5+.24+.1244} = .278$

P(2 def in batch | 0 def in sample) = $\dfrac{.1244}{.5+.24+.1244} = .144$

b P(0 def in batch | 1 def in sample) = 0

P(1 def in batch | 1 def in sample) = $\dfrac{.06}{.06+.0712} = .457$

P(2 def in batch | 1 def in sample) = $\dfrac{.0712}{.06+.0712} = .543$

62. Using a tree diagram, B = basic, D = deluxe, W = warranty purchase, W' = no warranty

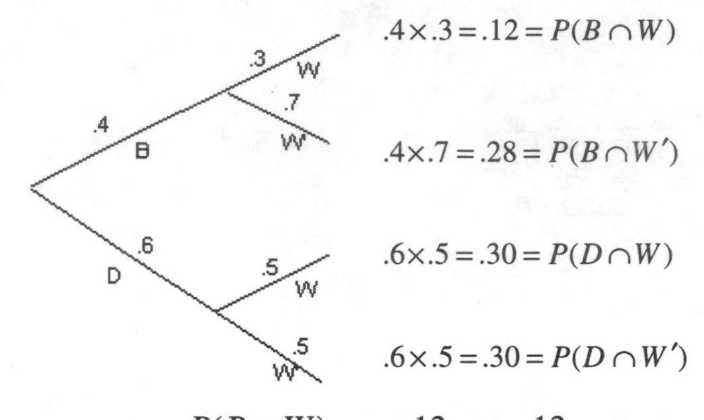

We want P(B|W) = $\dfrac{P(B \cap W)}{P(W)} = \dfrac{.12}{.30 + .12} = \dfrac{.12}{.42} = .2857$

63.

a

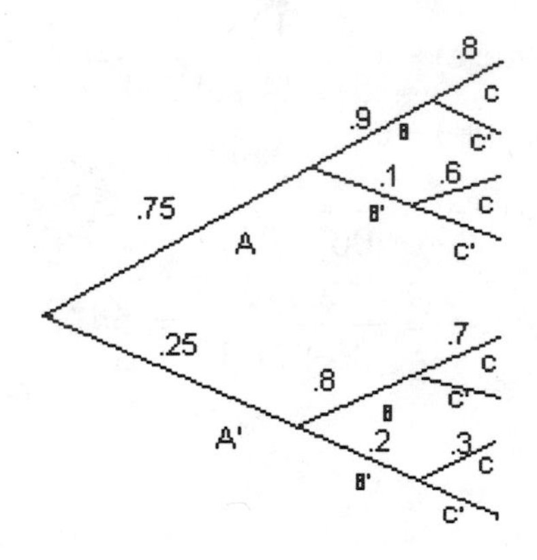

b P(A ∩ B ∩ C) = .75 × .9 × .8 = .5400

c P(B ∩ C) = P(A ∩ B ∩ C) + P(A′ ∩ B ∩ C)

.5400+.25×.8×.7 = .6800

d P(C) = P(A ∩ B ∩ C) + P(A′ ∩ B ∩ C) + P(A ∩ B′ ∩ C) + P(A′ ∩ B′ ∩ C)

= .54+.045+.14+.015 = .74

e P(A|B ∩ C) = $\dfrac{P(A \cap B \cap C)}{P(B \cap C)} = \dfrac{.54}{.68} = .7941$

Chapter 2

64.

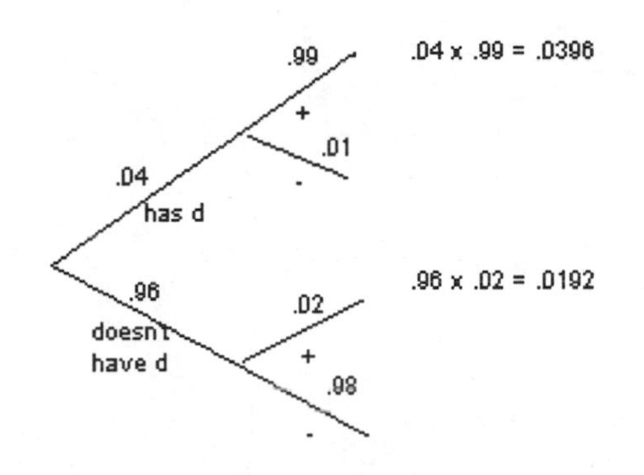

a P(+) = .0588

b P(has d | +) = $\frac{.0396}{.0588} = .6735$

c P(doesn't have d | -) = $\frac{.9408}{.9412} = .9996$

65.

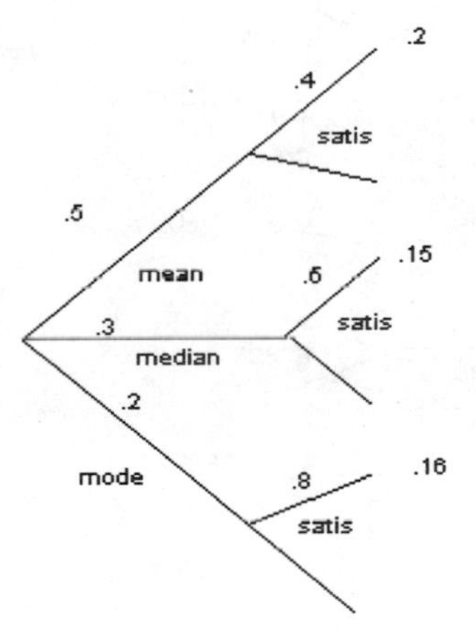

P(satis) = .51

P(mean | satis) = $\frac{.2}{.51} = .3922$

P(median | satis) = .2941

P(mode | satis) = .3137

So Mean (and not Mode!) is the most likely author, while Median is least.

66.

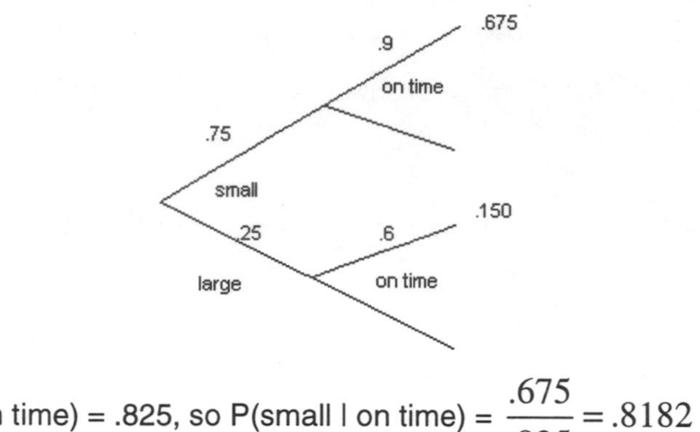

P(on time) = .825, so P(small | on time) = $\dfrac{.675}{.825} = .8182$

67.

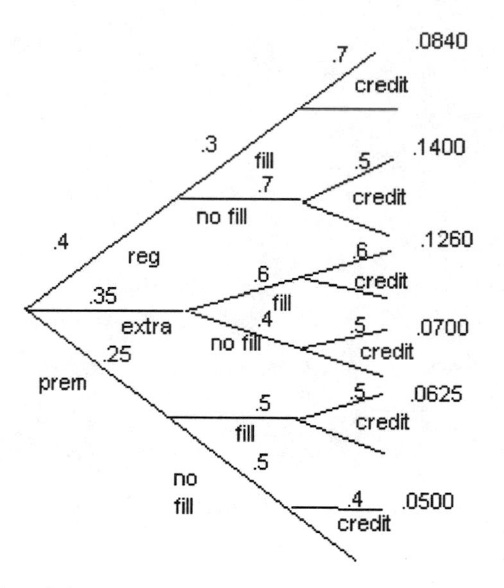

a P(U ∩ F ∩ Cr) = .1260

b P(Pr ∩ NF ∩ Cr) = .05

c P(Pr ∩ Cr) = .0625 + .05 = .1125

d P(F ∩ Cr) = .0840 + .1260 + .0625 = .2725

e P(Cr) = .5325

f P(PR | Cr) = $\dfrac{P(\Pr \cap Cr)}{P(Cr)} = \dfrac{.1125}{.5325} = .2113$

Chapter 2

Section 2.5

68. Using the definition, two events A and B are independent if $P(A|B) = P(A)$;

$P(A|B) = .6125$; $P(A) = .50$; $.6125 \neq .50$, so A and B are dependent.

Using the multiplication rule, the events are independent if $P(A \cap B) = P(A) \bullet P(B)$;

$P(A \cap B) = .25$; $P(A) \bullet P(B) = (.5)(.4) = .2$. $.25 \neq .2$, so A and B are dependent.

69.

a Because A and B are independent events, A′ and B′ are also independent. Then $P(B' \mid A') = P(B') = 1 - P(B) = 1 - .3 = .7$

b $P(A \cup B) = P(A) + P(B) - P(A \cap B)$

$= P(A) + P(B) - P(A)P(B) = .2 + .3 - (.2)(.3) = .44.$

c

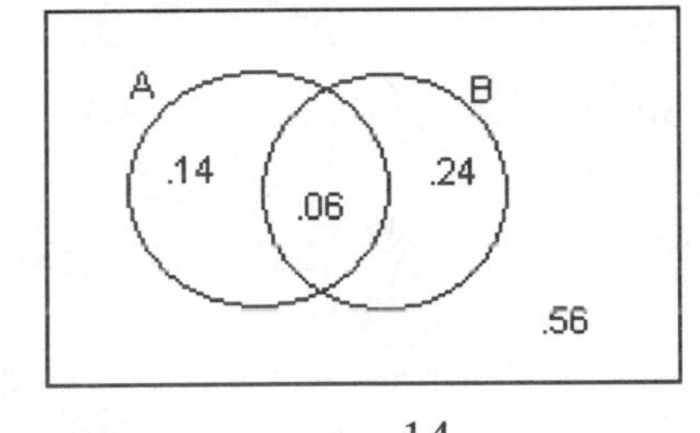

$$P[(A \cap B') \mid (A \cup B)] = \frac{.14}{.44} = .3182$$

70. $P(A_1 \cap A_2) = .11$, $P(A_1) \bullet P(A_2) = .055$. A_1 and A_2 are not independent.

$P(A_1 \cap A_3) = .05$, $P(A_1) \bullet P(A_3) = .0616$. A_1 and A_3 are not independent.

$P(A_2 \cap A_3) = .07$, $P(A_1) \bullet P(A_3) = .07$. A_2 and A_3 are independent.

71. $P(A' \cap B) = P(B) - P(A \cap B) = P(B) - P(A) \bullet P(B) = [1 - P(A)] \bullet P(B) = P(A') \bullet P(B)$.

Alternatively, $P(A' \mid B) = \dfrac{P(A' \cap B)}{P(B)} = \dfrac{P(B) - P(A \cap B)}{P(B)}$

$$= \frac{P(B) - P(A) \cdot P(B)}{P(B)} = 1 - P(A) = P(A').$$

72. Using subscripts to differentiate between the selected individuals,

$P(O_1 \cap O_2) = P(O_1) \bullet P(O_2) = (.44)(.44) = .1936$

$P(\text{two individuals match}) = P(A_1 \cap A_2) + P(B_1 \cap B_2) + P(AB_1 \cap AB_2) + P(O_1 \cap O_2)$

$= .42^2 + .10^2 + .04^2 + .44^2 = .3816$

Chapter 2

73. Let event E be the event that an error was signaled incorrectly. Then P(E) = .01.

For 10 independent points, $P(E_1 \cap E_2 \cap \ldots \cap E_{10}) = P(E_1)\, P(E_2) \ldots P(E_{10})$

$$= [P(E)]^{10} = (.01)^{10}$$

Similarly, for 25 points, the desired probability is $= [P(E)]^{25} = (.01)^{25}$

74. P(no error on any particular question) = .9, so P(no error on any of the 10 questions) $=(.9)^{10} = .3487$. Then P(at least one error) $= 1 - (.9)^{10} = .6513$. For **p** replacing .1, the two probabilities are $(1\text{-}\mathbf{p})^n$ and $1 - (1\text{-}\mathbf{p})^n$.

75. Let q denote the probability that a rivet is defective.

a P(seam need rework) = .14 = 1 – P(seam doesn't need rework)

= 1 – P(no rivets are defective)

= 1 – P(1st isn't def ∩ … ∩ 25th isn't def)

$= 1 - (1 - q)^{25}$, so $.86 = (1 - q)^{25}$, $1 - q = (.86)^{1/25}$, and thus

q = 1 = .99399 = .00601.

b The desired condition is $.10 = 1 - (1 - q)^{25}$, i.e. $(1 - q)^{25} = .90$, from which q = 1 - .99579 = .00421.

76. P(at least one opens) = 1 – P(none open) $= 1 - (.05)^5 = .99999969$

P(at least one fails to open) = 1 = P(all open) $= 1 - (.95)^5 = .2262$

77. Let A_1 = older pump fails, A_2 = newer pump fails, and $x = P(A_1 \cap A_2)$. Then $P(A_1) = .10 + x$, $P(A_2) = .05 + x$, and $x = P(A_1 \cap A_2) = P(A_1) \bullet P(A_2) = (.10 + x)(.05 + x)$. The resulting quadratic equation, $x^2 - .85x + .005 = 0$, has roots x = .0059 and x = .8441. Hopefully the smaller root is the actual probability of system failure.

78. P(system works) = P(1 – 2 works ∪ 3 – 4 works)

= P(1 – 2 works) + P(3 – 4 works) - P(1 – 2 works ∩ 3 – 4 works)

= P(1 works ∪ 2 works) + P(3 works ∩ 4 works) – P(1 – 2) • P(3 – 4)

= (.9+.9-.81) + (.9)(.9) – (.9+.9-.81)(.9)(.9)

= .99 + .81 - .8019 = .9981

79. P(both detect the defect) = 1 – P(at least one doesn't)

= 1 - .2 = .8

a P(1st detects ∩ 2nd doesn't) = P(1st detects) – P(1st does ∩ 2nd does)

= .9 - .8 = .1

Similarly, P(1st doesn't ∩ 2nd does) = .1, so P(exactly one does) = .1 + .1 = .2

b P(neither detects a defect) = 1 – [P(both do) + P(exactly 1 does)]

= 1 – [.8+.2] = 0

so P(all 3 escape) = (0)(0)(0) = 0.

80. P(pass) = .60

a (.60)(.60)(.60) = .216

b 1 – P(all pass) = 1 - .216 = .784

c P(exactly one passes) = (.60)(.40)(.40) + (.40)(.60)(.40) + (.40)(.40)(.60) = .288

d P(# pass $\leq$ 1) = P(0 pass) + P(exactly one passes) = $(.4)^3$ + .288 = .352

e P(3 pass I 1 or more pass) =

$$= \frac{P(3.pass \cap \geq 1.pass)}{P(\geq 1.pass)} = \frac{P(3.pass)}{P(\geq 1.pass)} = \frac{.216}{.936} = .231$$

81.

a Let D_1 = detection on 1st fixation, D_2 = detection on 2nd fixation.

P(detection in at most 2 fixations) = $P(D_1) + P(D_1' \cap D_2)$

$= P(D_1) + P(D2 \mid D1')P(D_1)$

$= p + p(1 - p) = p(2 - p)$.

b Define $D_1, D_2, \ldots, D_n$ as in **a**. Then P(at most n fixations)

$= P(D_1) + P(D_1' \cap D_2) + P(D_1' \cap D_2' \cap D_3) + \ldots + P(D_1' \cap D_2' \cap \ldots \cap D_{n-1}' \cap D_n)$

$= p + p(1 - p) + p(1 - p)^2 + \ldots + p(1 - p)^{n-1}$

$$= p[1 + (1 - p) + (1 - p)^2 + \ldots + (1 - p)^{n-1}] = p \bullet \frac{1-(1-p)^n}{1-(1-p)} = 1-(1-p)^n$$

Alternatively, P(at most n fixations) = 1 – P(at least n+1 are req'd)

= 1 – P(no detection in 1st n fixations)

$= 1 - P(D_1' \cap D_2' \cap \ldots \cap D_n')$

$= 1 - (1 - p)^n$

c P(no detection in 3 fixations) = $(1 - p)^3$

d P(passes inspection) = P({not flawed} $\cup$ {flawed and passes})

= P(not flawed) + P(flawed and passes)

= .9 + P(passes I flawed)$\bullet$ P(flawed) = $.9+(1 - p)^3(.1)$

e P(flawed I passed) = $\dfrac{P(flawed \cap passed)}{P(passed)} = \dfrac{.1(1-p)^3}{.9+.1(1-p)^3}$

For p = .5, P(flawed I passed) = $\dfrac{.1(.5)^3}{.9+.1(.5)^3} = .0137$

82.

a $P(A) = \frac{2000}{10,000} = .02$, P(B) = P(A ∩ B) + P(A′ ∩ B)

$= P(B|A)\,P(A) + P(B|A')\,P(A') = \frac{1999}{9999} \bullet (.2) + \frac{2000}{9999} \bullet (.8) = .2$

P(A ∩ B) = .039984; since P(A ∩ B) ≠ P(A)P(B), the events are not independent.

b P(A ∩ B) = .04. Very little difference. Yes.

c P(A) = P(B) = .2, P(A)P(B) = .04, but P(A ∩ B) = P(B|A)P(A) = $\frac{1}{9} \cdot \frac{2}{10} = .0222$, so the two numbers are quite different.

In **a**, the sample size is small relative to the "population" size, while here it is not.

83. P(system works) = P(1 – 2 works ∩ 3 – 4 – 5 – 6 works ∩ 7 works)
= P(1 – 2 works) • P(3 – 4 – 5 – 6 works) •P(7 works)
= (.99) (.9639) (.9) = .8588

With the subsystem in example 2.36 connected in parallel to this subsystem,
P(system works) = .8588+.9639 – (.8588)(.9639) = .9949

84.

a For route #1, P(late) = P(stopped at 2 or 3 or 4 crossings)
= 1 – P(stopped at 0 or 1) = $1 - [.9^4 + 4(.9)^3(.1)] = .0523$

For route #2, P(late) = P(stopped at 1 or 2 crossings)
= 1 – P(stopped at none) = 1 - .81 = .19

thus route #1 should be taken.

b P(4 crossing route | late) = $\frac{P(4crossing \cap late)}{P(late)}$

$$= \frac{(.5)(.0523)}{(.5)(.0523) + (.5)(.19)} = .216$$

85.

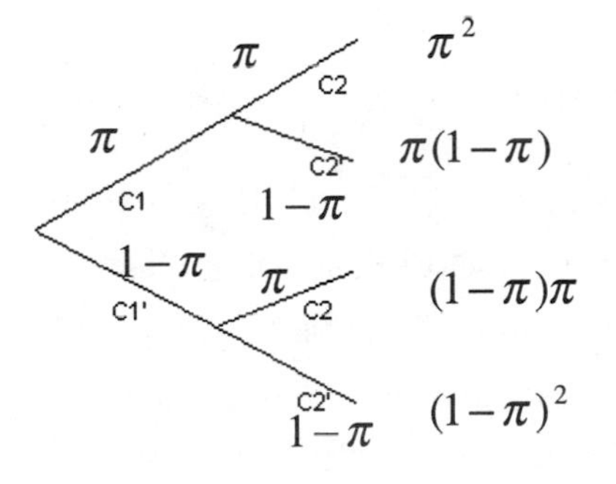

P(at most 1 is lost) = 1 – P(both lost)
$= 1 - \pi^2$
P(exactly 1 lost) = $2\pi(1 - \pi)$
P(exactly 1 | at most 1) = $\dfrac{P(exactly 1)}{P(at.most 1)} = \dfrac{2\pi(1-\pi)}{1-\pi^2}$

Supplementary Exercises

86.

a $\binom{20}{3} = 1140$ **b** $\binom{19}{3} = 969$

c # having at least 1 of the 10 best = 1140 - # of crews having none of 10 best
$= 1140 - \binom{10}{3} = 1140 - 120 = 1020$

d P(best will not work) = $\dfrac{969}{1140} = .85$

87.

a P(21 ft^3) = .17 + .13 = .30; P(Ice) = .12 + .17 + .24 = .53

b P(21 ft^3 | Ice) = $\dfrac{P(21 ft^3 \cap Ice)}{P(Ice)} = \dfrac{.17}{.53} = .3208 \neq .30$ = P(21 ft^3)
(The two events are not independent)

c P(at least 20 ft^3) = .17+.13+.24+.16=.70.
P(No Ice ∩ at least 20 ft^3) =.13+.16=.29,
so P(No Ice | at least 20 ft^3) = .29/.70 = .4143

88.

a The only way he will have one type of forms left is if they are all course substitution forms. He must choose all 6 of the withdrawal forms to pass to a subordinate. The desired probability is $\frac{\binom{6}{6}}{\binom{10}{6}} = .00476$

b He can start with the wd forms: W-C-W-C or with the cs forms: C-W-C-W:

# of ways: $6 \times 4 \times 5 \times 3 + 4 \times 6 \times 3 \times 5 = 2(360) = 720$;

The total # ways to arrange the four forms: $10 \times 9 \times 8 \times 7 = 5040$.

The desired probability is 720/5040 = .1429

89. $P(A)\,P(B) = .0002$; $P(A \cup B) = P(A) + P(B) - P(A \cap B) = P(A) + P(B) - .0002 = .03$
So $P(B) = .0302 - P(A)$, and $P(A)[\,.0302 - P(A)] = .0002$. Multiplying through, $[P(A)]^2 - .0302P(A) + .0002 = 0$, or $x^2 - .0302x + .0002 = 0$.

Solving yields $x \approx \frac{.0302 \pm .0106}{2}$, so $P(A) \approx .0204$, $P(B) \approx .0098$

90.

a $(.8)(.8)(.8) = .512$

b

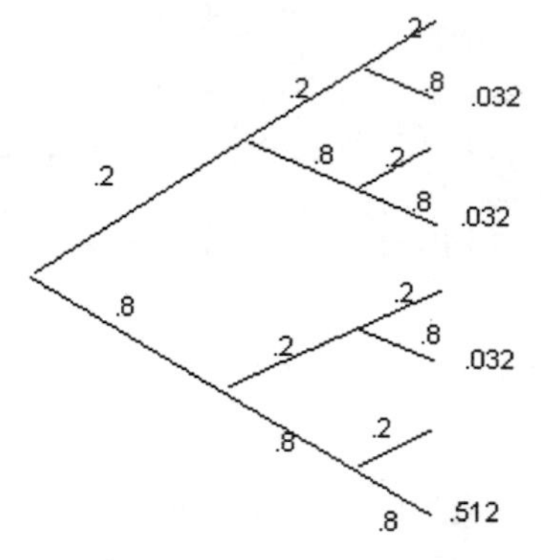

.512+.032+.023+.023 = .608

c $P(1 \text{ sent} \mid 1 \text{ received}) = \frac{P(1 sent \cap 1 received)}{P(1 received} = \frac{.4256}{.5432} = .7835$

91.

a There are 5×4×3×2×1 = 120 possible orderings, so P(BCDEF) = $\frac{1}{120} = .0083$

b # orderings in which F is 3rd = 4×3×1*×2×1 = 24, (* because F must be here), so P(F 3rd) = $\frac{24}{120} = .2$

c P(F last) = $\dfrac{4\times3\times2\times1\times1}{120} = .2$

92. P(F hasn't heard after 10 times) = P(not on #1 ∩ not on #2 ∩ … ∩ not on #10)

$$= \left(\frac{4}{5}\right)^{10} = .1074$$

93.

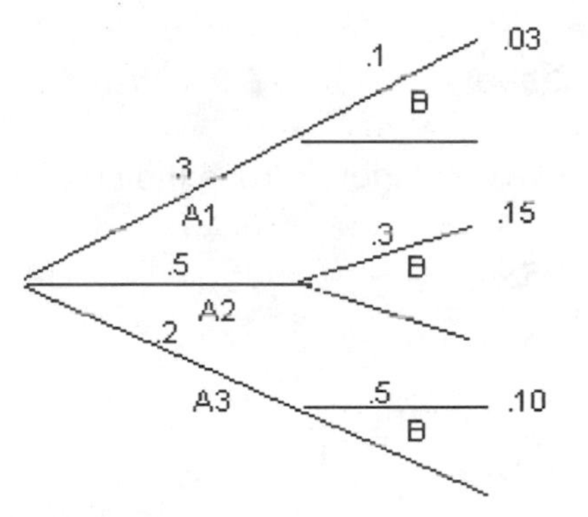

where B = at least 1 citation

P(at least 1 citation) = .03+.15+.10 = .28

P(good I at least 1 citation) = $\frac{.03}{.28} = .1071$

P(medium I at least 1 citation) = $\frac{.15}{.28} = .5357$

94. P(good I none in 3 yrs) = $\dfrac{P(good \cap none.in.3.yrs)}{P(none.in.3.yrs)}$

$$= \frac{P(good \cap none)}{P(good \cap none) + P(medium \cap none) + P(bad \cap none)}$$

$$= \frac{(.3)(.9)^3}{(.3)(.9)^3 + (.5)(.7)^3 + (.2)(59)^3)} = \frac{.2187}{.4152} = .5267$$

95. When three experiments are performed, there are 3 different ways in which detection can occur on exactly 2 of the experiments: (i) #1 and #2 and not #3 (ii) #1 and not #2 and #3; (iii) not#1 and #2 and #3. If the impurity is present, the probability of exactly 2 detections in three (independent) experiments is (.8)(.8)(.2) + (.8)(.2)(.8) + (.2)(.8)(.8) = .384. If the impurity is absent, the analogous probability is 3(.1)(.1)(.9) = .027. Thus

P(present | detected in exactly 2 out of 3) = $\dfrac{P(det\,ected.in.exactly.2 \cap present)}{P(det\,ected.in.exactly.2)}$

$$= \frac{(.384)(.4)}{(.384)(.4) + (.027)(.6)} = .905$$

96. P(exactly 1 selects category #1 | all 3 are different)

$$= \frac{P(exactly.1.selects\#1 \cap all.are.different)}{P(all.are.different)}$$

Denominator = $\dfrac{6\times5\times4}{6\times6\times6} = \dfrac{5}{9} = .5556$

Numerator = 3 P(contestant #1 selects category #1 and the other two select two different categories)

$$= 3\times\frac{1\times5\times4}{6\times6\times6} = \frac{5\times4\times3}{6\times6\times6}$$

The desired probability is then $\dfrac{5\times4\times3}{6\times5\times4} = \dfrac{1}{2} = .5$

97.

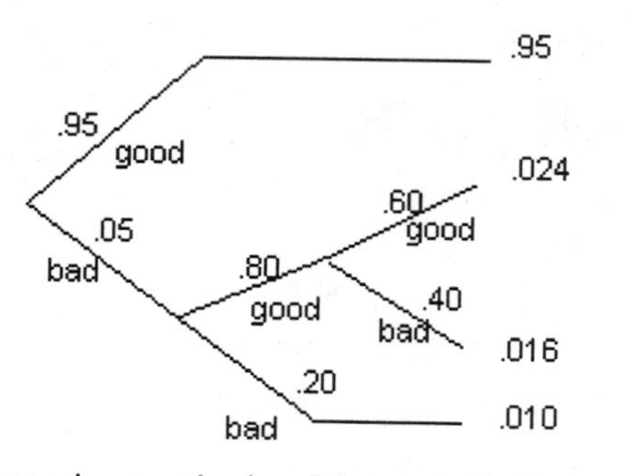

a P(pass inspection) = P(pass initially ∪ passes after recrimping)

= P(pass initially) + P(fails initially ∩ goes to recrimping ∩ is corrected after recrimping.

= .95 + (.05)(.80)(.60) (following path "bad-good-good" on tree diagram)

= .974

(continued)

b P(needed no recrimping | passed inspection) = $\dfrac{P(passed.initially)}{P(passed.inspection)}$

$$= \frac{.95}{.974} = .9754$$

98.

a P(both + ve) = P(carrier $\cap$ both + ve) + P(not a carrier $\cap$ both + ve)

=P(both + ve | carrier) x P(carrier)

+ P(both + ve | not a carrier) x P(not a carrier)

= $(.90)^2(.01) + (.05)^2(.99) = .01058$

P(both – ve) = $(.10)^2(.01) + (.95)^2(.99) = .89358$

P(tests agree) = .01058 + .89358 = .90416

b P(carrier | both + ve) = $\dfrac{P(carrier \cap both.positive)}{P(both.positive)} = \dfrac{(.90)^2(.01)}{.01058} = .7656$

99. Let A = 1st functions, B = 2nd functions, so P(B) = .9, P(A $\cup$ B) = .96, P(A $\cap$ B)=.75. Thus, P(A $\cup$ B) = P(A) + P(B) - P(A $\cap$ B) = P(A) + .9 - .75 = .96, implying P(A) = .81.

This give P(B | A) = $\dfrac{P(B \cap A)}{P(A)} = \dfrac{.75}{.81} = .926$

100. P(E_1 $\cap$ late) = P(late | E_1)P(E_1) = (.02)(.40) = .008

101.

a The law of total probability gives

P(late) = $\sum_{i=1}^{3} P(late \mid E_i) \cdot P(E_i)$

= (.02)(.40) + (.01)(.50) + (.05)(.10) = .018

b P(E_1' | on time) = 1 – P(E_1 | on time)

$$= 1 - \frac{P(E_1 \cap on.time)}{P(on.time)} = 1 - \frac{(.98)(.4)}{.982} = .601$$

102. Let B denote the event that a component needs rework. Then

$$P(B) = \sum_{i=1}^{3} P(B| A_i) \cdot P(A_i) = (.05)(.50) + (.08)(.30) + (.10)(.20) = .069$$

Thus $P(A_1 \mid B) = \dfrac{(.05)(.50)}{.069} = .362$

$P(A_2 \mid B) = \dfrac{(.08)(.30)}{.069} = .348$

$P(A_3 \mid B) = \dfrac{(.10)(.20)}{.069} = .290$

103.

a P(all different) = $\dfrac{(365)(364)...(356)}{(365)^{10}} = .883$

P(at least two the same) = 1 - .883 = .117

b P(at least two the same) = .476 for k=22, and = .507 for k=23

c P(at least two have the same SS number) = 1 – P(all different)

$$= 1 - \frac{(1000)(999)...(991)}{(1000)^{10}}$$

= 1 - .956 = .044

Thus P(at least one "coincidence") = P(BD coincidence $\cup$ SS coincidence)
= .117 + .044 – (.117)(.044) = .156

104.

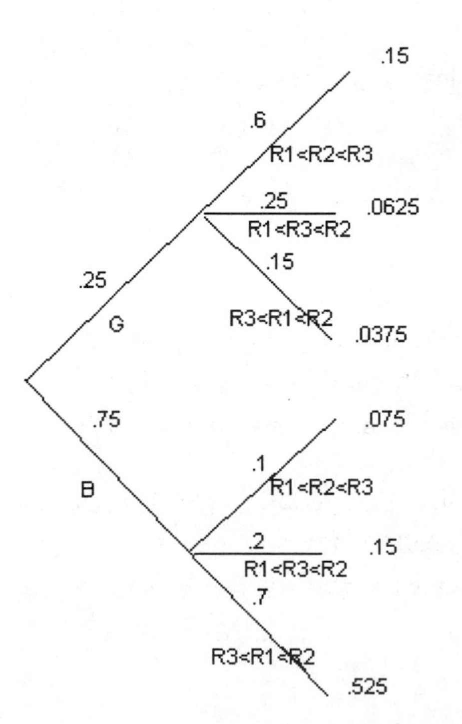

a $P(G \mid R_1 < R_2 < R_3) = \dfrac{.15}{.15+.075} = .67$, $P(B \mid R_1 < R_2 < R_3) = .33$, classify as granite.

b $P(G \mid R_1 < R_3 < R_2) = \dfrac{.0625}{.2125} = .2941 < .05$, so classify as basalt.

$P(G \mid R_3 < R_1 < R_2) = \dfrac{.0375}{.5625} = .0667$, so classify as basalt.

c P(erroneous classif) = P(B classif as G) + P(G classif as B)
= P(classif as G | B)P(B) + P(classif as B | G)P(G)
= $P(R_1 < R_2 < R_3 \mid B)(.75) + P(R_1 < R_3 < R_2 \text{ or } R_3 < R_1 < R_2 \mid G)(.25)$
= (.10)(.75) + (.25 + .15)(.25) = .175

d For what values of p will $P(G \mid R_1 < R_2 < R_3) > .5$, $P(G \mid R_1 < R_3 < R_2) > .5$, $P(G \mid R_3 < R_1 < R_2) > .5$?

$P(G \mid R_1 < R_2 < R_3) = \dfrac{.6p}{.6p+.1(1-p)} = \dfrac{.6p}{.1+.5p} > .5 \text{ iff } p > \dfrac{1}{7}$

$P(G \mid R_1 < R_3 < R_2) = \dfrac{.25p}{.25p+.2(1-p)} > .5 \text{ iff } p > \dfrac{4}{9}$

$P(G \mid R_3 < R_1 < R_2) = \dfrac{.15p}{.15p+.7(1-p)} > .5 \text{ iff } p > \dfrac{14}{17}$ (most restrictive)

If $p > \dfrac{14}{17}$ always classify as granite

105. P(detection by the end of the nth glimpse) = 1 – P(not detected in 1st n)
$= 1 - P(G_1' \cap G_2' \cap \ldots \cap G_n') = 1 - P(G_1')P(G_2') \ldots P(G_n')$
$= 1 - (1 - p_1)(1 - p_2) \ldots (1 - p_n) = 1 - \pi_{i=1}^{n}(1-p_i)$

106.

a P(walks on 4th pitch) = P(1st 4 pitches are balls) = $(.5)^4 = .0625$

b P(walks on 6th) = P(2 of the 1st 5 are strikes, #6 is a ball)
= P(2 of the 1st 5 are strikes)P(#6 is a ball)
= $[10(.5)^5](.5) = .15625$

c P(Batter walks) = P(walks on 4th) + P(walks on 5th) + P(walks on 6th)
= .0625 + .15625 + .15625 = .375

d P(first batter scores while no one is out) = P(first 4 batters walk)
$=(.375)^4 = .0198$

107.

a P(all in correct room) = $\frac{1}{4\times3\times2\times1} = \frac{1}{24} = .0417$

b The 9 outcomes which yield incorrect assignments are: 2143, 2341, 2413, 3142, 3412, 3421, 4123, 4321, and 4312, so P(all incorrect) = $\frac{9}{24} = .375$

108.

a P(all full) = $P(A \cap B \cap C)$ = (.6)(.5)(.4) = .12

P(at least one isn't full) = 1 − P(all full) = 1 - .12 = .88

b P(only NY is full) = $P(A \cap B' \cap C') = P(A)P(B')P(C')$ = .18

Similarly, P(only Atlanta is full) = .12 and P(only LA is full) = .08

So P(exactly one full) = .18 + .12 + .08 = .38

109. Note: s = 0 means that the very first candidate interviewed is hired. Each entry below is the candidate hired for the given policy and outcome.

Outcome	s=0	s=1	s=2	s=3	**Outcome**	s=0	s=1	s=2	s=3
1234	1	4	4	4	3124	3	1	4	4
1243	1	3	3	3	3142	3	1	4	2
1324	1	4	4	4	3214	3	2	1	4
1342	1	2	2	2	3241	3	2	1	1
1423	1	3	3	3	3412	3	1	1	2
1432	1	2	2	2	3421	3	2	2	1
2134	2	1	4	4	4123	4	1	3	3
2143	2	1	3	3	4132	4	1	2	2
2314	2	1	1	4	4213	4	2	1	3
2341	2	1	1	1	4231	4	2	1	1
2413	2	1	1	3	4312	4	3	1	2
2431	2	1	1	1	4321	4	3	2	1

s	0	1	2	3
P(hire#1)	$\frac{6}{24}$	$\frac{11}{24}$	$\frac{10}{24}$	$\frac{6}{24}$

So s = 1 is best.

Chapter 2

110. P(at least one occurs) = 1 – P(none occur)

$= 1 - (1 - p_1)(1 - p_2)(1 - p_3)(1 - p_4)$

$= p_1p_2(1 - p_3)(1 - p_4) + \ldots + (1 - p_1)(1 - p_2)p_3p_4$

$+ (1 - p_1)p_2p_3p_4 + \ldots + p_1p_2p_3(1 - p_4) + p_1p_2p_3p_4$

111. $P(A_1)$ = P(draw slip 1 or 4) = ½

$P(A_2)$ = P(draw slip 2 or 4) = ½

$P(A_3)$ = P(draw slip 3 or 4) = ½

$P(A_1 \cap A_2)$ = P(draw slip 4) = ¼

$P(A_2 \cap A_3)$ = P(draw slip 4) = ¼

$P(A_1 \cap A_3)$ = P(draw slip 4) = ¼

Hence $P(A_1 \cap A_2) = P(A_1)P(A_2)$ = ¼, $P(A_2 \cap A_3) = P(A_2)P(A_3)$ = ¼,

$P(A_1 \cap A_3) = P(A_1)P(A_3)$ = ¼, thus there exists pairwise independence

$P(A_1 \cap A_2 \cap A_3)$ = P(draw slip 4) = ¼ ≠ 1/8 = $P(A_1)p(A_2)P(A_3)$, so the events are not mutually independent.

Chapter 3

1.

S:	FFF	SFF	FSF	FFS	FSS	SFS	SSF	SSS
X:	0	1	1	1	2	2	2	3

2. X = 1 if a randomly selected book is non-fiction and X = 0 otherwise
X = 1 if a randomly selected executive is a female and X = 0 otherwise
X = 1 if a randomly selected driver has automobile insurance and X = 0 otherwise

3. M = the difference between the large and the smaller outcome with possible values 0, 1, 2, 3, 4, or 5; W = 1 if he sum of the two resulting numbers is even and W = 0 otherwise, a Bernoulli random variable.

4. In my perusal of a zip code directory, I found no 00000, nor did I find any zip codes with four zeros, a fact which was not obvious. Thus possible X values are 2, 3, 4, 5 (and not 0 or 1). X = % for the outcome 15213, X = 4 for the outcome 44074, and X = 3 for 94322.

5. No. In the experiment in which a coin is tossed repeatedly until a H results, let Y = 1 if the experiment terminates with at most 5 tosses and Y = 0 otherwise. The sample space is infinite, yet Y has only two possible values.

6. Possible X values are1, 2, 3, 4, … (all positive integers)

Outcome:	RL	AL	RAARL	RRRRL	AARRL
X:	2	2	5	5	5

7.

- **a** Possible values are 0, 1, 2, …, 12; discrete
- **b** With N = # on the list, values are 0, 1, 2, … , N; discrete
- **c** Possible values are 1, 2, 3, 4, … ; discrete
- **d** $\{ x: 0 < x < \infty \}$ if we assume that a rattlesnake can be arbitrarily short or long; not discrete
- **e** With c = amount earned per book sold, possible values are 0, c, 2c, 3c, … , 10,000c; discrete
- **f** $\{ y: 0 < y < 14 \}$ since 0 is the smallest possible pH and 14 is the largest possible pH; not discrete
- **g** With m and M denoting the minimum and maximum possible tension, respectively, possible values are $\{ x: m < x < M \}$; not discrete
- **h** Possible values are 3, 6, 9, 12, 15, … -- i.e. 3(1), 3(2), 3(3), 3(4), …giving a first element, etc,; discrete

8. Y = 3 : SSS; Y = 4: FSSS; Y = 5: FFSSS, SFSSS;
Y = 6: SSFSSS, SFFSSS, FSFSSS, FFFSSS;
Y = 7: SSFFS, SFSFSSS< SFFFSSS, FSSFSSS, FSFFSSS, FFSFSSS, FFFFSSS

Chapter 3

9.

a Returns to 0 can occur only after an even number of tosses; possible S values are 2, 4, 6, 8, ...(i.e. 2(1), 2(2), 2(3), 2(4),...) an infinite sequence, so x is discrete.

b Now a return to 0 is possible after any number of tosses greater than 1, so possible values are 2, 3, 4, 5, ... (1+1,1+2, 1+3, 1+4, ..., an infinite sequence) and X is discrete

10.

a T = total number of pumps in use at both stations. Possible values: 0, 1, 2, 3, 4, 5, 6, 7, 8, 9, 10

b X: -4, -3, -2, -1, 0, 1, 2, 3, 4, 5, 6

c U: 0, 1, 2, 3, 4, 5, 6

d Z: 0, 1, 2

Section 3.2

11.

a

x	4	6	8
P(x)	.45	.40	.15

b

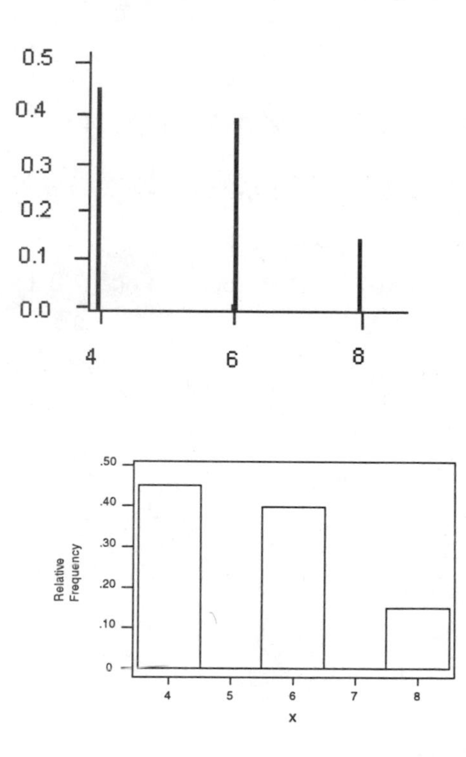

12.

a $\Sigma p(x) \neq 1$ for 1st and 3rd p(x)

b $P(2 \le X \le 4) = p(2) + p(3) + p(4) = .5$, $P(X \le 2) = p(0) + p(1) + p(2) = .6$,
$P(X \neq 0) = 1 - P(X = 0) = 1 - p(0) = .6$

c $\sum_{x=0}^{4} p(x) = c[(5-0) + (5-1) + \ldots + (5-4)] = 15c = 1 \Rightarrow c = \frac{1}{15}$

13.

a $P(X \le 3) = p(0) + p(1) + p(2) + p(3) = .10+.15+.20+.25 = .70$

b $P(X < 3) = P(X \le 2) = p(0) + p(1) + p(2) = .45$

c $P(3 \le X) = p(3) + p(4) + p(5) + p(6) = .55$

d $P(2 \le X \le 5) = p(2) + p(3) + p(4) + p(5) = .71$

e The number of lines not in use is $6 - X$, so $6 - X = 2$ is equivalent to $X = 4$, $6 - X = 3$ to $X = 3$, and $6 - X = 4$ to $X = 2$.
Thus we desire $P(2 \le X \le 4) = p(2) + p(3) + p(4) = .65$

f $6 - X \ge 4$ if $6 - 4 \ge X$, i.e. $2 \ge X$, or $X \le 2$, and $P(X \le 2) = .10+.15+.20 = .45$

14.

a $\sum_{y=1}^{5} p(y) = K[1 + 2 + 3 + 4 + 5] = 15K = 1 \Rightarrow K = \frac{1}{15}$

b $P(Y \le 3) = p(1) + p(2) + p(3) = \frac{6}{15} = .4$

c $P(2 \le Y \le 4) = p(2) + p(3) + p(4) = \frac{9}{15} = .6$

d $\sum_{y=1}^{5} \left(\frac{y^2}{50}\right) = \frac{1}{50}[1+4+9+16+25] = \frac{55}{50} \neq 1$; No

15.

a (1,2) (1,3) (1,4) (1,5) (2,3) (2,4) (2,5) (3,4) (3,5) (4,5)

b $P(X = 0) = p(0) = P[\{(3,4)\ (3,5)\ (4,5)\}] = \frac{3}{10} = .3$

$P(X = 2) = p(2) = P[\{(1,2)\}] = \frac{1}{10} = .1$

$P(X = 1) = p(1) = 1 - [p(0) + p(2)] = .60$, and $p(x) = 0$ if $x \neq 0, 1, 2$

c $F(0) = P(X \le 0) = P(X = 0) = .30$
$F(1) = P(X \le 1) = P(X = 0 \text{ or } 1) = .90$
$F(2) = P(X \le 2) = 1$

The c.d.f. is

$$F(x) = \begin{cases} 0 & x < 0 \\ .30 & 0 \le x < 1 \\ .90 & 1 \le x < 2 \\ 1 & 2 \le x \end{cases}$$

16.

a

x	Outcomes	p(x)	
0	FFFF	$(.7)^4$	=.2401
1	FFFS,FFSF,FSFF,SFFF	$4[(.7)^3(.3)]$	=.4116
2	FFSS,FSFS,SFFS,FSSF,SFSF,SSFF	$6[(.7)^2(.3)^2]$	=.2646
3	FSSS, SFSS,SSFS,SSSF	$4[(.7)(.3)^3]$	=.0756
4	SSSS	$(.3)^4$	=.0081

b

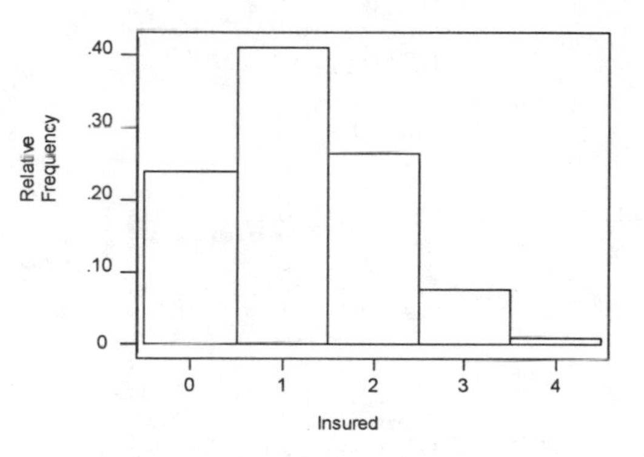

c p(x) is largest for X = 1

d $P(X \geq 2) = p(2) + p(3) + p(4) = .2646+.0756+.0081 = .3483$
This could also be done using the complement.

17.

a $P(2) = P(Y = 2) = P(1^{st}$ 2 batteries are acceptable)
$= P(AA) = (.9)(.9) = .81$

b $p(3) = P(Y = 3) = P(UAA \text{ or } AUA) = (.1)(.9)^2 + (.1)(.9)^2 = 2[(.1)(.9)^2] = .162$

c The fifth battery must be an A, and one of the first four must also be an A. Thus, $p(5) = P(AUUUA \text{ or } UAUUA \text{ or } UUAUA \text{ or } UUUAA) = 4[(.1)^3(.9)^2] = .00324$

d $P(Y = y) = p(y) = P($the yth is an A and so is exactly one of the first $y - 1)$
$=(y - 1)(.1)^{y-2}(.9)^2,\ y = 2,3,4,5,\ldots$

18.

a $p(1) = P(M = 1) = P[(1,1)] = \frac{1}{36}$

$p(2) = P(M = 2) = P[(1,2) \text{ or } (2,1) \text{ or } (2,2)] = \frac{3}{36}$

$p(3) = P(M = 3) = P[(1,3) \text{ or } (2,3) \text{ or } (3,1) \text{ or } (3,2) \text{ or } (3,3)] = \frac{5}{36}$

Similarly, $p(4) = \frac{7}{36}$, $p(5) = \frac{9}{36}$, and $p(6) = \frac{11}{36}$

b $F(m) =$ 0 for $m < 1$, $\frac{1}{36}$ for $1 \le m < 2$,

$$F(m) = \begin{cases} 0 & m < 1 \\ \frac{1}{36} & 1 \le m < 2 \\ \frac{4}{36} & 2 \le m < 3 \\ \frac{9}{36} & 3 \le m < 4 \\ \frac{16}{36} & 4 \le m < 5 \\ \frac{25}{36} & 5 \le m < 6 \\ 1 & m \ge 6 \end{cases}$$

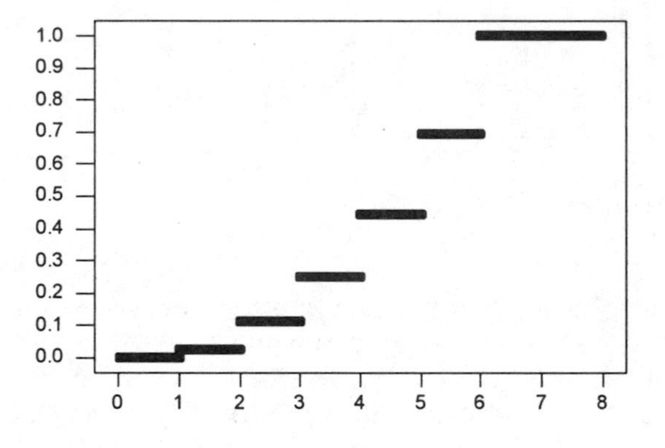

19. Let A denote the type O+ individual (type O positive blood) and B, C, D, the other 3 individuals. Then $p(1) – P(Y = 1) = P(\text{A first}) = \frac{1}{4} = .25$

$p(2) = P(Y = 2) = P(\text{B, C, or D first and A next}) = \frac{3}{4} \cdot \frac{1}{3} = \frac{1}{4} = .25$

$p(4) = P(Y = 3) = P(\text{A last}) = \frac{3}{4} \cdot \frac{2}{3} \cdot \frac{1}{2} = \frac{1}{4} = .25$

So $p(3) = 1 – (.25+.25+.25) = .25$

20. $P(0) = P(Y = 0) = P(\text{both arrive on Wed.}) = (.3)(.3) = .09$
$P(1) = P(Y = 1) = P[(W,Th) \text{or} (Th,W) \text{or} (Th,Th)] = (.3)(.4) + (.4)(.3) + (.4)(.4) = .40$
$P(2) = P(Y = 2) = P[(W,F) \text{or} (Th,F) \text{or} (F,W) \text{ or } (F,Th) \text{ or } (F,F)] = .32$
$P(3) = 1 – [.09 + .40 + .32] = .19$

21. The jumps in F(x) occur at x = 0, 1, 2, 3, 4, 5, and 6, so we first calculate F( ) at each of these values:

$F(0) = P(X \le 0) = P(X = 0) = .10$
$F(1) = P(X \le 1) = p(0) + p(1) = .25$
$F(2) = P(X \le 2) = p(0) + p(1) + p(2) = .45$
$F(3) = .70$, $F(4) = .90$, $F(5) = .96$, and $F(6) = 1$.

The c.d.f. is

$$F(x) = \begin{cases} .00 & x < 0 \\ .10 & 0 \le x < 1 \\ .25 & 1 \le x < 2 \\ .45 & 2 \le x < 3 \\ .70 & 3 \le x < 4 \\ .90 & 4 \le x < 5 \\ .96 & 5 \le x < 6 \\ 1.00 & 6 \le x \end{cases}$$

Then $P(X \le 3) = F(3) = .70$, $P(X < 3) = P(X \le 2) = F(2) = .45$, $P(3 \le X) = 1 - P(X \le 2)$ $= 1 - F(2) = 1 - .45 = .55$, and $P(2 \le X \le 5) = F(5) - F(1) = .96 - .25 = .71$

22.

a $P(X = 2) = .39 - .19 = .20$
b $P(X > 3) = 1 - .67 = .33$
c $P(2 \le X \le 5) = .92 - .19 = .78$
d $P(2 < X < 5) = .92 - .39 = .53$

23.

a Possible X values are those values at which F(x) jumps, and the probability of any particular value is the size of the jump at that value. Thus we have:

x	1	3	4	6	12
p(x)	.30	.10	.05	.15	.40

b $P(3 \le X \le 6) = F(6) - F(3-) = .60 - .30 = .30$
$P(4 \le X) = 1 - P(X < 4) = 1 - F(4-) = 1 - .40 = .60$

24. $P(0) = P(Y = 0) = P(\text{B first}) = p$
$P(1) = P(Y = 1) = P(\text{G first, then B}) = P(GB) = (1 - p)p$
$P(2) = P(Y = 2) = P(GGB) = (1 - p)^2p$
Continuing, $p(y) = P(Y=y) = P(\text{y G's and then a B}) = (1 - p)^y p$ for y = 0,1,2,3,…

25.

a Possible X values are 1, 2, 3, ...

$P(1) = P(X = 1) = P(\text{return home after just one visit}) = \frac{1}{3}$

$P(2) = P(X = 2) = P(\text{second visit and then return home}) = \frac{2}{3} \cdot \frac{1}{3}$

$P(3) = P(X = 3) = P(\text{three visits and then return home}) = \left(\frac{2}{3}\right)^2 \cdot \frac{1}{3}$

In general $p(x) = \left(\frac{2}{3}\right)^{x-1}\left(\frac{1}{3}\right)$ for x = 1, 2, 3, ...

b The number of straight line segments is Y = 1 + X (since the last segment traversed returns Alvie to O), so as in a, $p(y) = \left(\frac{2}{3}\right)^{y-2}\left(\frac{1}{3}\right)$ for y = 2, 3, ...

c Possible Z values are 0, 1, 2, 3 , ...

$p(0) = P(\text{male first and then home}) = \frac{1}{2} \cdot \frac{1}{3} = \frac{1}{6}$,

p(1) = P(exactly one visit to a female) = P(female 1st, then home) + P(F, M, home) + P(M, F, home) + P(M, F, M, home)

$= \left(\frac{1}{2}\right)\left(\frac{1}{3}\right) + \left(\frac{1}{2}\right)\left(\frac{2}{3}\right)\left(\frac{1}{3}\right) + \left(\frac{1}{2}\right)\left(\frac{2}{3}\right)\left(\frac{1}{3}\right) + \left(\frac{1}{2}\right)\left(\frac{2}{3}\right)\left(\frac{2}{3}\right)\left(\frac{1}{3}\right)$

$= \left(\frac{1}{2}\right)\left(1 + \frac{2}{3}\right)\left(\frac{1}{3}\right) + \left(\frac{1}{2}\right)\left(\frac{2}{3}\right)\left(\frac{2}{3} + 1\right)\left(\frac{1}{3}\right) = \left(\frac{1}{2}\right)\left(\frac{5}{3}\right)\left(\frac{1}{3}\right) + \left(\frac{1}{2}\right)\left(\frac{2}{3}\right)\left(\frac{5}{3}\right)\left(\frac{1}{3}\right)$

where the first term corresponds to initially visiting a female and the second term corresponds to initially visiting a male. Similarly,

$p(2) = \left(\frac{1}{2}\right)\left(\frac{2}{3}\right)^2\left(\frac{5}{3}\right)\left(\frac{1}{3}\right) + \left(\frac{1}{2}\right)\left(\frac{2}{3}\right)^2\left(\frac{5}{3}\right)\left(\frac{1}{3}\right)$. In general,

$p(z) = \left(\frac{1}{2}\right)\left(\frac{2}{3}\right)^{2z-2}\left(\frac{5}{3}\right)\left(\frac{1}{3}\right) + \left(\frac{1}{2}\right)\left(\frac{2}{3}\right)^{2z-2}\left(\frac{5}{3}\right)\left(\frac{1}{3}\right) = \left(\frac{24}{54}\right)\left(\frac{2}{3}\right)^{2z-2}$ for z = 1, 2, 3, ...

26.

a The sample space consists of all possible permutations of the four numbers 1, 2, 3, 4:

outcome	y value	outcome	y value	outcome	y value
1234	4	2314	1	3412	0
1243	2	2341	0	3421	0
1324	2	2413	0	4132	1
1342	1	2431	1	4123	0
1423	1	3124	1	4213	1
1432	2	3142	0	4231	2
2134	2	3214	2	4312	0
2143	0	3241	1	4321	0

b Thus $p(0) = P(Y = 0) = \frac{9}{24}$, $p(1) = P(Y = 1) = \frac{8}{24}$, $p(2) = P(Y = 2) = \frac{6}{24}$,

$p(3) = P(Y = 3) = 0$, $p(3) = P(Y = 3) = \frac{1}{24}$.

27. If $x_1 < x_2$, $F(x_2) = P(X \le x_2) = P(\{X \le x_1\} \cup \{x_1 < X \le x_2\})$

$= P(X \le x_1) + P(x_1 < X \le x_2) \ge P(X \le x_1) = F(x_1)$.

$F(x_1) = F(x_2)$ when $P(x_1 < X \le x_2) = 0$.

Chapter 3

Section 3.3

28.

a $\quad E(X) = \sum_{x=0}^{4} x \cdot p(x) = (0)(.08) + (1)(.15) + (2)(.45) + (3)(.27) + (4)(.05) = 2.06$

b $\quad V(X) = \sum_{x=0}^{4} (x-2.06)^2 \cdot p(x) = (0-2.06)^2(.08) + \ldots + (4-2.06)^2(.05)$

$= .339488 + .168540 + .001620 + .238572 + .188180 = .9364$

c $\quad \sigma_x = \sqrt{.9364} = .9677$

d $\quad V(X) = \left[\sum_{x=0}^{4} x^2 \cdot p(x)\right] - (2.06)^2 = 5.1800 - 4.2436 = .9364$

29.

a $\quad E(Y) = \sum_{x=0}^{4} y \cdot p(y) = (0)(.60) + (1)(.25) + (2)(.10) + (3)(.05) = .60$

b $\quad E(100Y^2) = \sum_{x=0}^{4} 100y^2 \cdot p(y) = (0)(.60) + (100)(.25)$

$+ (400)(.10) + (900)(.05) = 110$

30. $E(Y) = .60; \qquad E(Y^2) = 1.1$

$V(Y) = E(Y^2) - [E(Y)]^2 = 1.1 - (.60)^2 = .74$

$\sigma_y = \sqrt{.74} = .8602$

$E(Y) \pm \sigma_y = .60 \pm .8602 = (-.2602, 1.4602)$ or $(0, 1)$.

$P(Y = 0) + P(Y = 1) = .85$

31.

a $\quad E(X) = (13.5)(.2) + (15.9)(.5) + (19.1)(.3) = 16.38,$

$E(X^2) = (13.5)^2(.2) + (15.9)^2(.5) + (19.1)^2(.3) = 272.298,$

$V(X) = 272.298 - (16.38)^2 = 3.9936$

b $\quad E(25X - 8.5) = 25\,E(X) - 8.5 = (25)(16.38) - 8.5 = 401$

c $\quad V(25X - 8.5) = V(25X) = (25)^2V(X) = (625)(3.9936) = 2496$

d $\quad E[h(X)] = E[X - .01X^2] = E(X) - .01E(X^2) = 16.38 - 2.72 = 13.66$

32.

a $\quad E(X^2) = \sum_{x=0}^{1} x^2 \cdot p(x) = (0^2)((1-p) + (1^2)(p) = (1)(p) = p$

b $\quad V(X) = E(X^2) - [E(X)]^2 = p - p^2 = p(1-p)$

c $\quad E(x^{79}) = (0^{79})(1-p) + (1^{79})(p) = p$

33. $E(X) = \sum_{x=1}^{\infty} x \cdot p(x) = \sum_{x=1}^{\infty} x \cdot \frac{c}{x^3} = c\sum_{x=1}^{\infty} \frac{1}{x^2}$, but it is a well-known result from the theory of infinite series that $\sum_{x=1}^{\infty} \frac{1}{x^2} < \infty$, so E(X) is finite.

34. Let h(X) denote the net revenue (sales revenue – order cost) as a function of X. Then $h_3(X)$ and $h_4(X)$ are the net revenue for 3 and 4 copies purchased, respectively. For x = 1 or 2 , $h_3(X) = 2x - 3$, but at x = 3,4,5,6 the revenue plateaus. Following similar reasoning, $h_4(X) = 2x - 4$ for x=1,2,3, but plateaus at 4 for x = 4,5,6.

x	1	2	3	4	5	6
$h_3(x)$	-1	1	3	3	3	3
$h_4(x)$	-2	0	2	4	4	4
p(x)	$\frac{1}{15}$	$\frac{2}{15}$	$\frac{3}{15}$	$\frac{4}{15}$	$\frac{3}{15}$	$\frac{2}{15}$

$E[h_3(X)] = \sum_{x=1}^{6} h_3(x)\cdot p(x) = (-1)(\frac{1}{15}) + \ldots + (3)(\frac{2}{15}) = 2.4667$

Similarly, $E[h_4(X)] = \sum_{x=1}^{6} h_4(x)\cdot p(x) = (-2)(\frac{1}{15}) + \ldots + (4)(\frac{2}{15}) = 2.6667$

Ordering 4 copies gives a slightly higher revenue, on the average.

35.

P(x)	.8	.1	.08	.02
x	0	1,000	5,000	10,000
H(x)	0	500	4,500	9,500

E[h(X)] = 600. Premium should be $100 plus expected value of damage minus deductible or $700.

36. $E(X) = \sum_{x=1}^{n} x\cdot\left(\frac{1}{n}\right) = \left(\frac{1}{n}\right)\sum_{x=1}^{n} x = \frac{1}{n}\left[\frac{n(n+1)}{2}\right] = \frac{n+1}{2}$

$E(X^2) = \sum_{x=1}^{n} x^2\cdot\left(\frac{1}{n}\right) = \left(\frac{1}{n}\right)\sum_{x=1}^{n} x^2 = \frac{1}{n}\left[\frac{n(n+1)(2n+1)}{6}\right] = \frac{(n+1)(2n+1)}{6}$

So $V(X) = \frac{(n+1)(2n+1)}{6} - \left(\frac{n+1}{2}\right)^2 = \frac{n^2-1}{12}$

37. $E[h(X)] = E\left(\frac{1}{X}\right) = \sum_{x=1}^{6}\left(\frac{1}{x}\right)\cdot p(x) = \frac{1}{6}\sum_{x=1}^{6}\frac{1}{x} = .408$, whereas $\frac{1}{3.5} = .286$, so you expect to win more if you gamble.

38. E(X) = $\sum_{x=1}^{4} x \cdot p(x)$ = 2.3, E(X^2) = 6.1, so V(X) = 6.1 – $(2.3)^2$ = .81

Each lot weighs 5 lbs, so weight left = 100 – 5x.
Thus the expected weight left is 100 – 5E(X) = 88.5,
and the variance of the weight left is
V(100 – 5X) = V(-5X) = 25V(x) = 20.25.

39.

a The line graph of the p.m.f. of –X is just the line graph of the p.m.f. of X reflected about zero, but both have the same degree of spread about their respective means, suggesting V(-X) = V(X).

b With a - -1, b = 0, V(aX + b) = V(-X) = a^2V(X).

40. V(aX + b) = $\sum_x [aX + b - E(aX + b)]^2 \cdot p(x) = \sum_x [aX + b - (a\mu + b)]^2 p(x)$

$$= \sum_x [aX - (a\mu)]^2 p(x) = a^2 \sum_x [X - \mu]^2 p(x) = a^2 V(X).$$

41.

a E[X(X-1)] = E(X^2) – E(X), $\Rightarrow$ E(X^2) = E[X(X-1)] + E(X) = 32.5

b V(X) = 32.5 – $(5)^2$ = 7.5

c V(X) = E[X(X-1)] + E(X) – $[E(X)]^2$

42. With a = 1 and b = c, E(X – c) = E(aX + b) = aE(X) + b = E(X) – c. When c = μ, E(X - μ) = E(X) - μ = μ - μ = 0, so the expected deviation from the mean is zero.

43.

a

k	2	3	4	5	10
$\frac{1}{k^2}$	.25	.11	.06	.04	.01

b $\mu = \sum_{x=0}^{6} x \cdot p(x) = 2.64, \sigma^2 = \left[\sum_{x=0}^{6} x^2 \cdot p(x)\right] - \mu^2 = 2.37, \sigma = 1.54$

Thus μ - 2σ = -.44, and μ + 2σ = 5.72, so P(|x-μ| ≥ 2σ)
= P(X is lat least 2 s.d.'s from μ)
= P(x is either ≤-.44 or ≥ 5.72) = P(X = 6) = .04.
Chebyshev's bound of .025 is much too conservative. For K = 3,4,5, and 10, P(|x-μ| ≥ kσ) = 0, here again pointing to the very conservative nature of the bound $\frac{1}{k^2}$.

(continued)

c $\mu = 0$ and $\sigma = \frac{1}{3}$, so $P(|x-\mu| \geq 3\sigma) = P(|X| \geq 1)$

$= P(X = -1 \text{ or } +1) = \frac{1}{18} + \frac{1}{18} = \frac{1}{9}$, identical to the upper bound.

d Let $p(-1) = \frac{1}{50}, p(+1) = \frac{1}{50}, p(0) = \frac{24}{25}$.

Section 3.4

44.

a $b(3;8,.6) = \binom{8}{3}(.6)^3(.4)^5 = (56)(.00221184) = .124$

b $b(5;8,.6) = \binom{8}{5}(.6)^5(.4)^3 = (56)(.00497664) = .279$

c $P(3 \leq X \leq 5) = b(3;8,.6) + b(4;8,.6) + b(5;8,.6) = .635$

d $P(1 \leq X) = 1 - P(X = 0) = 1 - \binom{12}{0}(.1)^0(.9)^{12} = 1 - (.9)^{12} = .718$

45.

a $B(4;10,.3) = .850$
b $b(4;10,.3) = B(4;10,.3) - B(3;10,.3) = .200$
c $b(6;10,.7) = B(6;10,.7) - B(5;10,.7) = .200$
d $P(2 \leq X \leq 4) = B(4;10,.3) - B(1;10,.3) = .701$
e $P(2 < X) = 1 - P(X \leq 1) = 1 - B(1;10,.3) = .851$
f $P(X \leq 1) = B(1;10,.7) = .0000$
g $P(2 < X < 6) = P(3 \leq X \leq 5) = B(5;10,.3) - B(2;10,.3) = .570$

46. $X \sim Bin(25, .05)$

a $P(X \leq 2) = B(2;25,.05) = .873$
b $P(X \geq 5) = 1 - P(X \leq 4) = 1 - B(4;25,.05) = .1 - .993 = .007$
c $P(1 \leq X \leq 4) = P(X \leq 4) - P(X \leq 0) = .993 - .277 = .716$
d $P(X = 0) = P(X \leq 0) = .277$
e $E(X) = np = (25)(.05) = 1.25$
$V(X) = np(1 - p) = (25)(.05)(.95) = 1.1875$
$\sigma_x = 1.0897$

47. X ~ Bin(6, .10)

a $P(X = 1) = \binom{n}{x}(p)^x(1-p)^{n-x} = \binom{6}{1}(.1)^1(.9)^5 = .3543$

b $P(X \geq 2) = 1 - [P(X = 0) + P(X = 1)]$.

From **a** , we know P(X = 1) = .3543, and $P(X = 0) = \binom{6}{0}(.1)^0(.9)^6 = .5314$.

Hence $P(X \geq 2) = 1 - [.3543 + .5314] = .1143$

c Either 4 or 5 goblets must be selected

i) Select 4 goblets with zero defects: $P(X = 0) = \binom{4}{0}(.1)^0(.9)^4 = .6561$.

ii) Select 4 goblets, one of which has a defect, and the 5th is good:

$$\left[\binom{4}{1}(.1)^1(.9)^3\right] \times .9 = .26244$$

So the desired probability is .6561 + .26244 = .91854

48. Let S = comes to a complete stop, so p = .2 , n = 20

a $P(X \leq 6) = B(6;20,.20) = .913$

b $P(X = 6) = b(6;20,.20) = B(6;20,.20) - B(5;20,.20) = .913 - .804 = .109$

c $P(X \geq 6) = 1 - P(X \leq 5) = 1 - B(5;20,.20) = 1 - .804 = .196$

d E(X) = (20)(.20) = 4. We expect 4 of the next 20 to stop.

49. Let S = has at least one citation. Then p = .4, n = 15

a If at least 10 have no citations (Failure), then at most 5 have had at least one (Success): $P(X \leq 5) = B(5;15,.40) = .403$

b $P(X \geq 8) = 1 - P(X \leq 7) = 1 - B(7;15,.40) = 1 - .787 = .213$

c $P(5 \leq X \leq 10) = P(X \leq 10) - P(X \leq 4) = .991 - .217 = .774$

50. X ~ Bin(10, .60)

a $P(X \geq 6) = 1 - P(X \leq 5) = 1 - B(5;20,.60) = 1 - .367 = .633$

b E(X) = np = (10)(.6) = 6; V(X) = np(1 – p) = (10)(.6)(.4) = 2.4;

$\sigma_x = 1.55$

$E(X) \pm \sigma_x = (4.45, 7.55)$.

We desire $P(5 \leq X \leq 7) = P(X \leq 7) - P(X \leq 4) = .833 - .166 = .667$

c $P(3 \leq X \leq 7) = P(X \leq 7) - P(X \leq 2) = .833 - .012 = .821$

51. Let S represent a telephone that is submitted for service while under warranty and must be replaced. Then p = P(S) = P(replaced | submitted)·P(submitted) = (.40)(.20) = .08. Thus X, the number among the company's 10 phones that must be replaced, has a binomial distribution with n = 10, p = .08, so $p(2) = P(X=2) = \binom{10}{2}(.08)^2(.92)^8 = .1478$

52. Let p denote the actual proportion of defectives in the batch, and X denote the number of defectives in the sample.

a P(the batch is accepted) = $P(X \le 2) = B(2;10,p)$

p	.01	.05	.10	.20	.25
P(accept)	1.00	.988	.930	.678	.526

b

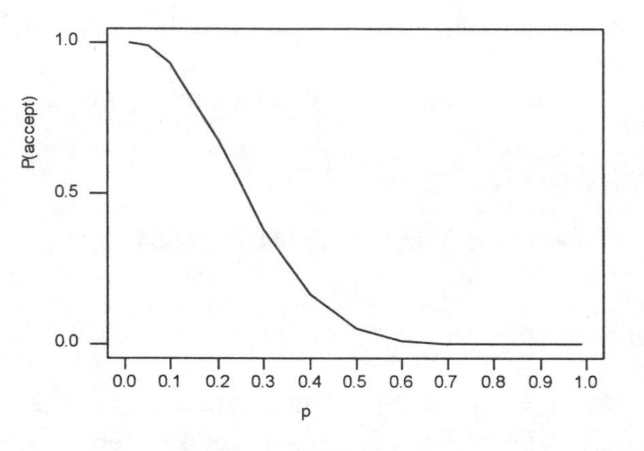

c P(the batch is accepted) = $P(X \le 1) = B(1;10,p)$

p	.01	.05	.10	.20	.25
P(accept)	.996	.914	.736	.376	.244

d P(the batch is accepted) = $P(X \le 2) = B(2;15,p)$

p	.01	.05	.10	.20	.25
P(accept)	1.00	.964	.816	.398	.236

e We want a plan for which P(accept) is high for $p \le .1$ and low for $p > .1$
The plan in **d** seems most satisfactory in these respects.

53.

a P(rejecting claim when p = .8) = B(15;25,.8) = .017

b P(not rejecting claim when p = .7) = $P(X \ge 16$ when p = .7)
= 1 - B(15;25,.7) = 1 - .189 = .811; for p = .6, this probability is
= 1 - B(15;25,.6) = 1 - .575 = .425.;

c The probability of rejecting the claim when p = .8 becomes B(14;25,.8) = .006, smaller than in **a** above. However, the probabilities of **b** above increase to .902 and .586, respectively.

54. h(x) = 1 · X + 2.25(25 – X) = 62.5 – 1.5X, so E(h(X)) = 62.5 – 1.5E(x)
= 62.5 – 1.5np – 62.5 – (1.5)(25)(.6) = $40.00

55. If topic A is chosen, when n = 2, P(at least half received)
= $P(X \ge 1) = 1 - P(X = 0) = 1 - (.1)^2 = .99$

If B is chosen, when n = 4, P(at least half received)
= $P(X \ge 2) = 1 - P(X \le 1) = 1 - (0.1)^4 - 4(.1)^3(.9) = .9963$

Thus topic B should be chosen. If p = .5, the probabilities are .75 for A and .6875 for B, so now A should be chosen.

56.

a np(1 – p) = 0 if either p = 0 (whence every trial is a failure, so there is no variability in X) or if p = 1 (whence every trial is a success and again there is no variability in X)

b $\frac{d}{dp}[np(1-p)]$ = n[(1 – p) + p(-1)] = n[1 – 2p = 0 ⇒ p = .5, which is easily seen to correspond to a maximum value of V(X).

57.

a $b(x; n, 1-p) = \binom{n}{x}(1-p)^x(p)^{n-x} = \binom{n}{n-x}(p)^{n-x}(1-p)^x$ = b(n-x; n, p)

Alternatively, P(x S's when P(S) = 1 – p) = P(n-x F's when P(F) = p), since the two events are identical), but the labels S and F are arbitrary so can be interchanged (if P(S) and P(F) are also interchanged), yielding P(n-x S's when P(S) = 1 – p) as desired.

b B(x;n,1 – p) = P(at most x S's when P(S) = 1 – p)
= P(at least n-x F's when P(F) = p)
= P(at least n-x S's when P(S) = p)
= 1 – P(at most n-x-1 S's when P(S) = p)
= 1 – B(n-x-1;n,p)

c Whenever p > .5, (1 – p) < .5 so probabilities involving X can be calculated using the results **a** and **b** in combination with tables giving probabilities only for $p \le .5$

58. Proof of E(X) = np:

$$E(X) = \sum_{x=0}^{n} x \cdot \binom{n}{x} p^x (1-p)^{n-x} = \sum_{x=1}^{n} x \cdot \frac{n!}{x!(n-x)!} p^x (1-p)^{n-x}$$

$$= \sum_{x=1}^{n} \frac{n!}{(x-1)!(n-x)!} p^x (1-p)^{n-x} = np \sum_{x=1}^{n} \frac{(n-1)!}{(x-1)!(n-x)!} p^{x-1} (1-p)^{n-x}$$

$$= np \sum_{y=0}^{n} \frac{(n-1)!}{(y)!(n-1-y)!} p^y (1-p)^{n-1-y} \text{ (y replaces x-1)}$$

$$= np \left\{ \sum_{y=0}^{n-1} \binom{n-1}{y} p^y (1-p)^{n-1-y} \right\}$$

The expression in braces is the sum over all possible values y = 0, 1, 2, … , n-1 of a binomial p.m.f. based on n-1 trials, so equals 1, leaving only np, as desired.

59.

a Although there are three payment methods, we are only concerned with S = uses a debit card and F = does not use a debit card. Thus we can use the binomial distribution. So n = 100 and p = .5. E(X) = np = 100(.5) = 50, and V(X) = 25.

b With S = doesn't pay with cash, n = 100 and p = .7, E(X) = np = 100(.7) = 70, and V(X) = 21.

60.

a Let X = the number with reservations who show, a binomial r.v. with n = 6 and p = .8. The desired probability is

P(X = 5 or 6) = b(5;6,.8) + b(6;6,.8) = .3932 + .2621 = .6553

b Let h(X) = the number of available spaces. Then

When x is:	0	1	2	3	4	5	6
H(x) is:	4	3	2	1	0	0	0

E[h(X)] = $\sum_{x=0}^{6} h(x) \cdot b(x;6,.8)$ = 4(.000) + 3(.002) = 2(.015 + 3(.082) = .277

c Possible X values are 0, 1, 2, 3, and 4. X = 0 if there are 3 reservations and none show or …or 6 reservations and none show, so

P(X = 0) = b(0;3,.8)(.1) + b(0;4,.8)(.2) + b(0;5,.8)(.3) + b(0;6,.8)(.4)
= .0080(.1) + .0016(.2) + .0003(.3) + .0001(.4) = .0013
P(X = 1) = b(1;3,.8)(.1) + … + b(1;6,.8)(.4) = .0172
P(X = 2) = .0906, P(X = 3) = .2273,
P(X = 4) = 1 – [.0013 + … + .2273] = .6636

61. When p = .5, μ = 10 and σ = 2.236, so 2σ = 4.472 and 3σ = 6.708.
The inequality |X – 10| ≥ 4.472 is satisfied if either X ≤ 5 or X ≥ 15, or P(|X - μ| ≥ 2σ) = P(X ≤ 5 or X ≥ 15) = .021 + .021 = .042.

In the case p = .75, μ = 15 and σ = 1.937, so 2σ = 3.874 and 3σ = 5.811. P(|X - 15| ≥ 3.874) = P(X ≤ 11 or X ≥ 19) = .041 + .024 = .065, whereas P(|X - 15| ≥ 5.811) = P(X ≤ 9) = .004. All these probabilities are considerably less than the upper bounds .25(for k = 2) and .11 (for k = 3) given by Chebyshev.

Section 3.5

62.

a P(X = 2) = h(2;5,6,15) = $\dfrac{\binom{6}{2}\binom{9}{3}}{\binom{15}{5}} = \dfrac{1260}{3003} = .420$

b P(X ≤ 2) = h(0;5,6,15) + h(1;5,6,15) + h(2;5,6,15) = .714

c P(X ≥ 1) = 1 – P(x = 0) = 1 - .042 = .958

d E(X) = $\dfrac{(5)(6)}{15}$ = 2, V(X) = $\dfrac{(15-5)}{(15-1)} \cdot \dfrac{(5)(6)}{15} \cdot \left(1 - \dfrac{6}{15}\right) = .857$

63.

a P(X = 1) = h(1;6,5,12) = $\dfrac{\binom{5}{1}\binom{7}{5}}{\binom{12}{6}} = .1136$

b P(X ≥ 4) = P(X = 4) + P(X = 5) = h(4;6,5,12) + h(5;6,5,12) = .0379 + .0011 = .039

c P(1 ≤ X ≤ 3) = P(X = 1) + P(X = 2) + P(X = 3) = .7955

64.

a P(X = 10) = h(10;15,30,50) = $\dfrac{\binom{30}{10}\binom{20}{5}}{\binom{50}{15}} = .2070$

b P(X ≥ 10) = h(10;15,30,50) + h(11;15,30,50) + ... + h(15;15,30,50)
= .2070+.1176+.0438+.0101+.0013+.0001 = .3799

c P(at least 10 from the same class) = P(at least 10 from second class [answer from **b**]) + P(at least 10 from first class). But "at least 10 from 1st class" is the same as "at most 5 from the second" or P(X ≤ 5).

P(X ≤ 5) = h(0;15,30,50) + h(1;15,30,50) + ... + h(5;15,30,50)
= .011697+.002045+.000227+.000150+.000001+.000000
= .01412

So the desired probability = P(x ≥ 10) + P(X ≤ 5) = .3799 + .01412 = .39402

(continued)

d E(X) = $n \cdot \frac{M}{N} = 15 \cdot \frac{30}{50} = 9$

V(X) = $\left(\frac{35}{49}\right)(9)\left(1 - \frac{30}{50}\right) = 2.5714$

σ_X = 1.6036

e Let Y = 15 – X. Then E(Y) = 15 – E(X) = 15 – 9 = 6
V(Y) = V(15 – X) – V(X) = 2.5714, so σ_Y = 1.6036

65.

a Possible values of X are 5, 6, 7, 8, 9, 10. (In order to have less than 5 of the granite, there would have to be more than 10 of the basaltic).

P(X = 5) = h(5; 15,10,20) = $\frac{\binom{10}{5}\binom{10}{10}}{\binom{20}{15}} = .0163$.

Following the same pattern for the other values, we arrive at the pmf, in table form below.

x	5	6	7	8	9	10
p(x)	.0163	.1354	.3483	.3483	.1354	.0163

b P(all 10 of one kind or the other) = P(X = 5) + P(X = 10) = .0163 + .0163 = .0326

c E(X) = $n \cdot \frac{M}{N} = 15 \cdot \frac{10}{20} = 7.5$; V(X) = $\left(\frac{5}{19}\right)(7.5)\left(1 - \frac{10}{20}\right) = .9868$; σ_X = .9934

$\mu \pm \sigma$ = 7.5 ± .9934 = (6.5066, 8.4934), so we want
P(X = 7) + P(X = 8) = .3483 + .3483 = .6966

66.

a h(x; 6,4,11)

b $6 \cdot \left(\frac{4}{11}\right) = 2.18$

67.

a h(x; 10,10,20) (the successes here are the top 10 pairs, and a sample of 10 pairs is drawn from among the 20)

b Let X = the number among the top 5 who play E-W. Then P(all of top 5 play the same direction) = P(X = 5) + P(X = 0) = h(5;10,5,20) + h(5;10,5,20)

$$= \frac{\binom{15}{5}}{\binom{20}{10}} + \frac{\binom{15}{10}}{\binom{20}{10}} = .033$$

c N = 2n; M = n; n = n
h(x;n,n,2n)

E(X) = $n \cdot \frac{n}{2n} = \frac{1}{2}n$;

V(X) =

$$\left(\frac{2n-n}{2n-1}\right) \cdot n \cdot \frac{n}{2n} \cdot \left(1 - \frac{n}{2n}\right) = \left(\frac{n}{2n-1}\right) \cdot \frac{n}{2} \cdot \left(1 - \frac{n}{2n}\right) = \left(\frac{n}{2n-1}\right) \cdot \frac{n}{2} \cdot \left(\frac{1}{2}\right)$$

68.

a h(x;10,15,50)

b When N is large relative to n, h(x;n,M,N) $\doteq b\left(x; n, \frac{M}{N}\right)$,

so h(x;10,150,500) $\doteq b(x;10,.3)$

c Using the hypergeometric model, E(X) = $10 \cdot \left(\frac{150}{500}\right) = 3$ and

V(X) = $\frac{490}{499}(10)(.3)(.7) = .982(2.1) = 2.06$

Using the binomial model, E(X) = (10)(.3) = 3, and V(X) = 10(.3)(.7) = 2.1

69.

a With S = a female child and F = a male child, let X = the number of F's before the 2nd S. Then P(X = x) = nb(x;2, .5)

b P(exactly 4 children) = P(exactly 2 males)
= nb(2;2,.5) = (3)(.0625) = .188

c P(at most 4 children) = P(X ≤ 2)

$$= \sum_{x=0}^{2} nb(x;2,.5) = .25+2(.25)(.5) + 3(.0625) = .688$$

d E(X) = $\frac{(2)(.5)}{.5} = 2$, expected number of children = E(X + 2) = E(X) + 2 = 4

70. The only possible values of X are 3, 4, and 5.

p(3) = P(X = 3) = P(first 3 are B's or first 3 are G's) = $2(.5)^3$ = .250

p(4) = P(two among the 1st three are B's and the 4th is a B) + P(two among the 1st three are G's and the 4th is a G) = $2 \cdot \binom{3}{2}(.5)^4 = .375$

p(5) = 1 – p(3) – p(4) = .375

71. This is identical to an experiment in which a single family has children until exactly 6 females have been born(since p = .5 for each of the three families), so p(x) = nb(x;6,.5) and E(X) = 6 (= 2+2+2, the sum of the expected number of males born to each one.)

72. The interpretation of "roll" here is a pair of tosses of a single player's die(two tosses by A or two by B). With S = doubles on a particular roll, p = $\frac{1}{6}$. Furthermore, A and B are really identical (each die is fair), so we can equivalently imagine A rolling until 10 doubles appear. The P(x rolls) = P(9 doubles among the first x – 1 rolls and a double on the xth roll = $\binom{x-1}{9}\left(\frac{5}{6}\right)^{x-10}\left(\frac{1}{6}\right)^{9} \cdot \left(\frac{1}{6}\right) = \binom{x-1}{9}\left(\frac{5}{6}\right)^{x-10}\left(\frac{1}{6}\right)^{10}$

Section 3.6

73.

a $P(X \le 8) = F(8;5) = .932$

b $P(X = 8) = F(8;5) - F(7;5) = .065$

c $P(X \ge 9) = 1 - P(X \le 8) = .068$

d $P(5 \le X \le 8) = F(8;5) - F(4;5) = .492$

74.

a $P(X \le 5) = F(5;8) = .191$

b $P(6 \le X \le 9) = F(9;8) - F(5;8) = .526$

c $P(X \ge 10) = 1 - P(X \le 9) = .283$

d $E(X) = \lambda = 10,\ \sigma_X = \sqrt{\lambda} = 2.83$

75.

a $P(X \le 10) = F(10;20) = .011$

b $P(X > 20) = 1 - F(20;20) = 1 - .559 = .441$

c $P(10 \le X \le 20) = F(20;20) - F(9;20) = .559 - .005 = .554$

$P(10 < X < 20) = F(19;20) - F(10;20) = .470 - .011 = .459$

d $E(X) = \lambda = 20,\ \sigma_X = \sqrt{\lambda} = 4.472$

$$\begin{aligned} P(X < \mu - 2\sigma \text{ or } X > \mu + 2\sigma) &= P(X < 20 - 8.944) + P(X > 20 + 8.944) \\ &= P(X < 11.056) + P(X > 28.944) \\ &= P(X \le 11) + P(X \ge 29) \\ &= F(11;20) + [1 - F(28;20)] \\ &= .021 + .004 = .025 \end{aligned}$$

76.

a $P(X = 1) = F(1;2) – F(0;2) = .982 - .819 = .163$

b $P(X \geq 2) = 1 – P(X \leq 1) = 1 – F(1;2) = 1 - .982 = .018$

c $P(1^{st}$ doesn't $\cap$ 2nd doesn't$) = P(1^{st}$ doesn't$) \cdot P(2^{nd}$ doesn't$) = (.819)(.819) = .671$

77. $p = \frac{1}{200}$; n = 1000; λ = np = 5

a $P(5 \leq X \leq 8) = F(8;5) – F(4;5) = .492$

b $P(X \geq 8) = 1 – P(X \leq 7) = 1 - .867 = .133$

78.

a The experiment is binomial with n = 10,000 and p = .005, so μ = np = 5 and $\sigma = \sqrt{npq} = 2.2355$.

b X has approximately a Poisson distribution with λ = 5, so $P(X > 10) = 1 – F(10;5) = 1 - .986 = .014$

79.

a λ = 8 when t = 1, so $P(X = 5) = F(5;8) – F(4;8) = .191 - .100 = .091$, $P(X \geq 5) = 1 - F(4;8) = .900$, and $P(X \geq 10) = 1 - F(9;8) = .283$

b t = 90 min = 1.5 hours, so λ = 12; thus the expected number of arrivals is 12 and the SD = $\sqrt{12} = 3.464$

c t = 2.5 hours implies that λ = 20; in this case, $P(X \geq 20) = 1 – F(19;20) = .530$ and $P(X \leq 10) = F(10;20) = .011$.

80.

a $P(X = 4) = F(4;5) – F(3;5) = .440 - .265 = .175$

b $P(X \geq 4) = 1 - P(X \leq 3) = 1 - .265 = .735$

c Tickets are given at the rate of 5 per hour, so for a 45 minute period the rate is λ = (5)(.75) = 3.75, which is also the expected number of tickets in a 45 minute period.

81.

a For a two hour period the parameter of the distribution is λt = (4)(2) = 8, so $P(X = 10) = F(10;8) – F(9;8) = .099$.

b For a 30 minute period, λt = (4)(.5) = 2, so $P(X = 0) = F(0,2) = .135$

c $E(X) = \lambda t = 2$

82. Let X = the number of diodes on a board that fail.

a E(X) = np = (200)(.01) = 2, V(X) = npq = (200)(.01)(.99) = 1.98, σ_X = 1.407

b X has approximately a Poisson distribution with λ = np = 2, so P(X ≥ 4) = 1 – P(X ≤ 3) = 1 – F(3;2) = 1 - .857 = .143

c P(board works properly) = P(all diodes work) = P(X = 0) = F(0;2) = .135
Let Y = the number among the five boards that work, a binomial r.v. with n = 5 and p = .135. Then P(Y ≥ 4) = P(Y = 4) + P(Y = 5) =

$$\binom{5}{4}(.135)^4(.865)+\binom{5}{5}(.135)^5(.865)^0 = .00144 + .00004 = .00148$$

83. α = 1/(mean time between occurrences) = $\frac{1}{.5}=2$

a αt = (2)(2) = 4

b P(X > 5) 1 – P(X ≤ 5) = 1 - .785 = .215

c Solve for t , given α = 2:
$.1 = e^{-\alpha t}$
$\ln(.1) = -\alpha t$
$t = \frac{2.3026}{2} \approx 1.15$ years

84.

$$E(X) = \sum_{x=0}^{\infty} x\frac{e^{-\lambda}\lambda^x}{x!} = \sum_{x=1}^{\infty} x\frac{e^{-\lambda}\lambda^x}{x!} = \lambda\sum_{x=1}^{\infty} x\frac{e^{-\lambda}\lambda^x}{x!} = \lambda\sum_{y=0}^{\infty} x\frac{e^{-\lambda}\lambda^y}{y!} = \lambda$$

85.

a For a one-quarter acre plot, the parameter is (80)(.25) = 20, so P(X ≤ 16) = F(16;20) = .221

b The expected number of trees is $\lambda\cdot$(area) = 80(85,000) = 6,800,000.

c The area of the circle is πr^2 = .031416 sq. miles or 20.106 acres. Thus X has a Poisson distribution with parameter 20.106

86.

a P(X = 10 and no violations) = P(no violations | X = 10) · P(X = 10)
= $(.5)^{10}$ · [F(10;10) – F(9;10)]
= (.000977)(.125) = .000122

b P(y arrive and exactly 10 have no violations)
= P(exactly 10 have no violations | y arrive) · P(y arrive)
= P(10 successes in y trials when p = .5) · $e^{-10}\frac{(10)^y}{y!}$

$$= \binom{y}{10}(.5)^{10}(.5)^{y-10}e^{-10}\frac{(10)^y}{y!} = \frac{e^{-10}(5)^y}{10!(y-10)!}$$

c P(exactly 10 without a violation) = $\sum_{y=10}^{\infty} \frac{e^{-10}(5)^y}{10!(y-10)!}$

$$= \frac{e^{-10}\cdot 5^{10}}{10!}\sum_{y=10}^{\infty}\frac{(5)^{y-10}}{(y-10)!} = \frac{e^{-10}\cdot 5^{10}}{10!}\sum_{u=0}^{\infty}\frac{(5)^u}{(u)!} = \frac{e^{-10}\cdot 5^{10}}{10!}\cdot e^5$$

$$= \frac{e^{-5}\cdot 5^{10}}{10!} = p(10;5).$$

In fact, generalizing this argument shows that the number of "no-violation" arrivals within the hour has a Poisson distribution with parameter 5; the 5 results from $\lambda p = 10(.5)$.

87.

a No events in $(0, t+\Delta t)$ if and only if no events in (o, t) and no events in $(t, t+\Delta t)$. Thus, $P_0(t+\Delta t) = P_0(t)\cdot P(\text{no events in } (t, t+\Delta t)) = P_0(t)[1 - \lambda\cdot\Delta t - o(\Delta t)]$

b $\frac{P_0(t+\Delta t)-P_0(t)}{\Delta t} = -\lambda P_0(t)\frac{\Delta' t}{\Delta' t} - P_0(t)\cdot\frac{o(\Delta t)}{\Delta t}$

c $\frac{d}{dt}\left[e^{-\lambda t}\right] = -\lambda e^{-\lambda t} = -\lambda P_0(t)$, as desired.

d $\frac{d}{dt}\left[\frac{e^{-\lambda t}(\lambda t)^k}{k!}\right] = \frac{-\lambda e^{-\lambda t}(\lambda t)^k}{k!} + \frac{k\lambda e^{-\lambda t}(\lambda t)^{k-1}}{k!}$

$$= -\lambda\frac{e^{-\lambda t}(\lambda t)^k}{k!} + \lambda\frac{e^{-\lambda t}(\lambda t)^{k-1}}{(k-1)!} = -\lambda P_k(t) + \lambda P_{k-1}(t) \text{ as desired.}$$

Supplementary Exercises

88. Outcomes are(1,2,3)(1,2,4) (1,2,5) … (5,6,7); there are 35 such outcomes. Each having probability $\frac{1}{35}$. The W values for these outcomes are 6 (=1+2+3), 7, 8, …, 18. Since there is just one outcome with W value 6, p(6) = P(W = 6) = $\frac{1}{35}$. Similarly, there are three outcomes with W value 9 [(1,2,6) (1,3,5) and 2,3,4)], so p(9) = $\frac{3}{35}$. Continuing in this manner yields the following distribution:

W	6	7	8	9	10	11	12	13	14	15	16	17	18
P(W)	$\frac{1}{35}$	$\frac{1}{35}$	$\frac{2}{35}$	$\frac{3}{35}$	$\frac{4}{35}$	$\frac{4}{35}$	$\frac{5}{35}$	$\frac{4}{35}$	$\frac{4}{35}$	$\frac{3}{35}$	$\frac{2}{35}$	$\frac{1}{35}$	$\frac{1}{35}$

Since the distribution is symmetric about 12, μ = 12, and $\sigma^2 = \sum_{w=6}^{18}(w-12)^2 p(w)$

$= \frac{1}{35}[(6)^2(1) + (5)^2(1) + \ldots + (5)^2(1) + (6)^2(1) = 8$

89.

a p(1) = P(exactly one suit) = P(all spades) + P(all hearts) + P(all diamonds)

$$+ \text{P(all clubs)} = 4\text{P(all spades)} = 4 \cdot \frac{\binom{13}{5}}{\binom{52}{5}} = .00198$$

p(2) = P(all hearts and spades with at least one of each) + …+ P(all diamonds and clubs with at least one of each)

= 6 P(all hearts and spades with at least one of each)

= 6 [P(1 h and 4 s) + P(2 h and 3 s) + P(3 h and 2 s) + P(4 h and 1 s)]

$$= 6 \cdot \left[2 \cdot \frac{\binom{13}{4}\binom{13}{1}}{\binom{52}{5}} + 2 \cdot \frac{\binom{13}{3}\binom{13}{2}}{\binom{52}{5}} \right] = 6 \left[\frac{18{,}590 + 44{,}616}{2{,}598{,}960} \right] = .14592$$

$$\text{p(4)} = \text{4P(2 spades, 1 h, 1 d, 1 c)} = \frac{4 \cdot \binom{13}{2}(13)(13)(13)}{\binom{52}{5}} = .26375$$

p(3) = 1 – [p(1) + p(2) + p(4)] = .58835

b $\mu = \sum_{x=1}^{4} x \cdot p(x) = 3.114,\ \sigma^2 = \left[\sum_{x=1}^{4} x^2 \cdot p(x) \right] - (3.114)^2 = .405, \sigma = .636$

90. p(y) = P(Y = y) = P(y trials to achieve r S's) = P(y-r F's before r^{th} S) = nb(y – r;r,p)

$$= \binom{y-1}{r-1} p^r (1-p)^{y-r}, \text{ y = r, r+1, r+2, …}$$

91.

a b(x;15,.75)

b B(10;15, .75) = .314

c B(12;15, .75) - B(7;15, .75) = .764 - .017 = .747

d μ = (15)(.75) = 11.75, σ^2= (15)(.75)(.25) = 2.81

e Requests can all be met if and only if X ≤ 10, and 15 – X ≤ 8, i.e. if 7 ≤ X ≤ 10, so P(all requests met) = B(10; 15,.75) - B(6; 15,.75) = .310

92. P(6-v light works) = P(at least one 6-v battery works) = 1 – P(neither works)
$= 1 - (1-p)^2$.
P(D light works) = P(at least 2 d batteries work) = 1 – P(at most 1 D battery works)
$= 1 - [(1-p)^4 + 4(1-p)^3]$
the 6-v should be taken if $1 - (1-p)^2 \geq 1 - [(1-p)^4 + 4(1-p)^3]$.
Simplifying, $1 \leq (1-p)^2 + 4p(1-p) \Rightarrow 0 \leq 2p - 3p^3 \Rightarrow p \leq \frac{2}{3}$.

93. Let X ~ Bin(5, .9). Then $P(X \geq 3) = 1 - P(X \leq 2) = 1 - B(2;5,.9) = .991$

94.

a $P(X \geq 5) = 1 - B(4;25,.05) = .007$
b $P(X \geq 5) = 1 - B(4;25,.10) = .098$
c $P(X \geq 5) = 1 - B(4;25,.20) = .579$
d All would decrease, which is bad if the % defective is large and good if the % is small.

95.

a N = 500, p = .005, so np = 2.5 and b(x; 500, .005) = p(x; 2.5), a Poisson p.m.f.
b P(X = 5) = p(5; 2.5) - p(4; 2.5) = .9580 - .8912 = .0668
c $P(X \geq 5) = 1 - p(4;2.5) = 1 - .8912 = .1088$

96.

a B(18;25,.5) - B(6;25,.5) = .986
b B(18;25,.8) - B(6;25,.8) = .220
c $P(X \leq 7 \text{ or } X \geq 18 \text{ when } p = .5) = 1 - P(8 \leq X \leq 17 \text{ when } p = .5)$
= 1 – [B(17;25,.5) - B(7;25,.5)] = 1 – [.978 - .022] = .044
d $P(X \leq 7 \text{ or } X \geq 18 \text{ when } p = .6) = 1 - [B(17;25,.6) - B(7;25,.6)] = .154$;
$P(X \leq 7 \text{ or } X \geq 18 \text{ when } p = .8) = .891$
e $\{X \leq 6 \text{ or } X \geq 19\}$ gives P(rejecting when p = .5) = .14, which is too large, so use $\{X \leq 5 \text{ or } X \geq 20\}$

97. Let Y denote the number of tests carried out. For n = 3, possible Y values are 1 and 4. P(Y = 1) = P(no one has the disease) = $(.9)^3$ = .729 and P(Y = 4) = .271, so E(Y) = (1)(.729) + (4)(.271) = 1.813, as contrasted with the 3 tests necessary without group testing.

98. Regarding any particular symbol being received as constituting a trial. Then p = P(S) = P(symbol is sent correctly or is sent incorrectly and subsequently corrected) $= 1 - p_1 + p_1p_2$. The block of n symbols gives a binomial experiment with n trials and $p = 1 - p_1 + p_1p_2$.

99. p(2) = P(X = 2) = P(S on #1 and S on #2) = p^2
p(3) = P(S on #3 and S on #2 and F on #1) = $(1 - p)p^2$
p(4) = P(S on #4 and S on #3 and F on #2) = $(1 - p)p^2$
p(5) = P(S on #5 and S on #4 and F on #3 and no 2 consecutive S's on trials prior to #3)
= $[\,1 - p(2)\,](1 - p)p^2$
p(6) = P(S on #6 and S on #5 and F on #4 and no 2 consecutive S's on trials prior to #4)
= $[\,1 - p(2) - p(3)](1 - p)p^2$
In general, for x = 5, 6, 7, ...: $p(x) = [\,1 - p(2) - \ldots - p(x - 3)](1 - p)p^2$
For p = .9,

x	2	3	4	5	6	7	8
p(x)	.81	.081	.081	.0154	.0088	.0023	.0010

So $P(X \le 8) = p(2) + \ldots + p(8) = .9995$

100.

a With X ~ Bin(25, .1),$P(2 \le X \le 6)$ = B(6;25,.1 – B(1;25,.1) = .991 - .271 = 720

b E(X) = np = 25(.1) = 2.5, $\sigma_X = \sqrt{npq} = \sqrt{25(.1)(.9)} = \sqrt{2.25} = 1.50$

c $P(X \ge 7$ when p = .1) = 1 – B(6;25,.1) = 1 - .991 = .009

d $P(X \le 6$ when p = .2) = B(6;25,.2) = = .780, which is quite large

101.

a Let event C = seed carries single spikelets, and event P = seed produces ears with single spikelets. Then $P(P \cap C) = P(P \mid C) \cdot P(C)$ = .29 (.40) = .116. Let X = the number of seeds out of the 10 selected that meet the condition $P \cap C$.

Then X ~ Bin(10, .116). P(X = 5) = $\binom{10}{5}(.116)^5(.884)^5 = .002857$

b For 1 seed, the event of interest is P = seed produces ears with single spikelets.
$P(P) = P(P \cap C) + P(P \cap C')$ = .116 (from **a**) + $P(P \mid C') \cdot P(C')$
= .116 + (.26)(.40) = .272.
Let Y = the number out of the 10 seeds that meet condition P.
Then Y ~ Bin(10, .272), and P(Y = 5) = .0767.
$P(Y \le 5)$ = b(0;10,.272) + ... + b(5;10,.272) = .041813 + ... + .076719 = .97024

102. With S = favored acquittal, the population size is N = 12, the number of population S's is M = 4, the sample size is n = 4, and the p.m.f. of the number of interviewed jurors who favor acquittal is the hypergeometric p.m.f. h(x;4,4,12). E(X) = $4 \cdot \left(\frac{4}{12}\right) = 1.33$

103.

a P(X = 0) = F(0;2) 0.135

b Let S = an operator who receives no requests. Then p = .135 and we wish P(4 S's in 5 trials) = b(4;5,..135) = $\binom{5}{4}(.135)^4(.884)^1$ = ..00144

c P(all receive x) = P(first receives x) · … · P(fifth receives x) = $\left[\frac{e^{-2}2^x}{x!}\right]^5$, and P(all receive the same number) is the sum from x = 0 to ∞.

104. P(at least one) = 1 – P(none) = 1 - $e^{-\lambda\pi R^2}\cdot\frac{(\lambda\pi R^2)^0}{0!}$ = 1 - $e^{-\lambda\pi R^2}$ = .99 ⇒ $e^{-\lambda\pi R^2}$ = .01

$\Rightarrow R^2 = \frac{-\ln(.01)}{\lambda\pi}$ = .7329 ⇒ R = .8561

105. The number sold is min (X, 5), so E[min(x, 5)] = $\sum^{\infty}\min(x,5)p(x;4)$

=(0)p(0;4) + (1) p(1;4) + (2) p(2;4) + (3) p(3;4) + (4) p(4;4) + $5\sum_{x=5}^{\infty}p(x;4)$

= 1.735 + 5[1 – F(4;4)] = 3.59

106.

a P(X = x) = P(A wins in x games) + P(B wins in x games)
= P(9 S's in 1st x-1 ∩ S on the xth) + P(9 F's in 1st x-1 ∩ F on the xth)

= $\binom{x-1}{9}p^9(1-p)^{x-10}p$ + $\binom{x-1}{9}(1-p)^9p^{x-10}(1-p)$

= $\binom{x-1}{9}\left[p^{10}(1-p)^{x-10}+(1-p)^{10}p^{x-10}\right]$

b Possible values of X are now 10, 11, 12, …(all positive integers ≥ 10). Now

P(X = x) = $\binom{x-1}{9}\left[p^{10}(1-p)^{x-10}+q^{10}(1-q)^{x-10}\right]$ for x = 10, … , 19,

So P(X ≥ 20) = 1 – P(X < 20) and P(X < 20) = $\sum_{x=10}^{19}P(X=x)$

107.

a No; probability of success is not the same for all tests

b There are four ways exactly three could have positive results. Let D represent those with the disease and D′ represent those without the disease.

Combination		Probability
D	D′	
0	3	$\left[\binom{5}{0}(.2)^0(.8)^5\right]\cdot\left[\binom{5}{3}(.9)^3(.1)^2\right]$ =(.32768)(.0729) = .02389
1	2	$\left[\binom{5}{1}(.2)^1(.8)^4\right]\cdot\left[\binom{5}{2}(.9)2(.1)^3\right]$ =(.4096)(.0081) = .00332
2	1	$\left[\binom{5}{2}(.2)^2(.8)^3\right]\cdot\left[\binom{5}{1}(.9)^1(.1)^4\right]$ =(.2048)(.00045) = .00009216
3	0	$\left[\binom{5}{3}(.2)^3(.8)^2\right]\cdot\left[\binom{5}{0}(.9)^0(.1)^5\right]$ =(.0512)(.00001) = .000000512

Adding up the probabilities associated with the four combinations yields 0.0273.

108. k(r,x) = $\dfrac{(x+r-1)(x+r-2)...(x+r-x)}{x!}$

With r = 2.5 and p = .3, p(4) = $\dfrac{(5.5)(4.5)(3.5)(2.5)}{4!}(.3)^{2.5}(.7)^4 = .1068$

Using k(r,0) = 1, P(X ≥ 1) = 1 – p(0) = 1 – $(.3)^{2.5}$ = .9507

109.

a p(x;λ,μ) = $\frac{1}{2}p(x;\lambda)+\frac{1}{2}p(x;\mu)$ where both p(x;λ) and p(x; μ) are Poisson p.m.f.'s and thus ≥ 0, so p(x;λ,μ) ≥ 0. Further,

$$\sum_{x=0}^{\infty}p(x;\lambda,\mu)=\frac{1}{2}\sum_{x=0}^{\infty}p(x;\lambda)+\frac{1}{2}\sum_{x=0}^{\infty}p(x;\mu)=\frac{1}{2}+\frac{1}{2}=1$$

b $.6p(x;\lambda)+.4p(x;\mu)$

(continued)

c $E(X) = \sum_{x=0}^{\infty} x[\frac{1}{2}p(x;\lambda) + \frac{1}{2}p(x;\mu)] = \frac{1}{2}\sum_{x=0}^{\infty} xp(x;\lambda) + \frac{1}{2}\sum_{x=0}^{\infty} xp(x;\mu)$

$= \frac{1}{2}\lambda + \frac{1}{2}\mu = \frac{\lambda+\mu}{2}$

d $E(X^2) = \frac{1}{2}\sum_{x=0}^{\infty} x^2 p(x;\lambda) + \frac{1}{2}\sum_{x=0}^{\infty} x^2 p(x;\mu) = \frac{1}{2}(\lambda^2+\lambda) + \frac{1}{2}(\mu^2+\mu)$ (since for a Poisson r.v., $E(X^2) = V(X) + [E(X)]^2 = \lambda + \lambda^2$),

so $V(X) = \frac{1}{2}\left[\lambda^2 + \lambda + \mu^2 + \mu\right] - \left[\frac{\lambda+\mu}{2}\right]^2 = \left(\frac{\lambda-\mu}{2}\right)^2 + \frac{\lambda+\mu}{2}$

110.

a $\frac{b(x+1;n,p)}{b(x;n,p)} = \frac{(n-x)}{(x+1)} \cdot \frac{p}{(1-p)} > 1$ if $np - (1-p) > x$, from which the stated conclusion follows.

b $\frac{p(x+1;\lambda)}{p(x;\lambda)} = \frac{\lambda}{(x+1)} > 1$ if $x < \lambda - 1$, from which the stated conclusion follows.
If λ is an integer, then $\lambda - 1$ is a mode, but $p(\lambda,\lambda) = p(1-\lambda, \lambda)$ so λ is also a mode[$p(x;\lambda)$ achieves its maximum for both $x = \lambda - 1$ and $x = \lambda$.

111. $P(X = j) = \sum_{i=1}^{10} P(\text{arm on track } i \cap X = j) = \sum_{i=1}^{10} P(X = j \mid \text{arm on } i) \cdot p_i$

$= \sum_{i=1}^{10} P(\text{next seek at I+j+1 or I-j-1}) \cdot p_i = \sum_{i=1}^{10} (p_{i+j+1} + P_{i-j-1})p_i$

where $p_k = 0$ if $k < 0$ or $k > 10$.

112. $E(X) = \sum_{x=0}^{n} x \cdot \frac{\binom{M}{x}\binom{N-M}{n-x}}{\binom{N}{n}} = \sum_{x=1}^{n} \frac{\frac{M!}{(x-1)!(M-x)!} \cdot \binom{N-M}{n-x}}{\binom{N}{n}}$

$n \cdot \frac{M}{N} \sum_{x=1}^{n} \binom{M-1}{x-1} \frac{\binom{N-M}{n-x}}{\binom{N-1}{n-1}} = n \cdot \frac{M}{N} \sum_{y=0}^{n-1} \binom{M-1}{y} \frac{\binom{N-1-(M-1)}{n-1-y}}{\binom{N-1}{n-1}}$

$n \cdot \frac{M}{N} \sum_{y=0}^{n-1} h(y; n-1, M-1, N-1) = n \cdot \frac{M}{N}$

113. Let A = {x: |x - μ| ≥ kσ}. Then $\sigma^2 = \sum_A (x-\mu)^2 p(x) \ge (k\sigma)^2 \sum_A p(x)$. But $\sum_A p(x)$ = P(X is in A) = P(|X - μ| ≥ kσ), so $\sigma^2 \ge k^2\sigma^2 \cdot$ P(|X - μ| ≥ kσ), as desired.

114.

a For [0,4], $\lambda = \int_0^4 e^{2+.6t}dt$ = 123.44, whereas for [2,6], $\lambda = \int_2^6 e^{2+.6t}dt$ = 409.82

b $\lambda = \int_0^{0.9907} e^{2+.6t}dt$ = 9.9996 ≈ 10, so the desired probability is F(15, 10) = .951.

Chapter 4

Section 4.1

1.

a P(x ≤ 1) = $\int_{-\infty}^{1} f(x)dx = \int_{0}^{1} \frac{1}{2}xdx = \frac{1}{4}x^2 \Big]_0^1 = .25$

b P(.5 ≤ X ≤ 1.5) = $\int_{.5}^{1.5} \frac{1}{2}xdx = \frac{1}{4}x^2 \Big]_{.5}^{1.5} = .5$

c P(x > 1.5) = $\int_{1.5}^{\infty} f(x)dx = \int_{1.5}^{2} \frac{1}{2}xdx = \frac{1}{4}x^2 \Big]_{1.5}^{2} = \frac{7}{16} \approx .438$

2. F(x) = $\frac{1}{10}$ for –5 ≤ x ≤ 5, and = 0 otherwise

a P(X < 0) = $\int_{-5}^{0} \frac{1}{10}dx = .5$

b P(-2.5 < X < 2.5) = $\int_{-2.5}^{2.5} \frac{1}{10}dx = .5$

c P(-2 ≤ X ≤ 3) = $\int_{-2}^{3} \frac{1}{10}dx = .5$

d P(k < X < k + 4) = $\int_{k}^{k+4} \frac{1}{10}dx = \frac{x}{10}\Big]_k^{k+4} = \frac{1}{10}[(k+4)-k] = .4$

3.

a Graph of f(x) = .09375(4 – x^2)

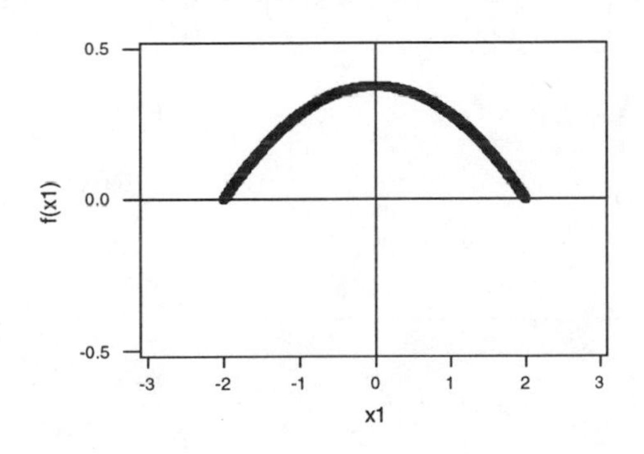

b P(X > 0) = $\int_{0}^{2} .09375(4-x^2)dx = .09375(4x - \frac{x^3}{3})\Big]_0^2 = .5$

c P(-1 < X < 1) = $\int_{-1}^{1} .09375(4-x^2)dx = .6875$

d P(x < -.5 OR x > .5) = 1 – P(-.5 ≤ X ≤ .5) = 1 - $\int_{-.5}^{.5} .09375(4-x^2)dx$

= 1 - .3672 = .6328

Chapter 4

4.

a $\int_{-\infty}^{\infty} f(x;\theta)dx = \int_0^{\infty} \frac{x}{\theta^2} e^{-x^2/2\theta^2} dx = -e^{-x^2/2\theta^2} \Big]_0^{\infty} = 0-(-1) = 1$

b P(X ≤ 200) = $\int_{-\infty}^{200} f(x;\theta)dx = \int_0^{200} \frac{x}{\theta^2} e^{-x^2/2\theta^2} dx$

$$= -e^{-x^2/2\theta^2} \Big]_0^{200} \approx -.1353+1 = .8647$$

P(X < 200) = P(X ≤ 200) ≈ .8647, since x is continuous.
P(X ≥ 200) = 1 - P(X ≤ 200) ≈ .1353

c P(100 ≤ X ≤ 200) = $\int_{100}^{200} f(x;\theta)dx = -e^{-x^2/20,000} \Big]_{100}^{200} \approx .4712$

d For x > 0, P(X ≤ x) =

$$\int_{-\infty}^{x} f(y;\theta)dy = \int_0^{x} \frac{y}{e^2} e^{-y^2/2\theta^2} dx = -e^{-y^2/2\theta^2} \Big]_0^{x} = 1-e^{-x^2/2\theta^2}$$

5.

a 1 = $\int_{-\infty}^{\infty} f(x)dx = \int_0^{2} kx^2 dx = k\left(\frac{x^3}{3}\right)\Big]_0^2 = k\left(\frac{8}{3}\right) \Rightarrow k = \frac{3}{8}$

b P(0 ≤ X ≤ 1) = $\int_0^{1} \frac{3}{8} x^2 dx = \frac{1}{8} x^3 \Big]_0^1 = \frac{1}{8} = .125$

c P(1 ≤ X ≤ 1.5) = $\int_1^{1.5} \frac{3}{8} x^2 dx = \frac{1}{8} x^3 \Big]_1^{1.5} = \frac{1}{8}\left(\frac{3}{2}\right)^3 - \frac{1}{8}(1)^3 = \frac{19}{64} \approx .2969$

d P(X ≥ 1.5) = 1 - $\int_0^{1.5} \frac{3}{8} x^2 dx = \frac{1}{8} x^3 \Big]_0^{1.5} = 1-\left[\frac{1}{8}\left(\frac{3}{2}\right)^3 - 0\right] = 1-\frac{27}{64} = \frac{37}{64} \approx .5781$

6.

a

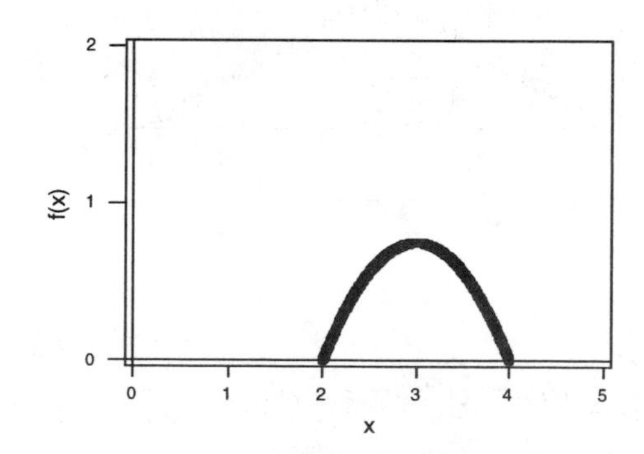

b 1 = $\int_2^{4} k[1-(x-3)^2]dx = \int_{-1}^{1} k[1-u^2]du = \frac{4}{3} \Rightarrow k = \frac{3}{4}$

c P(X > 3) = $\int_3^{4} \frac{3}{4}[1-(x-3)^2]dx = .5$ by symmetry of the p.d.f

d $P(\frac{11}{4} \le X \le \frac{13}{4}) = \int_{11/4}^{13/4} \frac{3}{4}[1-(x-3)^2]dx = \frac{3}{4}\int_{-1/4}^{1/4}[1-(u)^2]du = \frac{47}{128} \approx .367$

e P(|X-3| > .5) = 1 – P(|X-3| ≤ .5) = 1 – P(2.5 ≤ X ≤ 3.5)

$= 1 - \int_{-.5}^{.5} \frac{3}{4}[1-(u)^2]du = \frac{5}{16} \approx .313$

7.

a f(x) = $\frac{1}{10}$ for 25 ≤ x ≤ 35 and = 0 otherwise

b P(X > 33) = $\int_{33}^{35} \frac{1}{10} dx = .2$

c E(X) = $\int_{25}^{35} x \cdot \frac{1}{10} dx = \left.\frac{x^2}{20}\right]_{25}^{35} = 30$

30 ± 2 is from 28 to 32 minutes:

P(28 < X < 32) = $\int_{28}^{32} \frac{1}{10} dx = \left.\frac{1}{10}x\right]_{28}^{32} = .4$

d P(a ≤ x ≤ a+2) = $\int_{a}^{a+2} \frac{1}{10} dx = .2$, since the interval has length 2.

8.

a

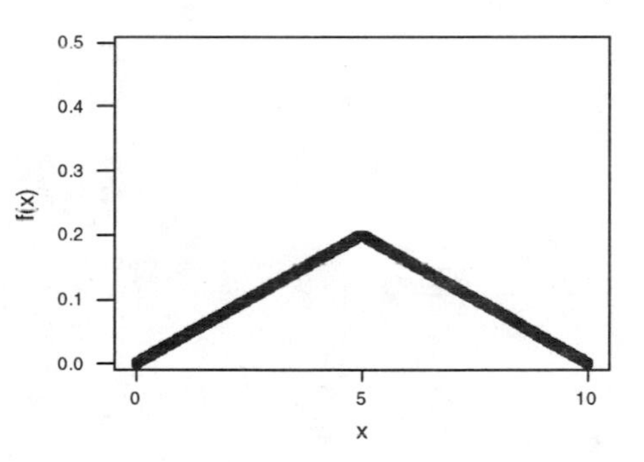

b $\int_{-\infty}^{\infty} f(y)dy = \int_0^5 \frac{1}{25} y dy + \int_5^{10} (\frac{2}{5} - \frac{1}{25} y) dy = \left.\frac{y^2}{50}\right]_0^5 + \left.\left(\frac{2}{5}y - \frac{1}{50}y^2\right)\right]_5^{10}$

$= \frac{1}{2} + \left[(4-2) - (2 - \frac{1}{2})\right] = \frac{1}{2} + \frac{1}{2} = 1$

c P(Y ≤ 3) = $\int_0^3 \frac{1}{25} y dy = \left.\frac{y^2}{50}\right]_0^5 = \frac{9}{50} \approx .18$

d P(Y ≤ 8) = $= \int_0^5 \frac{1}{25} y dy + \int_5^8 (\frac{2}{5} - \frac{1}{25} y) dy = \frac{23}{25} \approx .92$

e P(3 ≤ Y ≤ 8) = P(Y ≤ 8) - P(Y < 3) = $\frac{46}{50} - \frac{9}{50} = \frac{37}{50} = .74$

f P(Y < 2 or Y > 6) = $= \int_0^3 \frac{1}{25} y dy + \int_6^{10} (\frac{2}{5} - \frac{1}{25} y) dy = \frac{2}{5} = .4$

9.

a P(X ≤ 6) = $= \int_{.5}^{6} .15 e^{-.15(x-5)} dx = .15 \int_0^{5.5} e^{-.15u} du$ (after u = x - .5)

$= e^{-.15u} \Big]_0^{5.5} = 1 - e^{-.825} \approx .562$

b 1 - .562 = .438; .438

c P(5 ≤ Y ≤ 6) = P(Y ≤ 6) - P(Y ≤ 5) ≈ .562 - .491 = .071

10.

a

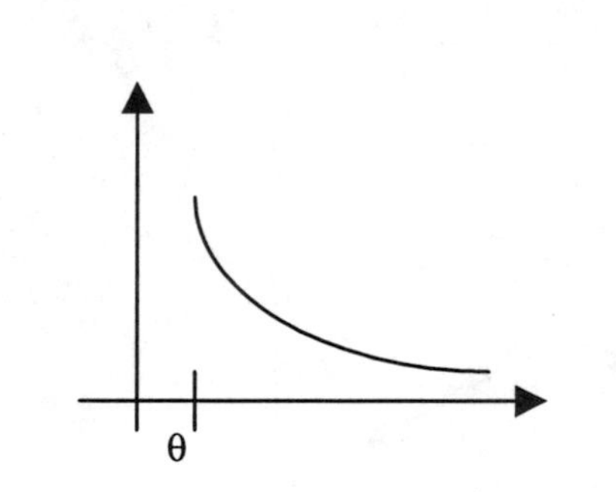

b $= \int_{-\infty}^{\infty} f(x;k,\theta) dx = \int_\theta^\infty \frac{k\theta^k}{x^{k+1}} dx = \theta^k \cdot \left(-\frac{1}{x^k}\right)\Big]_\theta^\infty = \frac{\theta^k}{\theta^k} = 1$

c P(X ≤ b) = $\int_\theta^b \frac{k\theta^k}{x^{k+1}} dx = \theta^k \cdot \left(-\frac{1}{x^k}\right)\Big]_\theta^b = 1 - \left(\frac{\theta}{b}\right)^k$

d P(a ≤ X ≤ b) = $\int_a^b \frac{k\theta^k}{x^{k+1}} dx = \theta^k \cdot \left(-\frac{1}{x^k}\right)\Big]_a^b = \left(\frac{\theta}{a}\right)^k - \left(\frac{\theta}{b}\right)^k$

Chapter 4

Section 4.2

11.

a P(X ≤ 1) = F(1) = $\frac{1}{4} = .25$

b P(.5 ≤ X ≤ 1) = F(1) – F(.5) = $\frac{3}{16} = .1875$

c P(X > .5) = 1 – P(X ≤ .5) = 1 – F(.5) = $\frac{15}{16} = .9375$

d $.5 = F(\tilde{\mu}) = \frac{\tilde{\mu}^2}{4} \Rightarrow \tilde{\mu}^2 = 2 \Rightarrow \tilde{\mu} = \sqrt{2} \approx 1.414$

e f(x) = F′(x) = $\frac{x}{2}$ for 0 ≤ x < 2, and = 0 otherwise

12.

a P(X < 0) = F(0) = .5

b P(-1 ≤ X ≤ 1) = F(1) – F(-1) = $\frac{11}{16} = .6875$

c P(X > .5) = 1 – P(X ≤ .5) = 1 – F(.5) = 1 - .6836 = .3164

d F(x) = F′(x) = $\frac{d}{dx}\left(\frac{1}{2} + \frac{3}{32}\left(4x - \frac{x^3}{3}\right)\right) = 0 + \frac{3}{32}\left(4 - \frac{3x^2}{3}\right) = .09375\left(4 - x^2\right)$

e $F(\tilde{\mu}) - .5$ by definition. F(0) = .5 from **a** above, which is as desired.

13.

a E(X) = $\int_{-\infty}^{\infty} x \cdot f(x)dx = \int_0^2 x \cdot \frac{1}{2}xdx = \frac{1}{2}\int_0^2 x^2 dx = \frac{x^3}{6}\bigg]_0^2 = \frac{8}{6} \approx 1.333$

b E(X²) = $\int_{-\infty}^{\infty} x^2 f(x)dx = \int_0^2 x^2 \frac{1}{2}xdx = \frac{1}{2}\int_0^2 x^3 dx = \frac{x^4}{8}\bigg]_0^2 = 2,$

So Var(X) = E(X²) – [E(X)]² = $2 - \left(\frac{8}{6}\right)^2 = \frac{8}{36} \approx .222$, $\sigma_X \approx .471$

c From **b** , E(X²) = 2

14.

a If X is uniformly distributed on the interval from A to B, then

$$E(X) = \int_A^B x \cdot \frac{1}{B-A}dx = \frac{A+B}{2}, E(X^2) = \frac{A^2 + AB + B^2}{3}$$

V(X) = E(X²) – [E(X)]² = $\frac{(B-A)^2}{2}$.

With A = 7.5 and B = 20, E(X) = 13.75, V(X) = 13.02

b F(X) = $\begin{cases} 0 & x < 7.5 \\ \frac{x-7.5}{12.5} & 7.5 \le x < 20 \\ 1 & x \ge 20 \end{cases}$

c $P(X \le 10) = F(10) = .200$; $P(10 \le X \le 15) = F(15) - F(10) = .4$

d $\sigma = 3.61$, so $\mu \pm \sigma = (10.14, 17.36)$
Thus, $P(\mu - \sigma \le X \le \mu + \sigma) = F(17.36) - F(10.14) = .5776$
Similarly, $P(\mu - \sigma \le X \le \mu + \sigma) = P(6.53 \le X \le 20.97) = 1$

15.

a $F(X) = 0$ for $x \le 0$, $= 1$ for $x \ge 1$, and for $0 < X < 1$,

$$F(X) = \int_{-\infty}^{x} f(y)dy = \int_0^x 90y^8(1-y)dy = 90\int_0^x (y^8 - y^9)dy$$

$$90\left(\tfrac{1}{9}y^9 - \tfrac{1}{10}y^{10}\right)\Big]_0^x = 10x^9 - 9x^{10}$$

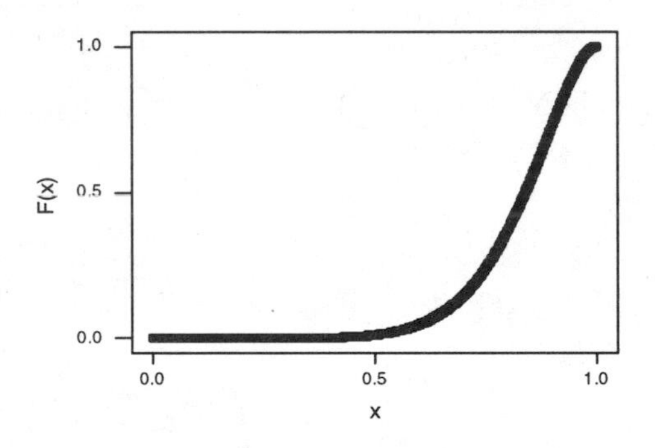

b $F(.5) = 10(.5)^9 - 9(.5)^{10} \approx .0107$

c $P(.25 \le X \le .5) = F(.5) - F(.25) \approx .0107 - [10(.25)^9 - 9(.25)^{10}]$
$\approx .0107 - .0000 \approx .0107$

d The 75th percentile is the value of x for which $F(x) = .75$
$\Rightarrow .75 = 10(x)^9 - 9(x)^{10} \Rightarrow x \approx .9036$

e $E(X) = \int_{-\infty}^{\infty} x \cdot f(x)dx = \int_0^1 x \cdot 90x^8(1-x)dx = 90\int_0^1 x^9(1-x)dx$

$= 9x^{10} - \frac{90}{11}x^{11}\Big]_0^1 = \frac{9}{11} \approx .8182$

$E(X^2) = \int_{-\infty}^{\infty} x^2 \cdot f(x)dx = \int_0^1 x^2 \cdot 90x^8(1-x)dx = 90\int_0^1 x^{10}(1-x)dx$

$= \frac{90}{11}x^{11} - \frac{90}{12}x^{12}\Big]_0^1 \approx .6818$

$V(X) \approx .6818 - (.8182)^2 = .0124$, $\sigma_X = .11134$.

f $\mu \pm \sigma = (.7068, .9295)$. Thus, $P(\mu - \sigma \le X \le \mu + \sigma) = F(.9295) - F(.7068)$
$= .8465 - .1602 = .6863$

Chapter 4

16.

a F(x) = 0 for x < 0 and F(x) = 1 for x > 2. For $0 \le x \le 2$,

F(x) = $\int_0^x \frac{3}{8} y^2 dy = \frac{1}{8} y^3 \Big]_0^x = \frac{1}{8} x^3$

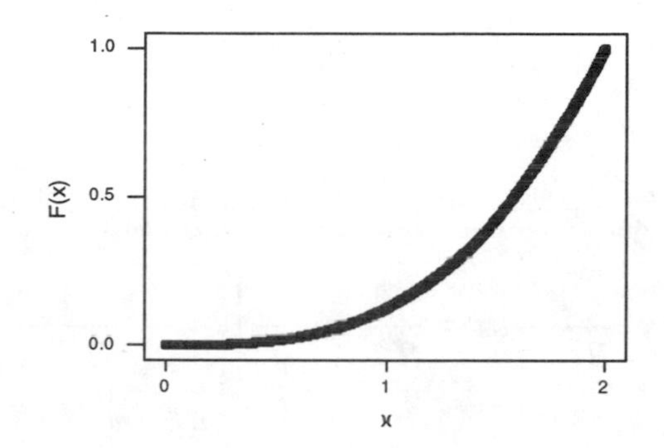

b $P(x \le .5) = F(.5) = \frac{1}{8}\left(\frac{1}{2}\right)^3 = \frac{1}{64}$

c $P(.25 \le X \le .5) = F(.5) - F(.25) = \frac{1}{64} - \frac{1}{8}\left(\frac{1}{4}\right)^3 = \frac{7}{512} \approx .0137$

d $.75 = F(x) = \frac{1}{8} x^3 \Rightarrow x^3 = 6 \Rightarrow x \approx 1.8171$

e $E(X) = \int_{-\infty}^{\infty} x \cdot f(x) dx = \int_0^2 x \cdot \left(\frac{3}{8} x^2\right) dx = \frac{3}{8} \int_0^1 x^3 dx = \frac{3}{8}\left(\frac{1}{4} x^4\right)\Big]_0^2 = \frac{3}{2} = 1.5$

$E(X^2) = \int_0^2 x \cdot \left(\frac{3}{8} x^2\right) dx = \frac{3}{8} \int_0^1 x^4 dx = \frac{3}{8}\left(\frac{1}{5} x5\right)\Big]_0^2 = \frac{12}{5} = 2.4$

$V(X) = \frac{12}{5} - \left(\frac{3}{2}\right)^2 = \frac{3}{20} = .15$ $\sigma_x = .3873$

f $\mu \pm \sigma = (1.1127, 1.8873)$. Thus, $P(\mu - \sigma \le X \le \mu + \sigma) = F(1.8873) - F(1.1127)$
$= .8403 - .1722 = .6681$

17.

a For $2 \le X \le 4$, $F(X) = \int_{-\infty}^{x} f(y) dy = \int_2^x \frac{3}{4}[1-(y-3)^2] dy$ (let u = y-3)

$= \int_{-1}^{x-3} \frac{3}{4}[1-u^2] du = \frac{3}{4}\left[u - \frac{u^3}{3}\right]_{-1}^{x-3} = \frac{3}{4}\left[x - \frac{7}{3} - \frac{(x-3)^3}{3}\right]$. Thus

$$F(x) = \begin{cases} 0 & x < 2 \\ \frac{1}{4}[3x - 7 - (x-3)^3] & 2 \le x \le 4 \\ 1 & x > 4 \end{cases}$$

b By symmetry of f(x), $\tilde{\mu} = 3$

c E(X) = $\int_2^4 x \cdot \frac{3}{4}[1-(x-3)^2]dx = \frac{3}{4}\int_{-1}^1 (y+3)(1-y^2)dx$

$$\frac{3}{4}\left[3y+\frac{y^2}{2}-y^3-\frac{y^4}{4}\right]_{-1}^1 = \frac{3}{4}\cdot 4 = 3$$

V(X) = $\int_{-\infty}^{\infty}(x-\mu)^2 f(x)dx = \frac{3}{4}\int_2^4 (x-3)^2 \cdot [1-(x-3)^2]dx$

$$= \frac{3}{4}\int_{-1}^1 y^2(1-y^2)dy = \frac{3}{4}\cdot\frac{4}{15} = \frac{1}{5} = .2$$

18.

a F(X) = $\frac{x-A}{B-A}$ = p $\Rightarrow$ x = (100p)th percentile = A + (B - A)p

b $E(X) = \int_A^B x \cdot \frac{1}{B-A}dx = \frac{1}{B-A}\cdot\left.\frac{x^2}{2}\right]_A^B = \frac{1}{2}\cdot\frac{1}{B-A}\cdot\left(B^2-A^2\right) = \frac{A+B}{2}$

$$E(X^2) = \frac{1}{3}\cdot\frac{1}{B-A}\cdot\left(B^3-A^3\right) = \frac{A^2+AB+B^2}{3}$$

$$V(X) = \left(\frac{A^2+AB+B^2}{3}\right) - \left(\frac{(A+B)}{2}\right)^2 = \frac{(B-A)^2}{12},\ \sigma_x = \frac{(B-A)}{\sqrt{12}}$$

c $E(X^n) = \int_A^B x^n \cdot \frac{1}{B-A}dx = \frac{B^{n+1}-A^{n+1}}{(n+1)(B-A)}$

19.

a P(X ≤ 1) = F(1) = .25[1 + ln(4)] ≈ .597

b P(1 ≤ X ≤ 3) = F(3) – F(1) ≈ .966 - .597 ≈ .369

c f(x) = F′(x) = .25 ln(4) - .25 ln(x) for o < x < 4

20.

a for 0 ≤ y ≤ 5, F(y) = $\int_0^y \frac{1}{25}u\,du = \frac{y^2}{50}$

for 5 ≤ y ≤ 10, F(y) = $\int_0^y f(u)du = \int_0^5 f(u)du + \int_5^y f(u)du$

$$= \frac{1}{2} + \int_0^y\left(\frac{2}{5}-\frac{u}{25}\right)du = \frac{2}{5}y - \frac{y^2}{50} - 1$$

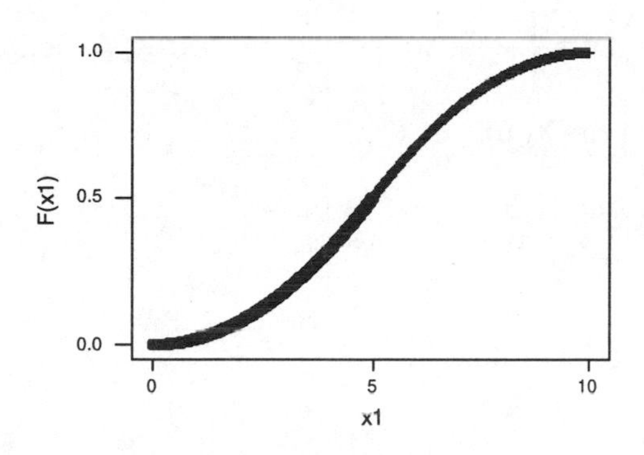

b for $0 < p \le .5$, p = F(y_p) = $\dfrac{y_p^2}{50} \Rightarrow y_p = (50p)^{1/2}$

For $.5 < p \le 1$, p = $\dfrac{2}{5}y_p - \dfrac{y_p^2}{50} - 1 \Rightarrow y_p = 10 - 5\sqrt{2(1-p)}$

c E(Y) = 5 by straightforward integration (or by symmetry of f(y)), and similarly V(Y)= $\dfrac{50}{12} = 4.1667$. For the waiting time X for a single bus,

E(X) = 2.5 and V(X) = $\dfrac{25}{12}$ ≤≥·≈⇒∩∪±′α λσΓ

21. E(area) = E(πR^2) = $\int_{-\infty}^{\infty} \pi r^2 f(r)dr = \int_9^{11} \pi r^2 \left(\frac{3}{4}\right)\left(1-(10-r)^2\right)dr$

$$= \left(\frac{3}{4}\right)\pi \int_9^{11} r^2\left(1-(100-20r+r^2)\right)dr = \frac{3}{4}\pi \int_9^{11} -99r^2 + 20r^3 - r^4 dr = 100 \cdot 2\pi$$

22.

a For $1 \le x \le 2$, F(x) = $\int_1^x 2\left(1-\frac{1}{y^2}\right)dy = 2\left(y+\frac{1}{y}\right)\Big]_1^x = 2\left(x+\frac{1}{x}\right) - 4$, so

$$F(x) = \begin{cases} 0 & x < 1 \\ 2\left(x+\frac{1}{x}\right) - 4 & 1 \le x \le 2 \\ 1 & x > 2 \end{cases}$$

b $2\left(x_p + \dfrac{1}{x_p}\right) - 4 = p \Rightarrow 2x_p^2 - (4-p)x_p + 2 = 0 \Rightarrow x_p = \frac{1}{4}[4 + p + \sqrt{p^2 + 8p}]$

To find $\tilde{\mu}$, set p = .5 ⇒ $\tilde{\mu}$ = 1.64

c E(X) = $\int_1^2 x \cdot 2\left(1-\frac{1}{x^2}\right)dx = 2\int_1^2\left(x-\frac{1}{x}\right)dx = 2\left(\frac{x^2}{2}-\ln(x)\right)\Big]_1^2 = 1.614$

$E(X^2) = 2\int_1^2 \left(x^2-1\right)dx = 2\left(\frac{x^3}{3}-x\right)\Big]_1^2 = \frac{8}{3} \Rightarrow$ Var(X) = .0626

d Amount left = max(1.5 – X, 0), so

E(amount left) = $\int_1^2 \max(1.5-x,0) f(x)dx = 2\int_1^{1.5}(1.5-x)\left(1-\frac{1}{x^2}\right)dx = .061$

23. With X = temperature in °C, temperature in °F = $\frac{9}{5}X+32$, so

$E\left[\frac{9}{5}X+32\right] = \frac{9}{5}(120)+32 = 248, \quad Var\left[\frac{9}{5}X+32\right] = \left(\frac{9}{5}\right)^2 \cdot (2)^2 = 12.96$,

so σ = 3.6

24.

a E(X) = $\int_\theta^\infty x \cdot \frac{k\theta^k}{x^{k+1}}dx = k\theta^k \int_\theta^\infty \frac{1}{x^k}dx = \frac{k\theta^k x^{-k+1}}{-k+1}\Big]_\theta^\infty = \frac{k\theta}{k-1}$

b E(X) = ∞

c $E(X^2) = k\theta^k \int_\theta^\infty \frac{1}{x^{k-1}}dx = \frac{k\theta^2}{k-2}$, so

Var(X) = $\left(\frac{k\theta^2}{k-2}\right) - \left(\frac{k\theta}{k-1}\right)^2 = \frac{k\theta^2}{(k-2)(k-1)^2}$

d Var(x) = ∞, since $E(X^2) = \infty$.

e $E(X^n) = k\theta^k \int_\theta^\infty x^{n-(k+1)}dx$, which will be finite if n – (k+1) < -1, i.e. if n < k.

25.

a P(Y ≤ 1.8 $\tilde{\mu}$ + 32) = P(1.8X + 32 ≤ 1.8 $\tilde{\mu}$ + 32) = P(X ≤ $\tilde{\mu}$) = .5

b 90th for Y = 1.8η(.9) + 32 where η(.9) is the 90th percentile for X, since
P(Y ≤ 1.8η(.9) + 32) = P(1.8X + 32 ≤ 1.8η(.9) + 32)
= (X ≤ η(.9)) = .9 as desired.

c The (100p)th percentile for Y is 1.8η(p) + 32, verified by substituting p for .9 in the argument of **b**. When Y = aX + b, (i.e. a linear transformation of X), and the (100p)th percentile of the X distribution is η(p), then the corresponding (100p)th percentile of the Y distribution is a·η(p) + b. (same linear transformation applied to X's percentile)

Chapter 4

Section 4.3

26.

- **a** $P(0 \le Z \le 2.17) = \Phi(2.17) - \Phi(0) = .4850$
- **b** $\Phi(1) - \Phi(0) = .3413$
- **c** $\Phi(0) - \Phi(-2.50) = .4938$
- **d** $\Phi(2.50) - \Phi(-2.50) = .9876$
- **e** $\Phi(1.37) = .9147$
- **f** $P(-1.75 < Z) + [1 - P(Z < -1.75)] = 1 - \Phi(-1.75) = .9599$
- **g** $\Phi(2) - \Phi(-1.50) = .9104$
- **h** $\Phi(2.50) - \Phi(1.37) = .0791$
- **i** $1 - \Phi(1.50) = .0668$
- **j** $P(|Z| \le 2.50) = P(-2.50 \le Z \le 2.50) = \Phi(2.50) - \Phi(-2.50) = .9876$

27.

- **a** .9838 is found in the 2.1 row and the .04 column of the standard normal table so $c = 2.14$.
- **b** $P(0 \le Z \le c) = .291 \Rightarrow \Phi(c) = .7910 \Rightarrow c = .81$
- **c** $P(c \le Z) = .121 \Rightarrow 1 - P(c \le Z) = P(Z < c) = \Phi(c) = 1 - .121 = .8790 \Rightarrow c = 1.17$
- **d** $P(-c \le Z \le c) = \Phi(c) - \Phi(-c) = \Phi(c) - (1 - \Phi(c)) = 2\Phi(c) - 1$
  $\Rightarrow \Phi(c) = .9920 \Rightarrow c = .97$
- **e** $P(c \le | Z |) = .016 \Rightarrow 1 - .016 = .9840 = 1 - P(c \le | Z |) = P(| Z | < c)$
  $= P(-c < Z < c) = \Phi(c) - \Phi(-c) = 2\Phi(c) - 1$
  $\Rightarrow \Phi(c) = .9920 \Rightarrow c = 2.41$

28.

- **a** $\Phi(c) = .9100 \Rightarrow c \approx 1.34$ (.9099 is the entry in the 1.3 row, .04 column)
- **b** 9th percentile = -91st percentile = -1.34
- **c** $\Phi(c) = .7500 \Rightarrow c \approx .675$ since .7486 and .7517 are in the .67 and .68 entries, respectively.
- **d** $25^{th} = -75^{th} = -.675$
- **e** $\Phi(c) = .06 \Rightarrow c \approx .-1.555$ (both .0594 and .0606 appear as the –1.56 and –1.55 entries, respectively).

29.

- **a** Area under Z curve above $z_{.0055}$ is .0055, which implies that
  $\Phi(z_{.0055}) = 1 - .0055 = .9945$, so $z_{.0055} = 2.54$
- **b** $\Phi(z_{.09}) = .9100 \Rightarrow z = 1.34$ (since .9099 appears as the 1.34 entry).
- **c** $\Phi(z_{.633})$ = area below $z_{.633} = .3370 \Rightarrow z_{.633} \approx -.42$

30.

a $P(X \le 100) = P\left(z \le \frac{100-80}{10}\right) = P(Z \le 2) = \Phi(2.00) = .9772$

b $P(X \le 80) = P\left(z \le \frac{80-80}{10}\right) = P(Z \le 0) = \Phi(0.00) = .5$

c $P(65 \le X \le 100) = P\left(\frac{65-80}{10} \le z \le \frac{100-80}{10}\right) = P(-1.50 \le Z \le 2)$

$= \Phi(2.00) - \Phi(-1.50) = .9772 - .0668 = .9104$

d $P(70 \le X) = P(-1.00 \le Z) = 1 - \Phi(-1.00) = .8413$

e $P(85 \le X \le 95) = P(.50 \le Z \le 1.50) = \Phi(1.50) - \Phi(.50) = .2417$

f $P(|X - 80| \le 10) = P(-10 \le X - 80 \le 10) = P(70 \le X \le 90)$

$P(-1.00 \le Z \le 1.00) = .6826$

31.

a $P(X \le 17) = P\left(z \le \frac{17-15}{1.25}\right) = P(Z \le 1.60) = \Phi(1.60 = .9452)$

b $P(10 \le X \le 12) = P(-4.00 \le Z \le -2.40) \approx P(Z \le -2.40) = \Phi(-2.40) = .0082$

c $P(|X - 10| \le 2(1.25)) = P(-2.50 \le X-15 \le 2.50) = P(12.5 \le X \le 17.5)$

$P(-2.00 \le Z \le 2.00) = .9544$

32.

a $P(X > .25) = P(Z > -.83) = 1 - .2033 = .7967$

b $P(X \le .10) = \Phi(-3.33) = .0004$

c We want the value of the distribution, c, that is the 95th percentile (5% of the values are higher). The 95th percentile of the standard normal distribution = 1.645. So c = .30 + (1.645)(.06) = .3987. The largest 5% of all concentration values are above .3987 mg/cm^3.

33.

a $P(X \ge 10) = P(Z \ge .43) = 1 - \Phi(.43) = 1 - .6664 = .3336$.

$P(X > 10) = P(X \ge 10) = .3336$, since for any continuous distribution, $P(x = a) = 0$.

b $P(X > 20) = P(Z > 4) \approx 0$

c $P(5 \le X \le 10) = P(-1.36 \le Z \le .43) = \Phi(.43) - \Phi(-1.36) = .6664 - .0869 = .5795$

d $P(8.8 - c \le X \le 8.8 + c) = .98$, so $8.8 - c$ and $8.8 + c$ are at the 1st and the 99th percentile of the given distribution, respectively. The 1st percentile of the standard normal distribution has the value –2.33, so

$8.8 - c = \mu + (-2.33)\sigma = 8.8 - 2.33(2.8) \Rightarrow c = 2.33(2.8) = 6.524$.

34. Let X denote the diameter of a randomly selected cork made by the first machine, and let Y be defined analogously for the second machine.

$P(2.9 \le X \le 3.1) = P(-1.00 \le Z \le 1.00) = .6826$

$P(2.9 \le Y \le 3.1) = P(-7.00 \le Z \le 3.00) = .9987$

So the second machine wins handily.

35.

a $\mu + \sigma \cdot$(91st percentile from std normal) = 25 + 5(1.34) = 31.7
b 25 + 5(-1.555) = 17.225
c μ = 3.000 μm; σ = 0.150. We desire the 90th percentile: 30 + 1.28(0.15) = 3.192

36. $\mu = 43;\ \sigma = 4.5$

a $P(X < 40) = P\left(z \le \frac{40-43}{4.5}\right) = P(Z < -0.667) = .2514$

$P(X > 60) = P\left(z > \frac{60-43}{4.5}\right) = P(Z > 3.778) \approx 0$

b 43 + (-0.67)(4.5) = 39.985

37. $P(\text{damage}) = P(X < 100) = P\left(z < \frac{100-200}{300}\right) = P(Z < -3.33) = .0004$

P(at least one among five is damaged) = 1 – P(none damaged)
= 1 – $(.9996)^5$ = 1 - .998 = .002

38. From Table A.3, $P(-1.96 \le Z \le 1.96) = .95$. Then $P(\mu - .1 \le X \le \mu + .1) =$
$P\left(\frac{-.1}{\sigma} < z < \frac{.1}{\sigma}\right)$ implies that $\frac{.1}{\sigma} = 1.96$, and thus that $\sigma = \frac{.1}{1.96} = .0510$

39. Since 1.28 is the 90th z percentile ($z_{.1}$ = 1.28) and –1.645 is the 5th z percentile ($z_{.05}$ = 1.645), the given information implies that $\mu + \sigma(1.28) = 10.256$ and $\mu + \sigma(-1.645) = 9.671$, from which $\sigma(-2.925) = -.585$, $\sigma = .2000$, and $\mu = 10$.

40.

a $P(\mu - 1.5\sigma \le X \le \mu + 1.5\sigma) = P(-1.5 \le Z \le 1.5) = \Phi(1.50) - \Phi(-1.50) = .8664$
b $P(X < \mu - 2.5\sigma \text{ or } X > \mu + 2.5\sigma) = 1 - P(\mu - 2.5\sigma \le X \le \mu + 2.5\sigma)$
$= 1 - P(-2.5 \le Z \le 2.5) = 1 - .9876 = .0124$
c $P(\mu - 2\sigma \le X \le \mu - \sigma \text{ or } \mu + \sigma \le X \le \mu + 2\sigma)$ = P(within 2 sd's) – P(within 1 sd)
$= P(\mu - 2\sigma \le X \le \mu + 2\sigma) - P(\mu - \sigma \le X \le \mu + \sigma) = .9544 - .6826 = .2718$

41.

a $P(X > 30.5) = P\left(z > \frac{30.5-31}{.2}\right) = P(Z > -2.5)$
b $P(30.5 < X < 31.5) = P(-2.5 < Z < 2.5) = .9876$
c $P(X < 30.4) = P(Z < -3) = .0013$, so P(at least one) = 1 – $(.9987)^4$ = .0052

42.

a $P(67 \le X \le 75) = P(-1.00 \le Z \le 1.67) = .7938$

b $P(70 - c \le X \le 70 + c) = P\left(\frac{-c}{3} \le Z \le \frac{c}{3}\right) = 2\Phi(\frac{c}{3}) - 1 = .95 \Rightarrow \Phi(\frac{c}{3}) = .9750$

$\frac{c}{3} = 1.96 \Rightarrow c = 5.88$

c $10 \cdot P(\text{a single one is acceptable}) = 9.05$

d $p = P(X < 73.84) = P(Z < 1.28) = .9$, so $P(Y \le 8) = B(8;10,.9) = .264$

43. The stated condition implies that 99% of the area under the normal curve with $\mu = 10$ and $\sigma = 2$ is to the left of $c - 1$, so $c - 1$ is the 99th percentile of the distribution. Thus $c - 1 = \mu + \sigma(2.33) = 14.66$, and $c = 15.66$.

44.

a By symmetry, $P(-1.72 \le Z \le -.55) = P(.55 \le Z \le 1.72) = \Phi(1.72) - \Phi(.55)$

b $P(-1.72 \le Z \le .55) = \Phi(.55) - \Phi(-1.72) = \Phi(.55) - [1 - \Phi(1.72)]$

No, symmetry of the Z curve about 0.

45. $P(|X - \mu| \ge \sigma) = P(X \le \mu - \sigma \text{ or } X \ge \mu + \sigma)$

$= 1 - P(\mu - \sigma \le X \le \mu + \sigma) = 1 - P(-1 \le Z \le 1) = .3174$

Similarly, $P(|X - \mu| \ge 2\sigma) = 1 - P(-2 \le Z \le 2) = .0456$

And $P(|X - \mu| \ge 3\sigma) = 1 - P(-3 \le Z \le 3) = .0026$

46.

a $P(20 - .5 \le X \le 30 + .5) = P(19.5 \le X \le 30.5) = P(-1.1 \le Z \le 1.1) = .7286$

b $P(\text{at most } 30) = P(X \le 30 + .5) = P(Z \le 1.1) = .8643.$

$P(\text{less than } 30) = P(X < 30 - .5) = P(Z < .9) = .8159$

47.

P:	.5	.6	.8
μ:	12.5	15	20
σ:	2.50	2.45	2.00

a

	$P(15 \le X \le 20)$	$P(14.5 \le \text{normal} \le 20.5)$
.5	.212	$P(.80 \le Z \le 3.20) = .2112$
.6	.577	$P(-.20 \le Z \le 2.24) = .5668$
.8	.573	$P(-2.75 \le Z \le .25) = .5957$

b

P(X ≤15)	P(normal ≤ 15.5)
.885	P(Z ≤ 1.20) = .8849
.575	P(Z ≤ .20) = .5793
.017	P(Z ≤ -2.25) = .0122

c

P(20 ≤X)	P(19.5 ≤ normal)
.002	.0026
.029	.0329
.617	.5987

48. P = .10; n = 200; np = 20, npq = 18

a $P(X \le 30) = \Phi\left(\frac{30+.5-20}{\sqrt{18}}\right) = \Phi(2.47) = .9932$

b $P(X < 30) = P(X \le 29) = \Phi\left(\frac{29+.5-20}{\sqrt{18}}\right) = \Phi(2.24) = .9875$

c $P(15 \le X \le 25) = P(X \le 25) - P(X \le 14) = \Phi\left(\frac{25+.5-20}{\sqrt{18}}\right) - \Phi\left(\frac{14+.5-20}{\sqrt{18}}\right)$

$\Phi(1.30) - \Phi(-1.30) = .9032 - .0968 = .8064$

49. N = 500, p = .4, $\mu = 200$, $\sigma = 10.9545$

a $P(180 \le X \le 230) = P(179.5 \le \text{normal} \le 230.5) = P(-1.87 \le Z \le 2.78) = .9666$

b $P(X < 175) = P(X \le 174) = P(\text{normal} \le 174.5) = P(Z \le -2.33) = .0099$

50. $P(X \le \mu + \sigma$[(100p)th percentile for std normal])

$P\left(\frac{X-\mu}{\sigma} \le [...]\right) = P(Z \le [...]) = p$ as desired

51.

a $F_Y(y) = P(Y \le y) = P(aX + b \le y) = P\left(X \le \frac{(y-b)}{a}\right)$ (for a > 0).

Now differentiate with respect to y to obtain

$f_Y(y) = F_y'(y) = \frac{1}{\sqrt{2\pi}a\sigma} e^{-\frac{1}{2a^2\sigma^2}[y-(a\mu+b)]^2}$ so Y is normal with mean $a\mu + b$ and variance $a^2\sigma^2$.

b Normal, mean $\frac{9}{5}(115)+32 = 239$, variance = 12.96

52.

a $P(Z \ge 1) \approx .5 \cdot \exp\left(\frac{83+351+562}{703+165}\right) = .1587$

b $P(Z > 3) \approx .5 \cdot \exp\left(\frac{-2362}{399.3333}\right) = .0013$

c $P(Z > 4) \approx .5 \cdot \exp\left(\frac{-3294}{340.75}\right) = .0000317$, so

$P(-4 < Z < 4) \approx 1 - 2(.0000317) = .999937$

d $P(Z > 5) \approx .5 \cdot \exp\left(\frac{-4392}{305.6}\right) = .00000029$

53.

a $\Gamma(6) = 5! = 120$

b $\Gamma\left(\frac{5}{2}\right) = \frac{3}{2}\Gamma\left(\frac{1}{2}\right) = \frac{3}{2}\cdot\frac{1}{2}\cdot\Gamma\left(\frac{1}{2}\right) = \left(\frac{3}{4}\right)\sqrt{\pi} \approx 1.329$

c $F(4;5) = .371$ from row 4, column 5 of Table A.4

d $F(5;4) = .735$

e $F(0;4) = P(X \le 0; \alpha = 4) = 0$

54.

a $P(X \le 5) = F(5;7) = .238$

b $P(X < 5) = P(X \le 5) = .238$

c $P(X > 8) = 1 - P(X < 8) = 1 - F(8;7) = .313$

d $P(3 \le X \le 8) = F(8;7) - F(3;7) = .653$

e $P(3 < X < 8) = .653$

f $P(X < 4 \text{ or } X > 6) = 1 - P(4 \le X \le 6) = 1 - [F(6;7) - F(4;7)] = .713$

55.

a $P(X \le 1) = F\left(\frac{1}{½};2\right) = F(2;2) = .594$

b $P(2 < X) = 1 - P(X \le 2) = 1 - F\left(\frac{2}{½};2\right) = 1 - F(4;2) = .092$

c $P(.5 \le X \le 1.5) = F(3;2) - F(1;2) = .537$

56.

a $\mu = 20,\ \sigma^2 = 80 \Rightarrow \alpha\beta = 20,\ \alpha\beta^2 = 80 \Rightarrow \beta = \frac{80}{20},\ \alpha = 5$

b $P(X \le 24) = F\left(\frac{24}{4};5\right) = F(6;5) = .715$

c $P(20 \le X \le 40) = F(10;5) - F(5;5) = .411$

57. $\mu = 24,\ \sigma^2 = 144 \Rightarrow \alpha\beta = 24,\ \alpha\beta^2 = 144 \Rightarrow \beta = 6,\ \alpha = 4$

a $P(12 \le X \le 24) = F(4;4) - F(2;4) = .424$

b $P(X \le 24) = F(4;4) = .567$, so while the mean is 24, the median is less than 24. ($P(X \le \tilde{\mu}) = .5$); This is a result of the positive skew of the gamma distribution.

58.

a $E(X) = \frac{1}{\lambda} = 1$

b $\sigma = \frac{1}{\lambda} = 1$

c $P(X \le 4) = 1 - e^{-(1)(4)} = 1 - e^{-4} = .982$

d $P(2 \le X \le 5) = 1 - e^{-(1)(5)} - \left[1 - e^{-(1)(2)}\right] = e^{-2} - e^{-5} = .129$

59.

a $P(X \le 100) = 1 - e^{-(100)(.01386)} = 1 - e^{-1.386} = .7499$

$P(X \le 200) = 1 - e^{-(200)(.01386)} = 1 - e^{-2.772} = .9375$

$P(100 \le X \le 200) = P(X \le 200) - P(X \le 100) = .9375 - .7499 = .1876$

b $\mu = \frac{1}{.01386} = 72.15,\ \sigma = 72.15$

$P(X > \mu + 2\sigma) = P(X > 72.15 + 2(72.15)) = P(X > 216.45) =$

$1 - \left[1 - e^{-(216.45)(.01386)}\right] = e^{-2.9999} = .0498$

c $.5 = P(X \le \tilde{\mu}) \Rightarrow 1 - e^{-(\tilde{\mu})(.01386)} = .5 \Rightarrow e^{-(\tilde{\mu})(.01386)} = .5$

$-\tilde{\mu}(.01386) = \ln(.5) = .693 \Rightarrow \tilde{\mu} = 50$

60. Mean $= \frac{1}{\lambda} = 25{,}000$ implies $\lambda = .00004$

a $P(X > 20{,}000) = 1 - P(X \le 20{,}000) = 1 - F(20{,}000;\ .00004)$

$= e^{-(.00004)(20{,}000)} = .449$

$P(X \le 30{,}000) = F(30{,}000;\ .00004) = e^{-1.2} = .699$

$P(20{,}000 \le X \le 30{,}000) = .699 - .551 = .148$

b $\sigma = \frac{1}{\lambda} = 25{,}000$, so $P(X > \mu + 2\sigma) = P(x > 75{,}000) =$

$1 - F(75{,}000;.00004) = .05.$

Similarly, $P(X > \mu + 3\sigma) = P(x > 100{,}000) = .018$

61.

a $E(X) = \alpha\beta = n\frac{1}{\lambda} = \frac{n}{\lambda}$; for λ = .5, n = 10, E(X) = 20

b $P(X \le 30) = F\left(\frac{30}{2};10\right) = F(15;10) = .930$

c P(X ≤ t) = P(at least n events in time t) = P(Y ≥ n) when Y ~ Poisson with parameter λt. Thus P(X ≤ t) = 1 – P(Y < n) = 1 – P(Y ≤ n – 1)

$$= 1 - \sum_{k=0}^{n-1} \frac{e^{-\lambda t}(\lambda t)^k}{k!}.$$

62.

a $\{X \ge t\} = A_1 \cap A_2 \cap A_3 \cap A_4 \cap A_5$

b $P(X \ge t) = P(A_1) \cdot P(A_2) \cdot P(A_3) \cdot P(A_4) \cdot P(A_5) = \left(e^{-\lambda t}\right)^5 = e^{-.05t}$, so $F_x(t) = P(X \le t) = 1 - e^{-.05t}$, $f_x(t) = .05e^{-.05t}$ for t ≥ 0. Thus X also ha an exponential distribution , but with parameter λ = .05.

c By the same reasoning, $P(X \le t) = 1 - e^{-n\lambda t}$, so X has an exponential distribution with parameter nλ.

63. With x_p = (100p)th percentile, $p = F(x_p) = 1 - e^{-\lambda x_p} \Rightarrow e^{-\lambda x_p} = 1 - p$,
$\Rightarrow -\lambda x_p = \ln(1-p) \Rightarrow x_p = \frac{-[\ln(1-p)]}{\lambda}$. For p = .5, $x_{.5} = \tilde{\mu} = \frac{.693}{\lambda}$.

64. $\int_0^{\infty} \frac{1}{\beta^{\alpha}\Gamma(\alpha)} x^{\alpha-1} e^{-x/\beta} dx = \int \frac{1}{\Gamma(\alpha)} y^{\alpha-1} e^{-y} dy$ (after $y = \frac{x}{\beta}$, $dy = \frac{dx}{\beta}$), and this integral is 1 since the integrand is a standard gamma density.

65.

a $\{X^2 \le y\} = \{-\sqrt{y} \le X \le \sqrt{y}\}$

b $P(X^2 \le y) = \int_{-\sqrt{y}}^{\sqrt{y}} \frac{1}{\sqrt{2\pi}} e^{-z^2/2} dz$. Now differentiate with respect to y to obtain the chi-squared p.d.f. with ν = 1.

Chapter 4

Section 4.5

66.

a $E(X) = 3\Gamma\left(1+\frac{1}{2}\right) = 3\cdot\frac{1}{2}\cdot\Gamma\left(\frac{1}{2}\right) = 2.66,$

$Var(X) = 9\left[\Gamma(1+1) - \Gamma^2\left(1+\frac{1}{2}\right)\right] = 1.926$

b $P(X \le 6) = 1 - e^{-(6/\beta)^\alpha} = 1 - e^{-(6/3)^2} = 1 - e^{-4} = .982$

c $P(1.5 \le X \le 6) = 1 - e^{-(6/3)^2} - \left[1 - e^{-(1.5/3)^2}\right] = e^{-.25} - e^{-4} = .760$

67.

a $P(X \le 200) = F(200;2.5, 200) = 1 - e^{-(200/200)^{2.5}} = 1 - e^{-1} \approx .632$

$P(X < 200) = P(X \le 200) \approx .632$

$P(X > 300) = 1 - F(300; 2.5, 200) = e^{-(1.5)^{2.5}} = .0636$

b $P(100 \le X \le 200) = F(200;2.5, 200) - F(100;2.5, 200) \approx .632 - .162 = .470$

c The median $\tilde{\mu}$ is requested. The equation $F(\tilde{\mu}) = .5$ reduces to

$.5 = e^{-(\tilde{\mu}/200)^{2.5}}$, i.e., $\ln(.5) \approx -\left(\frac{\tilde{\mu}}{200}\right)^{2.5}$, so $\tilde{\mu} = (.6931)^{.4}(200) = 172.727$.

68.

a For $x > 3.5$, $F(x) = P(X \le x) = P(X - 3.5 \le x - 3.5) = 1 - e^{-\left[\frac{(x-3.5)}{1.5}\right]^2}$

b $E(X - 3.5) = 1.5\Gamma\left(\frac{3}{2}\right) = 1.329$ so $E(X) = 4.829$

$Var(X) = Var(X - 3.5) = (1.5)^2\left[\Gamma(2) - \Gamma^2\left(\frac{3}{2}\right)\right] = .483$

c $P(X > 5) = 1 - P(X \le 5) = 1 - \left[1 - e^{-1}\right] = e^{-1} = .368$

d $P(5 \le X \le 8) = 1 - e^{-9} - \left[1 - e^{-1}\right] = e^{-1} - e^{-9} = .3679 - .0001 = .3678$

69. $\mu = \int_0^\infty x \cdot \frac{\alpha}{\beta^\alpha} x^{\alpha-1} e^{-(x/\beta)^\alpha} dx$ = (after $y = \left(\frac{x}{\beta}\right)^\alpha$, $dy = \frac{\alpha x^{\alpha-1}}{\beta^\alpha} dx$)

$\beta \int_0^\infty y^{1/\alpha} e^{-y} dy = \beta \cdot \Gamma\left(1 + \frac{1}{\alpha}\right)$ by definition of the gamma function.

70.

a $.5 = F(\tilde{\mu}) = 1 - e^{-(\mu/3)^2} \Rightarrow$

$e^{-\mu/9} = .5 \Rightarrow \tilde{\mu}^2 = -9\ln(.5) = 6.2383 \Rightarrow \tilde{\mu} = 2.50$

b $1 - e^{-[(\tilde{\mu}-3.5)/1.5]^2} = .5 \Rightarrow$ $(\tilde{\mu} - 3.5)^2 = -2.25\ln(.5) = 1.5596 \Rightarrow \tilde{\mu} = 4.75$

c $P = F(x_p) = 1 - e^{-(x_p/\beta)^\alpha} \Rightarrow (x_p/\beta)^\alpha = -\ln(1-p) \Rightarrow x_p = \beta[-\ln(1-p)]^{1/\alpha}$

d The desired value of t is the 90th percentile (since 90% will not be refused and 10% will be). From **c,** the 90th percentile of the distribution of X – 3.5 is $1.5[-\ln(.1)]^{1/2} = 2.27661$, so t = 3.5 + 2.2761 = 5.7761

71.

a $E(X) = e^{3.5+(1.2)^2/2} = 68.0335$; $V(X) = e^{2(3.5)+(1.2)^2} \cdot \left(e^{(1.2)^2} - 1\right) = 14907.168$;
$\sigma_x = 122.0949$

b $P(50 \le X \le 250) = P\left(z \le \frac{\ln(250) - 3.5}{1.2}\right) - P\left(z \le \frac{\ln(50) - 3.5}{1.2}\right)$

$P(Z \le 1.68) - P(Z \le .34) = .9535 - .6331 = .3204$.

c $P(X \le 68.0335) = P\left(z \le \frac{\ln(68.0335) - 3.5}{1.2}\right) = P(Z \le .60) = .7257$. The lognormal distribution is not a symmetric distribution.

72.

a $.5 = F(\tilde{\mu}) = \Phi\left(\frac{\ln(\tilde{\mu}) - \mu}{\sigma}\right)$, (where $\tilde{\mu}$ refers to the lognormal distribution and μ and σ to the normal distribution). Since the median of the standard normal distribution is 0, $\frac{\ln(\tilde{\mu}) - \mu}{\sigma} = 0$, so $\ln(\tilde{\mu}) = \mu \Rightarrow \tilde{\mu} = e^\mu$. For the power distribution, $\tilde{\mu} = e^{3.5} = 33.12$

b $1 - \alpha = \Phi(z_\alpha) = P(Z \le z_\alpha) = \left(\frac{\ln(X) - \mu}{\sigma} \le z_\alpha\right) = P(\ln(X) \le \mu + \sigma z_\alpha)$

$= P(X \le e^{\mu + \sigma z_\alpha})$, so the 100(1 - α)th percentile is $e^{\mu + \sigma z_\alpha}$. For the power distribution, the 95th percentile is $e^{3.5+(1.645)(1.2)} = e^{5.474} = 238.41$

73.

a $E(X) = e^{5+(.01)/2} = e^{5.005} = 149.157$; $Var(X) = e^{10+(.01)} \cdot \left(e^{.01} - 1\right) = 223.594$

b $P(X > 120) = 1 - P(X \le 120) = 1 - P(\ln(X) \le 4.788) = 1 - \Phi(-2.12) = .9830$

c $P(110 \le X \le 130) = P(4.7005 \le \ln(X) \le 4.8675) = \Phi(-1.32) - \Phi(-3) = .0921$

d $\tilde{\mu} = e^5 = 148.41$

e P(any particular one has X > 120) = .983 $\Rightarrow$ expected # = 10(.983) = 9.83

f We wish the 5th percentile, which is $e^{5+(-1.645)(.1)} = 125.90$

Chapter 4

74.

a $E(X) = e^{1.9+9^2/2} = 10.024$; Var(X) = $e^{3.8+(.81)} \cdot \left(e^{.81} - 1\right) = 125.395$, σ_x = 11.20

b P(X ≤ 10) = P(ln(X) ≤ 2.3026) = P(Z ≤ .45) = .6736
P(5 ≤ X ≤ 10) = P(1.6094 ≤ ln(X) ≤2.3026)
= P(-.32 ≤ Z ≤ .45) = .6736 - .3745 = .2991

75. The point of symmetry must be $\frac{1}{2}$, so we require that $f\left(\frac{1}{2} - \mu\right) = f\left(\frac{1}{2} + \mu\right)$, i.e., $\left(\frac{1}{2} - \mu\right)^{\alpha-1}\left(\frac{1}{2} + \mu\right)^{\beta-1} = \left(\frac{1}{2} + \mu\right)^{\alpha-1}\left(\frac{1}{2} - \mu\right)^{\beta-1}$, which in turn implies that α = β.

76.

a E(X) = $\frac{5}{(5+2)} = \frac{5}{7} = .714$, V(X) = $\frac{10}{(49)(8)} = .0255$

b f(x) = $\frac{\Gamma(7)}{\Gamma(5)\Gamma(2)} \cdot x^4 \cdot (1-x) = 30\left(x^4 - x^5\right)$ for 0 ≤ X ≤ 1,

so P(X ≤ .2) = $\int_0^{.2} 30\left(x^4 - x^5\right)dx = .0016$

c P(.2 ≤ X ≤ .4) = $\int_{.2}^{.4} 30\left(x^4 - x^5\right)dx = .03936$

d E(1 – X) = 1 – E(X) = 1 - $\frac{5}{7} = \frac{2}{7} = .286$

77.

a E(X) = $\int_0^1 x \cdot \frac{\Gamma(\alpha+\beta)}{\Gamma(\alpha)\Gamma(\beta)} x^{\alpha-1}(1-x)^{\beta-1}dx = \frac{\Gamma(\alpha+\beta)}{\Gamma(\alpha)\Gamma(\beta)}\int_0^1 x^{\alpha}(1-x)^{\beta-1}dx$

$$\frac{\Gamma(\alpha+\beta)}{\Gamma(\alpha)\Gamma(\beta)} \frac{\Gamma(\alpha+1)\Gamma(\beta)}{\Gamma(\alpha+\beta+1)} = \frac{\alpha\Gamma(\alpha)}{\Gamma(\alpha)\Gamma(\beta)} \cdot \frac{\Gamma(\alpha+\beta)}{(\alpha+\beta)\Gamma(\alpha+\beta)} = \frac{\alpha}{\alpha+\beta}$$

b E[(1 – X)^m] = $\int_0^1 (1-x)^m \cdot \frac{\Gamma(\alpha+\beta)}{\Gamma(\alpha)\Gamma(\beta)} x^{\alpha-1}(1-x)^{\beta-1}dx$

$$= \frac{\Gamma(\alpha+\beta)}{\Gamma(\alpha)\Gamma(\beta)}\int_0^1 x^{\alpha-1}(1-x)^{m+\beta-1}dx = \frac{\Gamma(\alpha+\beta)\cdot\Gamma(m+\beta)}{\Gamma(\alpha+\beta+m)\Gamma(\beta)}$$

For m = 1, E(1 – X) = $\frac{\beta}{\alpha+\beta}$.

78.

a E(Y) = 10 $\Rightarrow E\left(\frac{Y}{20}\right) = \frac{1}{2} = \frac{\alpha}{\alpha+\beta}$; Var(Y) = $\frac{100}{7} \Rightarrow Var\left(\frac{Y}{20}\right) = \frac{100}{2800} = \frac{1}{28}$

$\frac{\alpha\beta}{(\alpha+\beta)^2(\alpha+\beta+1)} \Rightarrow \alpha = 3, \beta = 3$, after some algebra.

b $P(8 \le X \le 12) = F\left(\frac{12}{20};3,3\right) - F\left(\frac{8}{20};3,3\right) = F(.6;3,3) - F(.4;3,3)$.

The standard density function here is $30y^2(1-y)^2$,

so $P(8 \le X \le 12) = \int_{.4}^{.6} 30y^2(1-y)^2\,dy = .365$.

c We expect it to snap at 10, so $P(Y < 8 \text{ or } Y > 12) = 1 - P(8 \le X \le 12)$ $= 1 - .365 = .665$.

Section 4.6

79. The given probability plot is quite linear, and thus it is quite plausible that the tension distribution is normal.

80. The z percentiles and observations are as follows:

percentile	observation
-1.645	152.7
-1.040	172.0
-0.670	172.5
-0.390	173.3
-0.130	193.0
0.130	204.7
0.390	216.5
0.670	234.9
1.040	262.6
1.645	422.6

The accompanying plot is quite straight except for the point corresponding to the largest observation. This observation is clearly much larger than what would be expected in a normal random sample. Because of this outlier, it would be inadvisable to analyze the data using any inferential method that depended on assuming a normal population distribution.

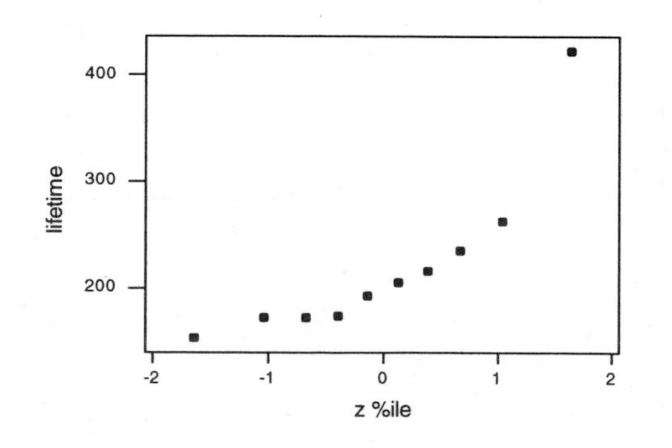

81. The z percentile values are as follows: -1.86, -1.32, -1.01, -0.78, -0.58, -0.40, -0.24, -0.08, 0.08, 0.24, 0.40, 0.58, 0.78, 1.01, 1.30, and 1.86. The accompanying probability plot is reasonably straight, and thus it would be reasonable to use estimating methods that assume a normal population distribution.

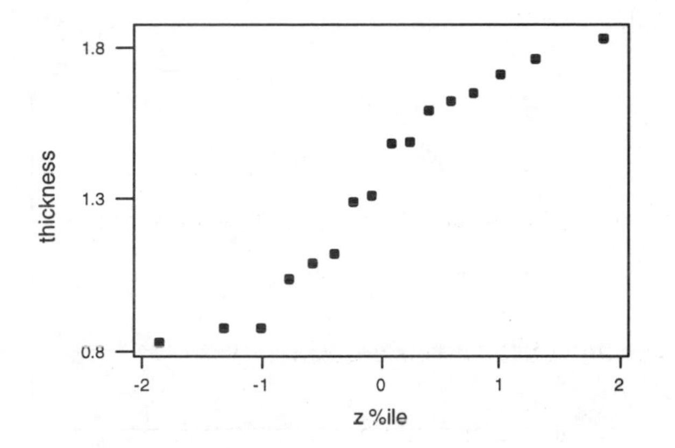

82. The Weibull plot uses ln(observations) and the z percentiles of the p_i values given. The accompanying probability plot appears sufficiently straight to lead us to agree with the argument that the distribution of fracture toughness in concrete specimens could well be modeled by a Weibull distribution.

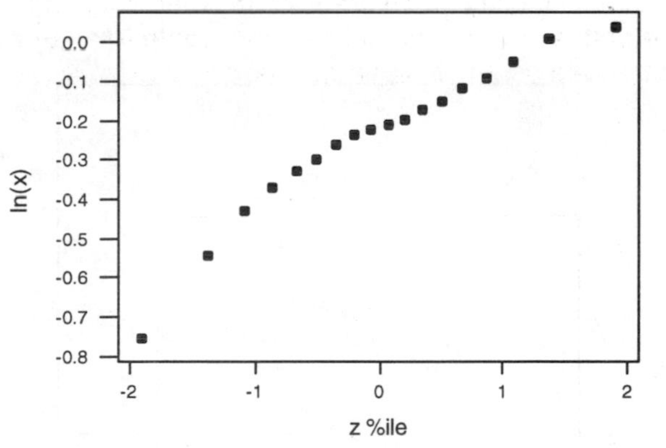

83. The (z percentile,observation) pairs are (-1.66, .736), (-1.32, .863), (-1.01, .865), (-.78, .913), (-.58, .915), (-.40, .937), (-.24, .983), (-.08, 1.007), (.08, 1.011), (.24, 1.064), (.40, 1.109), (.58, 1.132), (.78, 1.140), (1.01, 1.153), (1.32, 1.253), (1.86, 1.394). The accompanying probability plot is very straight, suggesting that an assumption of population normality is extremely plausible.

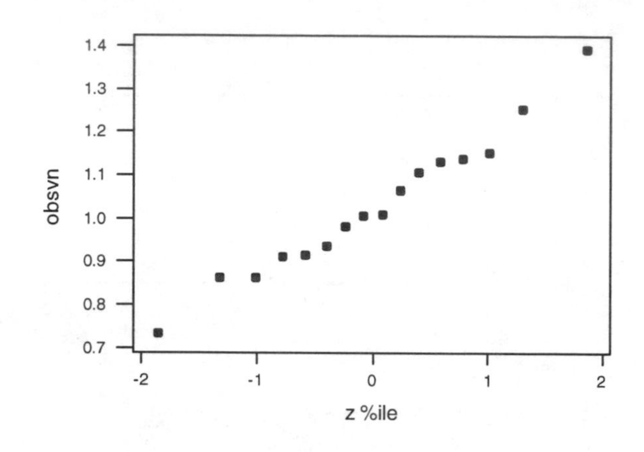

84.

a The 10 largest z percentiles are 1.96, 1.44, 1.15, .93, .76, .60, .45, .32, .19 and .06; the remaining 10 are the negatives of these values. The accompanying normal probability plot is reasonably straight. An assumption of population distribution normality is plausible.

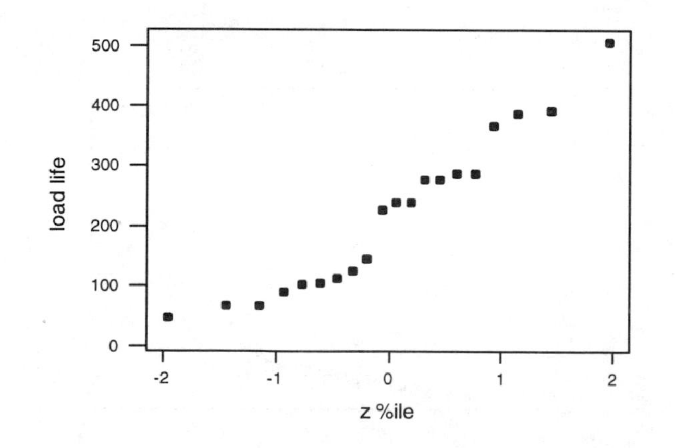

b For a Weibull probability plot, the natural logs of the observations are plotted against extreme value percentiles; these percentiles are -3.68, -2.55, -2.01, -1.65, -1.37, -1.13, -.93, -.76, -.59, -.44, -.30, -.16, -.02, .12, .26, .40, .56, .73, .95, and 1.31. The accompanying probability plot is roughly as straight as the one for checking normality (a plot of ln(x) versus the z percentiles, appropriate for checking the plausibility of a lognormal distribution, is also reasonably straight - any of 3 different families of population distributions seems plausible!).

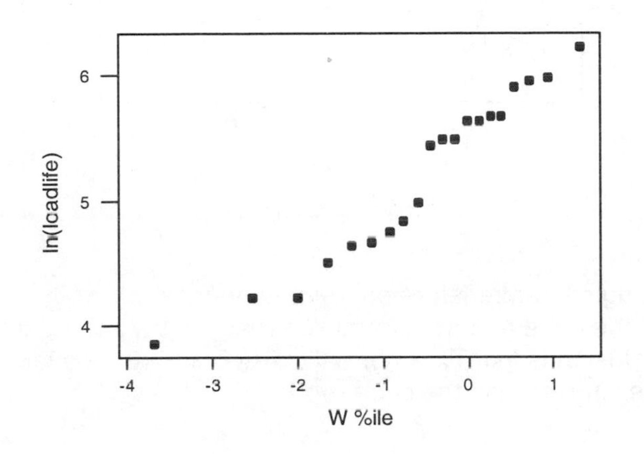

85. To check for plausibility of a lognormal population distribution for the rainfall data of Exercise 81 in Chapter 1, take the natural logs and construct a normal probability plot. This plot and a normal probability plot for the original data appear below. Clearly the log transformation gives quite a straight plot, so lognormality is plausible. The curvature in the plot for the original data implies a positively skewed population distribution - like the lognormal distribution.

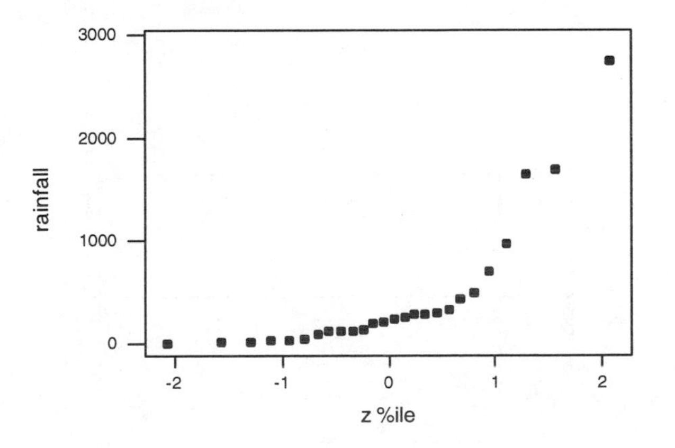

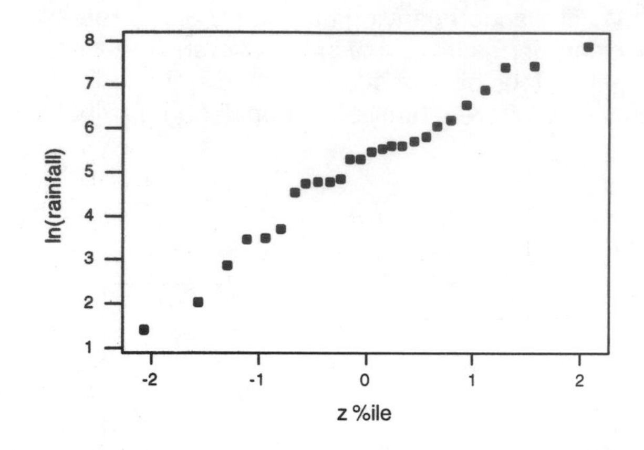

86. The plot of the original (untransformed) data appears somewhat curved. Both the square root and the cube root transformation result in very straight plots. A reasonable conclusion is that Tx is normally distributed (the square root transformation is simpler than the cube root).

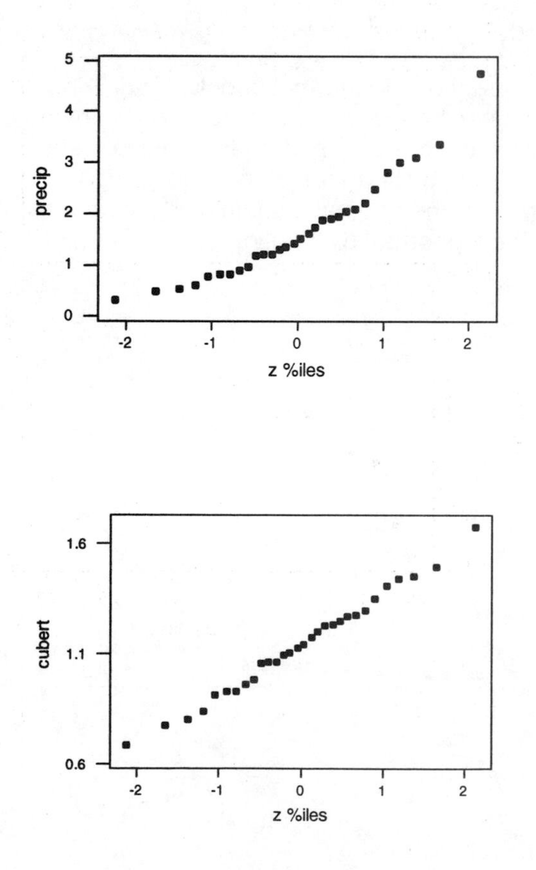

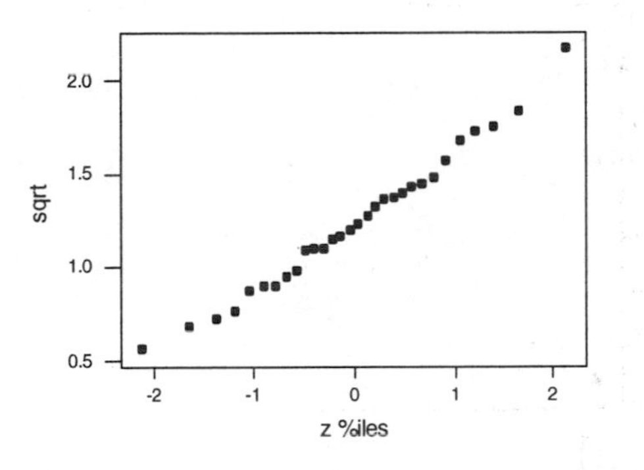

87. The $(100p)^{th}$ percentile $\eta(p)$ for the exponential distribution with $\lambda = 1$ satisfies $F(\eta(p)) = 1 - \exp[-\eta(p)] = p$, i.e., $\eta(p) = -\ln(1 - p)$. With n = 16, we need $\eta(p)$ for $p = \frac{5}{16}, \frac{1.5}{16}, \ldots, \frac{15.5}{16}$. These are .032, .398, .170, .247, .330, .421, .521, .633, .758, .901, 1.068, 1.269, 1.520, 1.856, 2.367, 3.466. this plot exhibits substantial curvature, casting doubt on the assumption of an exponential population distribution. Because λ is a scale parameter (as is σ for the normal family), $\lambda = 1$ can be used to assess the plausibility of the entire exponential family.

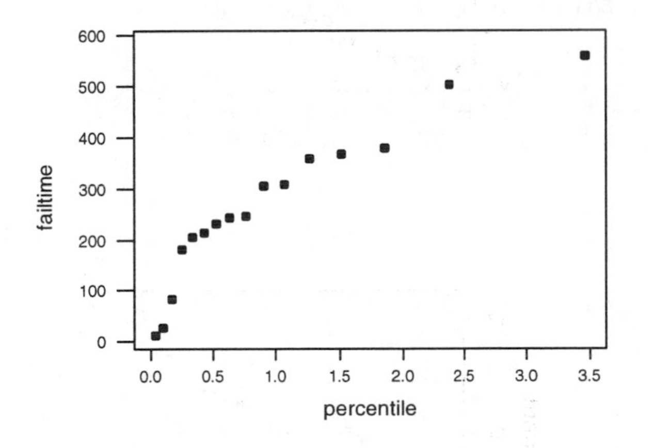

Chapter 4

Supplementary Exercises

88.

a $\quad P(10 \le X \le 20) = \dfrac{10}{25} = .4$

b $\quad P(X \ge 10) = P(10 \le X \le 25) = \dfrac{15}{25} = .6$

c $\quad$ For $0 \le X \le 25$, $F(x) = \int_0^x \frac{1}{25}\,dy = \frac{x}{25}$. $F(x) = 0$ for $x < 0$ and $= 1$ for $x > 25$.

d $\quad E(X) = \dfrac{(A+B)}{2} = \dfrac{(0+25)}{2} = 12.5$

$Var(X) = \dfrac{(B-A)^2}{12} = \dfrac{625}{12} = 52.083$

89.

a $\quad$ For $0 \le Y \le 25$, $F(y) = \frac{1}{24}\int_0^y \left(u - \frac{u^2}{12}\right) = \frac{1}{24}\left[\left(\frac{u^2}{2} - \frac{u^3}{36}\right)\right]_0^y$. Thus

$$F(y) = \begin{cases} 0 & y < 0 \\ \frac{1}{48}\left(y^2 - \frac{y^3}{18}\right) & 0 \le y \le 12 \\ 1 & y > 12 \end{cases}$$

b $\quad P(Y \le 4) = F(4) = .259$, $P(Y > 6) = 1 - F(6) = .5$
$P(4 \le X \le 6) = F(6) - F(4) = .5 - .259 = .241$

c $\quad E(Y) = \dfrac{1}{24}\int_0^{12} y^2\left(1 - \dfrac{y}{12}\right)dy = \dfrac{1}{24}\left[\dfrac{y^3}{3} - \dfrac{y^4}{48}\right]_0^{12} = 6$

$E(Y^2) = \dfrac{1}{24}\int_0^{12} y^3\left(1 - \dfrac{y}{12}\right)dy = 43.2$, so $V(Y) = 43.2 - 36 = 7.2$

d $\quad P(Y < 4 \text{ or } Y > 8) = 1 - P(4 \le X \le 8) = .518$

e $\quad$ the shorter segment has length $\min(Y, 12 - Y)$ so

$E[\min(Y, 12 - Y)] = \int_0^{12} \min(y, 12 - y)\cdot f(y)dy = \int_0^6 \min(y, 12 - y)\cdot f(y)dy$

$+ \int_6^{12} \min(y, 12 - y)\cdot f(y)dy = \int_0^6 y\cdot f(y)dy + \int_6^{12} (12 - y)\cdot f(y)dy = \dfrac{90}{24} = .3.75$

90.

a and **b** Clearly f(x) ≥ 0. The c.d.f. is , for x > 0,

$$F(x)=\int_{-\infty}^{x} f(y)dy=\int_{0}^{x}\frac{32}{(y+4)^3}dy=-\frac{1}{2}\cdot\frac{32}{(y+4)^2}\Bigg]_0^x=1-\frac{16}{(x+4)^2}$$

(F(x) = 0 for x ≤ 0.)

Since F(∞) = $\int_{-\infty}^{\infty} f(y)dy=1$, f(x) is a legitimate pdf.

c P(2 ≤ X ≤ 5) = F(5) – F(2) = $1-\frac{16}{81}-\left(1-\frac{16}{36}\right)=.247$

d $E(x)=\int_{-\infty}^{\infty} x\cdot f(x)dx=\int_{-\infty}^{\infty} x\cdot\frac{32}{(x+4)^3}dx=\int_{0}^{\infty}(x+4-4)\cdot\frac{32}{(x+4)^3}dx$

$$=\int_{0}^{\infty}\frac{32}{(x+4)^2}dx-4\int_{0}^{\infty}\frac{32}{(x+4)^3}dx=8-4=4$$

e E(salvage value) =

$$=\int_{0}^{\infty}\frac{100}{x+4}\cdot\frac{32}{(y+4)^3}dx=3200\int_{0}^{\infty}\frac{1}{(y+4)^4}dx=\frac{3200}{(3)(64)}=16.67$$

91.

a By differentiation,

$$f(x)=\begin{cases} x^2 & 0\le x<1 \\ \frac{7}{4}-\frac{3}{4}x & 1\le y\le\frac{7}{3} \\ 0 & otherwise \end{cases}$$

b P(.5 ≤ X ≤ 2) = F(2) – F(.5) = $1-\frac{1}{2}\left(\frac{7}{3}-2\right)\left(\frac{7}{4}-\frac{3}{4}\cdot 2\right)-\frac{(.5)^3}{3}=\frac{11}{12}=.917$

c E(X) = $\int_{0}^{1} x\cdot x^2dx+\int_{1}^{7/3} x\cdot\left(\frac{7}{4}-\frac{3}{4}x\right)dx=\frac{131}{108}=1.213$

92. μ = 40 V; σ = 1.5 V

a P(39 < X < 42) = $\Phi\left(\frac{42-40}{1.5}\right)-\Phi\left(\frac{39-40}{1.5}\right)$

= Φ(1.33) - Φ(-.67) = .9082 - .2514 = .6568

b We desire the 85th percentile: 40 + (1.04)(1.5) = 41.56

c P(X > 42) = 1 – P(X ≤ 42) = 1 $-\Phi\left(\frac{42-40}{1.5}\right)$ = 1 - Φ(1.33) = .0918

Let D represent the number of diodes out of 4 with voltage exceeding 42.

P(D ≥ 1) = 1 – P(D = 0) = $1-\binom{4}{0}(.0918)^0(.9082)^4$ =1 - .6803 = .3197

93. $\mu = 137.2$ oz.; $\sigma = 1.6$ oz

a $P(X > 135) = 1 - \Phi\left(\frac{135-137.2}{1.6}\right) = 1 - \Phi(-1.38) = 1 - .0838 = .9162$

b With Y = the number among ten that contain more than 135 oz, Y ~ Bin(10, .9162, so $P(Y \geq 8) = b(8; 10, .9162) + b(9; 10, .9162) + b(10; 10, .9162) = .9549$.

c $\mu = 137.2$; $\frac{135-137.2}{\sigma} = -1.65 \Rightarrow \sigma = 1.33$

94.

a Let S = defective. Then p = P(S) = .05; n = 250 $\Rightarrow \mu = np = 12.5$, $\sigma = 3.446$. The random variable X = the number of defectives in the batch of 250. X ~ Binomial. Since np = 12.5 ≥ 10, and nq = 237.5 ≥ 10, we can use the normal approximation.

$P(X_{bin} \geq 25) \approx 1 - \Phi\left(\frac{24.5-12.5}{3.446}\right) = 1 - \Phi(3.48) = 1 - .9997 = .0003$

b $P(X_{bin} = 10) \approx P(X_{norm} \leq 10.5) - P(X_{norm} \leq 9.5)$
$= \Phi(-.58) - \Phi(-.87) = .2810 - .1922 = .0888$

95.

a $P(X > 100) = 1 - \Phi\left(\frac{100-96}{14}\right) = 1 - \Phi(.29) = 1 - .6141 = .3859$

b $P(50 < X < 75) = \Phi\left(\frac{75-96}{14}\right) - \Phi\left(\frac{50-96}{14}\right)$
$= \Phi(-1.5) - \Phi(-3.29) = .0668 - .0005 = .0663$.

c a = 5th percentile = 96 + (-1.645)(14) = 72.97.
b = 95th percentile = 96 + (1.645)(14) = 119.03. The interval (72.97, 119.03) contains the central 90% of all grain sizes.

96.

a F(X) = 0 for x < 1 and = 1 for x > 3. For 1 ≤ x ≤ 3, $F(x) = \int_{-\infty}^{x} f(y)dy$
$= \int_{-\infty}^{1} 0dy + \int_{1}^{x} \frac{3}{2} \cdot \frac{1}{y^2} dy = 1.51\left(1 - \frac{1}{x}\right)$

b $P(X \leq 2.5) = F(2.5) = 1.5(1 - .4) = .9$; $P(1.5 \leq x \leq 2.5) = F(2.5) - F(1.5) = .4$

c $E(X) = = \int_{1}^{3} x \cdot \frac{3}{2} \cdot \frac{1}{x^2} dx = \frac{3}{2}\int_{1}^{3} \frac{1}{x} dx = 1.5\ln(x)\Big]_1^3 = 1.648$

d $E(X^2) = = \int_{1}^{3} x^2 \cdot \frac{3}{2} \cdot \frac{1}{x^2} dx = \frac{3}{2}\int_{1}^{3} dx = 3$, so $V(X) = E(X^2) - [E(X)]^2 = .284$, $\sigma = .553$

e $h(x) = \begin{cases} 0 & 1 \leq x \leq 1.5 \\ x - 1.5 & 1.5 \leq x \leq 2.5 \\ 1 & 2.5 \leq x \leq 3 \end{cases}$

so $E[h(X)] = = \int_{1.5}^{2.5} (x-1.5) \cdot \frac{3}{2} \cdot \frac{1}{x^2} dx + \int_{2.5}^{3} 1 \cdot \frac{3}{2} \cdot \frac{1}{x^2} dx = .267$

97.

a

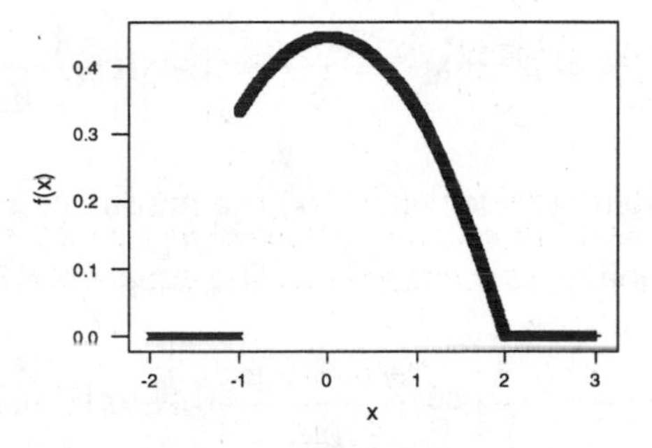

b F(x) = 0 for x < -1 or == 1 for x > 2. For $-1 \le x \le 2$,

$$F(x) = \int_{-1}^{x} \frac{1}{9}\left(4 - y^2\right)dy = \frac{1}{9}\left(4x - \frac{x^3}{3}\right) + \frac{11}{27}$$

c The median is 0 iff F(0) = .5. Since F(0) = $\frac{11}{27}$, this is not the case. Because $\frac{11}{27} < .5$, the median must be greater than 0.

d Y is a binomial r.v. with n = 10 and p = P(X > 1) = 1 – F(1) = $\frac{5}{27}$

98.

a E(X) = $\frac{1}{\lambda}$ = 1.075, $\sigma = \frac{1}{\lambda}$ = 1.075

b P(3.0 < X) = 1 – P(X ≤ 3.0) = 1 – F(3.0) = $3^{-.93(3.0)}$ = .0614
P(1.0 ≤ X ≤ 3.0) = F(3.0) – F(1.0) = .333

c The 90th percentile is requested; denoting it by c, we have
.9 = F(c) = 1 – $e^{-(.93)c}$, whence c = $\frac{\ln(.1)}{(-.93)} = 2.476$

99.

a P(X ≤ 150) = $\exp\left[-\exp\left(\frac{-(150-150)}{90}\right)\right] = \exp[-\exp(0)] = \exp(-1) = .368$, where exp(u) = e^u. P(X ≤ 300) = $\exp[-\exp(-1.6667)] = .828$,
and P(150 ≤ X ≤ 300) = .828 - .368 = .460.

b The desired value c is the 90th percentile, so c satisfies
$.9 = \exp\left[-\exp\left(\frac{-(c-150)}{90}\right)\right]$. Taking the natural log of each side twice in succession yields ln[ln(.9)] = $\frac{-(c-150)}{90}$, so c = 90(2.250367) + 150 = 352.53.

c $f(x) = F'(X) = \frac{1}{\beta}\cdot\exp\left[-\exp\left(\frac{-(x-\alpha)}{\beta}\right)\right]\cdot\exp\left(\frac{-(x-\alpha)}{\beta}\right)$

d We wish the value of x for which f(x) is a maximum; this is the same as the value of x for which ln[f(x)] is a maximum. The equation of $\frac{d[\ln(f(x))]}{dx} = 0$ gives $\exp\left(\frac{-(x-\alpha)}{\beta}\right) = 1$, so $\frac{-(x-\alpha)}{\beta} = 0$, which implies that $x = \alpha$. Thus the mode is α.

e $E(X) = .5772\beta + \alpha = 201.95$, whereas the mode is 150 and the median is $-(90)\ln[-\ln(.5)] + 150 = 182.99$. The distribution is positively skewed.

100.

a $E(cX) = cE(X) = \frac{c}{\lambda}$

b $E[c(1 - .5e^{ax})] = \int_0^\infty c(1-.5e^{ax})\cdot\lambda e^{-\lambda x}dx = \frac{c[.5\lambda - a]}{\lambda - a}$

101.

a From a graph of $f(x;\mu,\sigma)$ or by differentiation, $x^* = \mu$.

b No; the density function has constant height for $A \le X \le B$.

c $F(x;\lambda)$ is largest for x = 0 (the derivative at 0 does not exist since f is not continuous there) so $x^* = 0$.

d $\ln[f(x;\alpha,\beta)] = -\ln(\beta^\alpha) - \ln(\Gamma(\alpha)) + (\alpha-1)\ln(x) - \frac{x}{\beta};$

$$\frac{d}{dx}\ln[f(x;\alpha,\beta)] = \frac{\alpha-1}{x} - \frac{1}{\beta} \Rightarrow x = x^* = (\alpha-1)\beta$$

e From **d** $x^* = \left(\frac{\nu}{2} - 1\right)(2) = \nu - 2.$

102.

a $\int_{-\infty}^{\infty} f(x)dx = \int_{-\infty}^{0} .1e^{.2x}dx + \int_{0}^{\infty} .1e^{-.2x}dx = .5 + .5 = 1$

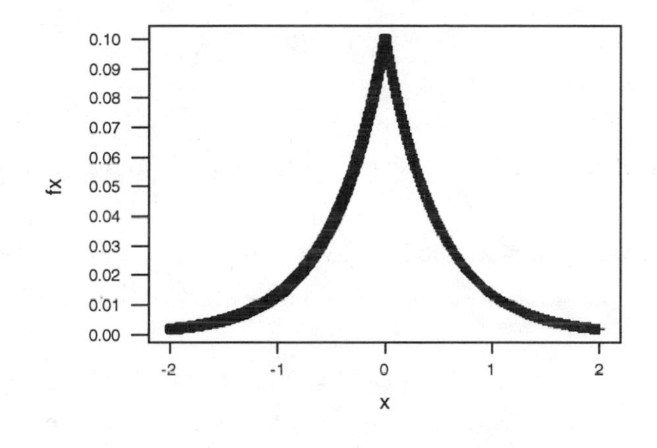

b For x < 0, F(x) = $\int_{-\infty}^{x} .1e^{.2y}dy = \frac{1}{2}e^{.2x}$.

For x ≥ 0, F(x) = $\frac{1}{2} + \int_{0}^{x} .1e^{-.2y}dy = 1 - \frac{1}{2}e^{-.2x}$.

c P(X < 0) = F(0) = $\frac{1}{2} = .5$, P(X < 2) = F(2) = 1 - .5e$^{-.4}$ = .665,

P(-1 ≤ X ≤ 2) – F(2) – F(-1) = .256, 1 - (-2 ≤ X ≤ 2) = .670

103.

a Clearly f(x; λ_1, λ_2, p) ≥ 0 for all x, and $\int_{-\infty}^{\infty} f(x;\lambda_1,\lambda_2,p)dx$

$= \int_0^\infty \left[p\lambda_1 e^{-\lambda_1 x} + (1-p)\lambda_2 e^{-\lambda_2 x}\right]dx = p\int_0^\infty \lambda_1 e^{-\lambda_1 x}dx + (1-p)\int_0^\infty \lambda_2 e^{-\lambda_2 x}dx$

= p + (1 – p) = 1

b For x > 0, F(x; λ_1, λ_2, p) = $\int_0^x f(y;\lambda_1,\lambda_2,p)dy = p(1-e^{-\lambda_1 x}) + (1-p)(1-e^{-\lambda_2 x})$.

c E(X) = $\int_0^\infty x\cdot\left[p\lambda_1 e^{-\lambda_1 x}) + (1-p)\lambda_2 e^{-\lambda_2 x})\right]dx$

$= p\int_0^\infty x\lambda_1 e^{-\lambda_1 x}dx + (1-p)\int_0^\infty x\lambda_2 e^{-\lambda_2 x}dx = \frac{p}{\lambda_1} + \frac{(1-p)}{\lambda_2}$

d $E(X^2) = \frac{2p}{\lambda_1^2} + \frac{2(1-p)}{\lambda_2^2}$, so $Var(X) = \frac{2p}{\lambda_1^2} + \frac{2(1-p)}{\lambda_2^2} - \left[\frac{p}{\lambda_1} + \frac{(1-p)}{\lambda_2}\right]^2$

e For an exponential r.v., CV = $\frac{1/\lambda}{1/\lambda} = 1$. For X hyperexponential,

$$CV = \left[\frac{\frac{2p}{\lambda_1^2}+\frac{2(1-p)}{\lambda_2^2}}{\left[\frac{p}{\lambda_1}+\frac{(1-p)}{\lambda_2}\right]^2} - 1\right]^{1/2} = \left[\frac{2(p\lambda_2^2 + (1-p)\lambda_1^2)}{(p\lambda_2 + (1-p)\lambda_1)^2} - 1\right]^{1/2}$$

= $[2r - 1]^{1/2}$ where r = $\frac{(p\lambda_2^2 + (1-p)\lambda_1^2)}{(p\lambda_2 + (1-p)\lambda_1)^2}$. But straightforward algebra shows that r > 1 provided $\lambda_1 \neq \lambda_2$, so that CV > 1.

f $\mu = \frac{n}{\lambda}$, $\sigma^2 = \frac{n}{\lambda^2}$, so $\sigma = \frac{\sqrt{n}}{\lambda}$ and CV = $\frac{1}{\sqrt{n}} < 1$ if n > 1.

104.

a $1 = \int_5^\infty \frac{k}{x^\alpha} dx = k \cdot \frac{5^{1-\alpha}}{\alpha - 1} \Rightarrow k = (\alpha - 1)5^{1-\alpha}$ where we must have $\alpha > 1$.

b For $x \geq 5$, F(x) = $\int_5^x \frac{k}{y^\alpha} dy = 5^{1-\alpha}\left[\frac{1}{5^{1-\alpha}} - \frac{1}{x^{\alpha-1}}\right] = 1 - \left(\frac{5}{x}\right)^{\alpha-1}$.

c E(X) = $\int_5^\infty x \cdot \frac{k}{x^\alpha} dx = \int_5^\infty x \cdot \frac{k}{x^{\alpha-1}} dx = \frac{k}{5^{\alpha-2} \cdot (\alpha - 2)}$, provided $\alpha > 2$.

d $P\left(\ln\left(\frac{X}{5}\right) \leq y\right) = P\left(\frac{X}{5} \leq e^y\right) = P(X \leq 5e^y) = F(5e^y) = 1 - \left(\frac{5}{5e^y}\right)^{\alpha-1}$

$1 - e^{-(\alpha-1)y}$, the cdf of an exponential r.v. with parameter α - 1.

105.

a A lognormal distribution, since $\ln\left(\frac{I_o}{I_i}\right)$ is a normal r.v.

b $P(I_o > 2I_i) = P\left(\frac{I_o}{I_i} > 2\right) = P\left(\ln\left(\frac{I_o}{I_i}\right) > \ln 2\right) = 1 - P\left(\ln\left(\frac{I_o}{I_i}\right) \leq \ln 2\right)$

$1 - \Phi\left(\frac{\ln 2 - 1}{.05}\right) = 1 - \Phi(-6.14) = 1$

c $E\left(\frac{I_o}{I_i}\right) = e^{1+.0025/2} = 2.72,\ Var\left(\frac{I_o}{I_i}\right) = e^{2+.0025} \cdot (e^{.0025} - 1) = .0185$

106.

a

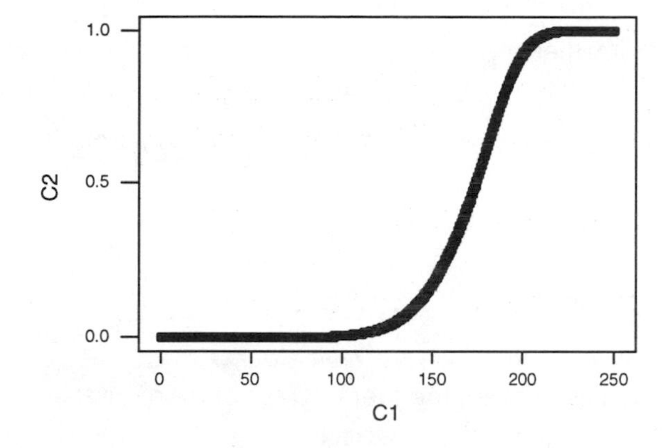

b $P(X > 175) = 1 - F(175; 9, 180) = e^{-\left(\frac{175}{180}\right)^9} = .4602$

$P(150 \le X \le 175) = F(175; 9, 180) - F(150; 9, 180) = .5398 - .1762 = .3636$

c $P(\text{at least one}) = 1 - P(\text{none}) = 1 - (1 - .3636)^2 = .5950$

d We want the 10th percentile: $.10 = F(x; 9, 180) = 1 - e^{-\left(\frac{x}{180}\right)^9}$. A small bit of algebra leads us to x = 140.178. Thus 10% of all tensile strengths will be less than 140.178 MPa.

107. $F(y) = P(Y \le y) = P(\sigma Z + \mu \le y) = P\left(Z \le \frac{(y-\mu)}{\sigma}\right) = \int_{-\infty}^{\frac{(y-\mu)}{\sigma}} \frac{1}{\sqrt{2\pi}} e^{-\frac{1}{2}z^2} dz$. Now differentiate with respect to y to obtain a normal pdf with parameters μ and σ.

108.

a $F_Y(y) = P(Y \le y) = P(60X \le y) = P\left(X \le \frac{y}{60}\right) = F\left(\frac{y}{60\beta}; \alpha\right)$. Thus $f_Y(y)$

$$= f\left(\frac{y}{60\beta}; \alpha\right) \cdot \frac{1}{60\beta} = \frac{y^{\alpha-1} e^{\frac{-y}{60\beta}}}{(60\beta)^\alpha \Gamma(\alpha)},$$

which shows that Y has a gamma distribution with parameters α and 60β.

b With c replacing 60 in **a**, the same argument shows that cX has a gamma distribution with parameters α and $c\beta$.

109.

a $Y = -\ln(X) \Rightarrow x = e^{-y} = k(y)$, so $k'(y) = -e^{-y}$. Thus since $f(x) = 1$, $g(y) = 1 \cdot |-e^{-y}| = e^{-y}$ for $0 < y < \infty$, so y has an exponential distribution with parameter $\lambda = 1$.

b $y = \sigma Z + \mu \Rightarrow y = h(z) = \sigma Z + \mu \Rightarrow z = k(y) = \dfrac{(y-\mu)}{\sigma}$ and $k'(y) = \dfrac{1}{\sigma}$, from which the result follows easily.

c $y = h(x) = cx \Rightarrow x = k(y) = \dfrac{y}{c}$ and $k'(y) = \dfrac{1}{c}$, from which the result follows easily.

110.

a $F(x) = \lambda e^{-\lambda x}$ and $F(x) = 1 - e^{-\lambda x}$, so $r(x) = \dfrac{\lambda e^{-\lambda x}}{e^{-\lambda x}} = \lambda$, a constant (independent of X); this is consistent with the memoryless property of the exponential distribution.

b $r(x) = \left(\dfrac{\alpha}{\beta^{\alpha}}\right)x^{\alpha-1}$; for $\alpha > 1$ this is increasing, while for $\alpha < 1$ it is a decreasing function.

c $\ln(1 - F(x)) = -\int \alpha\left(1 - \dfrac{x}{\beta}\right)dx = -\alpha\left[x - \dfrac{x^2}{2\beta}\right] \Rightarrow F(x) = 1 - e^{-\alpha\left(x - \frac{x^2}{2\beta}\right)}$,

$f(x) = \alpha\left(1 - \dfrac{x}{\beta}\right)e^{-\alpha\left(x - \frac{x^2}{2\beta}\right)}$ $\quad 0 \le x \le \beta$

111.

a $F_X(x) = P\left(-\dfrac{1}{\lambda}\ln(1-U) \le x\right) = P(\ln(1-U) \ge -\lambda x) = P(1 - U \ge e^{-\lambda x})$

$= P(U \le 1 - e^{-\lambda x}) = 1 - e^{-\lambda x}$ since $F_U(u) = u$ (U is uniform on [0, 1]). Thus X has an exponential distribution with parameter λ.

b By taking successive random numbers $u_1, u_2, u_3, \ldots$ and computing $x_i = -\dfrac{1}{10}\ln(1 - u_i), \ldots$ we obtain a sequence of values generated from an exponential distribution with parameter $\lambda = 10$.

Chapter 4

112.

a $E(g(X)) \approx E[g(\mu) + g'(\mu)(X - \mu)] = E(g(\mu)) + g'(\mu)\cdot E(X - \mu)$, but $E(X) - \mu = 0$ and $E(g(\mu)) = g(\mu)$ (since $g(\mu)$ is constant), giving $E(g(X)) \approx g(\mu)$.
$V(g(X)) \approx V[g(\mu) + g'(\mu)(X - \mu)] = V[g'(\mu)(X - \mu)] = (g'(\mu))^2\cdot V(X - \mu) = (g'(\mu))^2\cdot V(X)$.

b $g(I) = \frac{v}{I}, g'(I) = \frac{-v}{I^2}$, so $E(g(I)) = \mu_R \approx \frac{v}{\mu_I} = \frac{v}{20}$

$$V(g(I)) \approx \left(\frac{-v}{\mu_I^2}\right)^2 \cdot V(I), \sigma_{g(I)} \approx \frac{v}{20^2}\cdot\sigma_I = \frac{v}{800}$$

113. $g(\mu) + g'(\mu)(X - \mu) \le g(X)$ implies that $E[g(\mu) + g'(\mu)(X - \mu)] = E(g(\mu)) = g(\mu) \le E(g(X))$, i.e. that $g(E(X)) \le E(g(X))$.

114. For $y > 0$, $F(y) = P(Y \le y) = P\left(\frac{2X^2}{\beta^2} \le y\right) = P\left(X^2 \le \frac{\beta^2 y}{2}\right) = P\left(X \le \frac{\beta\sqrt{y}}{\sqrt{2}}\right)$. Now take the cdf of X (Weibull), replace x by $\frac{\beta\sqrt{y}}{\sqrt{2}}$, and then differentiate with respect to y to obtain the desired result $f_Y(y)$.

Chapter 5

Section 5.1

1.

a $P(X = 1, Y = 1) = p(1,1) = .20$

b $P(X \le 1 \text{ and } Y \le 1) = p(0,0) + p(0,1) + p(1,0) + p(1,1) = .42$

c At least one hose is in use at both islands. $P(X \ne 0 \text{ and } Y \ne 0) = p(1,1) + p(1,2) + p(2,1) + p(2,2) = .70$

d By summing row probabilities, $p_x(x) = .16, .34, .50$ for $x = 0, 1, 2$, and by summing column probabilities, $p_y(y) = .24, .38, .38$ for $y = 0, 1, 2$. $P(X \le 1) = p_x(0) + p_x(1) = .50$

e $P(0,0) = .10$, but $p_x(0) \cdot p_y(0) = (.16)(.24) = .0384 \ne .10$, so X and Y are not independent.

2.

a

	p(x,y)	y: 0	1	2	3	4	
	0	.30	.05	.025	.025	.10	.5
x	1	.18	.03	.015	.015	.06	.3
	2	.12	.02	.01	.01	.04	.2
		.6	.1	.05	.05	.2	

b $P(X \le 1 \text{ and } Y \le 1) = p(0,0) + p(0,1) + p(1,0) + p(1,1) = .56$
$= (.8)(.7) = P(X \le 1) \cdot P(Y \le 1)$

c $P(X + Y = 0) = P(X = 0 \text{ and } Y = 0) = p(0,0) = .30$

d $P(X + Y \le 1) = p(0,0) + p(0,1) + p(1,0) = .53$

3.

a $p(1,1) = .15$, the entry in the 1st row and 1st column of the joint probability table.

b $P(X_1 = X_2) = p(0,0) + p(1,1) + p(2,2) + p(3,3) = .08+.15+.10+.07 = .40$

c $A = \{ (x_1, x_2): x_1 \ge 2 + x_2 \} \cup \{ (x_1, x_2): x_2 \ge 2 + x_1 \}$
$P(A) = p(2,0) + p(3,0) + p(4,0) + p(3,1) + p(4,1) + p(4,2) + p(0,2) + p(0,3) + p(1,3) = .22$

d $P(\text{exactly } 4) = p(1,3) + p(2,2) + p(3,1) + p(4,0) = .17$
$P(\text{at least } 4) = P(\text{exactly } 4) + p(4,1) + p(4,2) + p(4,3) + p(3,2) + p(3,3) + p(2,3) = .46$

4.

a $P_1(0) = P(X_1 = 0) = p(0,0) + p(0,1) + p(0,2) + p(0,3) = .19$
$P_1(1) = P(X_1 = 1) = p(1,0) + p(1,1) + p(1,2) + p(1,3) = .30$, etc.

x_1	0	1	2	3	4
$p_1(x_1)$	.19	.30	.25	.14	.12

b $P_2(0) = P(X_2 = 0) = p(0,0) + p(1,0) + p(2,0) + p(3,0) + p(4,0) = .19$, etc

x_2	0	1	2	3
$p_2(x_2)$	.19	.30	.28	.23

c $p(4,0) = 0$, yet $p_1(4) = .12 > 0$ and $p_2(0) = .19 > 0$, so $p(x_1, x_2) \neq p_1(x_1) \cdot p_2(x_2)$ for every (x_1, x_2), and the two variables are not independent.

5.

a $P(X = 3, Y = 3) = P(3 \text{ customers, each with 1 package})$
$= P(\text{each has 1 package} \mid 3 \text{ customers}) \cdot P(3 \text{ customers})$
$= (.6)^3 \cdot (.25) = .054$

b $P(X = 4, Y = 11) = P(\text{total of 11 packages} \mid 4 \text{ customers}) \cdot P(4 \text{ customers})$
Given that there are 4 customers, there are 4 different ways to have a total of 11 packages: 3, 3, 3, 2 or 3, 3, 2, 3 or 3, 2, 3 ,3 or 2, 3, 3, 3. Each way has probability $(.1)^3(.3)$, so $p(4, 11) = 4(.1)^3(.3)(.15) = .00018$

6.

a $p(4,2) = P(Y = 2 \mid X = 4) \cdot P(X = 4) = \left[\binom{4}{2}(.6)^2(.4)^2\right] \cdot (.15) = .0518$

b $P(X = Y) = p(0,0) + p(1,1) + p(2,2) + p(3,3) + p(4,4) = .1+(.2)(.6) + (.3)(.6)^2 + (.25)(.6)^3 + (.15)(.6)^4 = .4014$

c $p(x,y) = 0$ unless $y = 0, 1, \ldots, x$; $x = 0, 1, 2, 3, 4$. For any such pair,

$$p(x,y) = P(Y = y \mid X = x) \cdot P(X = x) = \binom{x}{y}(.6)^y(.4)^{x-y} \cdot p_x(x)$$

$p_y(4) = p(y = 4) = p(x = 4, y = 4) = p(4,4) = (.6)^4 \cdot (.15) = .0194$

$$p_y(3) = p(3,3) + p(4,3) = (.6)^3(.25) + \binom{4}{3}(.6)^3(.4)(.15) = .1058$$

$$p_y(2) = p(2,2) + p(3,2) + p(4,2) = (.6)^2(.3) + \binom{3}{2}(.6)^2(.4)(.25)$$

$$+\binom{4}{2}(.6)^2(.4)^2(.15) = .2678$$

$$p_y(1) = p(1,1) + p(2,1) + p(3,1) + p(4,1) = (.6)(.2) + \binom{2}{1}(.6)(.4)(.3)$$

$$\binom{3}{1}(.6)(.4)^2(.25) + \binom{4}{1}(.6)(.4)^3(.15) = .3590$$

$p_y(0) = 1 - [.3590+.2678+.1058+.0194] = .2480$

7.

a $p(1,1) = .030$

b $P(X \le 1$ and $Y \le 1 = p(0,0) + p(0,1) + p(1,0) + p(1,1) = .120$

c $P(X = 1) = p(1,0) + p(1,1) + p(1,2) = .100$; $P(Y = 1) = p(0,1) + \ldots + p(5,1) = .300$

d $P(\text{overflow}) = P(X + 3Y > 5) = 1 - P(X + 3Y \le 5) = 1 - P[(X,Y)=(0,0)$ or …or $(5,0)$ or $(0,1)$ or $(1,1)$ or $(2,1)] = 1 - .620 = .380$

e The marginal probabilities for X (row sums from the joint probability table) are $p_x(0) = .05$, $p_x(1) = .10$, $p_x(2) = .25$, $p_x(3) = .30$, $p_x(4) = .20$, $p_x(5) = .10$; those for Y (column sums) are $p_y(0) = .5$, $p_y(1) = .3$, $p_y(2) = .2$. It is now easily verified that for every (x,y), $p(x,y) = p_x(x) \cdot p_y(y)$, so X and Y are independent.

8.

a

$$\text{numerator} = \binom{8}{3}\binom{10}{2}\binom{12}{1} = (56)(45)(12) = 30{,}240$$

$$\text{denominator} = \binom{30}{6} = 593{,}775;\ p(3,2) = \frac{30{,}240}{593{,}775} = .0509$$

b

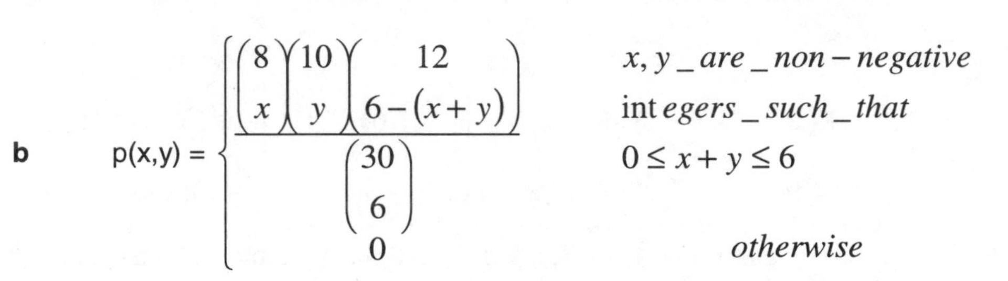

$$p(x,y) = \begin{cases} \dfrac{\binom{8}{x}\binom{10}{y}\binom{12}{6-(x+y)}}{\binom{30}{6}} & x, y\_are\_non-negative\ \ int\,egers\_such\_that\ \ 0 \le x+y \le 6 \\ 0 & otherwise \end{cases}$$

9.

a

$$1 = \int_{-\infty}^{\infty}\int_{-\infty}^{\infty} f(x,y)\,dxdy = \int_{20}^{30}\int_{20}^{30} K(x^2+y^2)\,dxdy$$

$$= K\int_{20}^{30}\int_{20}^{30} x^2\,dydx + K\int_{20}^{30}\int_{20}^{30} y^2\,dxdy = 10K\int_{20}^{30} x^2\,dx + 10K\int_{20}^{30} y^2\,dy$$

$$= 20K \cdot \left(\frac{19{,}000}{3}\right) \Rightarrow K = \frac{3}{380{,}000}$$

b

$$P(X < 26 \text{ and } Y < 26) = \int_{20}^{26}\int_{20}^{26} K(x^2+y^2)\,dxdy = 12K\int_{20}^{26} x^2\,dx$$

$$4Kx^3\Big|_{20}^{26} = 38{,}304K = .3024$$

c

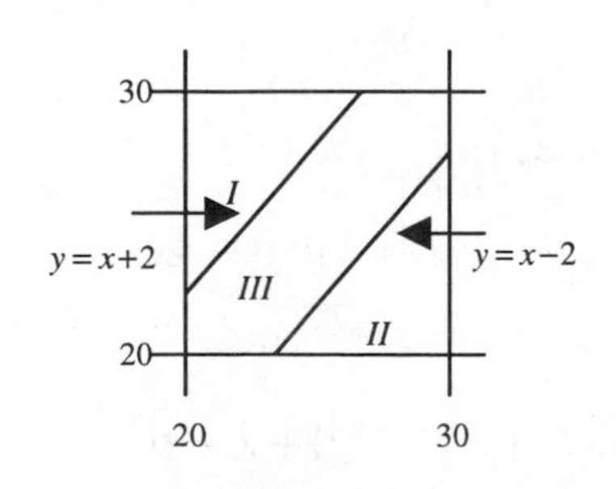

$$P(\,|X-Y|\le 2\,) = \iint_{\substack{region\\III}} f(x,y)dxdy$$

$$1-\iint_{I} f(x,y)dxdy - \iint_{II} f(x,y)dxdy$$

$$1-\int_{20}^{28}\int_{x+2}^{30} f(x,y)dydx - \int_{22}^{30}\int_{20}^{x-2} f(x,y)dydx$$

= (after much algebra) .3593

d $\quad f_x(x) = \int_{-\infty}^{\infty} f(x,y)dy = \int_{20}^{30} K(x^2+y^2)dy = 10Kx^2 + K\left.\frac{y^3}{3}\right|_{20}^{30}$

$= 10Kx^2 + .05, \qquad 20 \le x \le 30$

e $\quad f_y(y)$ is obtained by substituting y for x in (d); clearly $f(x,y) \ne f_x(x) \cdot f_y(y)$, so X and Y are not independent.

10.

a $\quad f(x,y) = \begin{cases} 1 & 5 \le x \le 6, 5 \le y \le 6 \\ 0 & otherwise \end{cases}$

since $f_x(x) = 1$, $f_y(y) = 1$ for $5 \le x \le 6$, $5 \le y \le 6$

b $\quad P(5.25 \le X \le 5.75, 5.25 \le Y \le 5.75) = P(5.25 \le X \le 5.75) \cdot P(5.25 \le Y \le 5.75) =$ (by independence) (.5)(.5) = .25

c

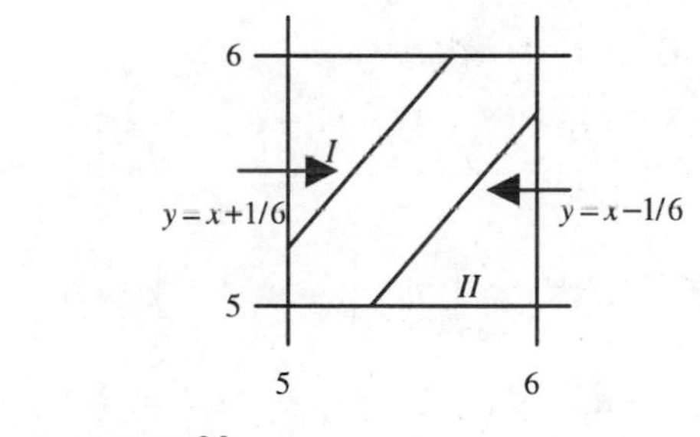

$$P((X,Y) \in A) = \iint_A 1dxdy$$

= area of A = 1 – (area of I + area of II)

$$= 1-\frac{25}{36} = \frac{11}{36} = .306$$

11.

a $\quad p(x,y) = \frac{e^{-\lambda}\lambda^x}{x!} \cdot \frac{e^{-\mu}\mu^y}{y!}$ for x = 0, 1, 2, …; y = 0, 1, 2, …

b $\quad p(0,0) + p(0,1) + p(1,0) = e^{-\lambda-\mu}\left[1+\lambda+\mu\right]$

c P(X+Y=m) = $\sum_{k=0}^{m} P(X=k, Y=m-k) = \sum_{k=0}^{m} e^{-\lambda-\mu} \frac{\lambda^k}{k!} \frac{\mu^{m=k}}{(m-k)!}$

$\frac{e^{-(\lambda+\mu)}}{m!} \sum_{k=0}^{m} \binom{m}{k} \lambda^k \mu^{m-k} = \frac{e^{-(\lambda+\mu)}(\lambda+\mu)^m}{m!}$, so the total # of errors X+Y also has a Poisson distribution with parameter $\lambda+\mu$.

12.

a P(X> 3) = $\int_3^\infty \int_0^\infty xe^{-x(1+y)} dydx = \int_3^\infty e^{-x} dx = .050$

b The marginal pdf of X is $\int_0^\infty xe^{-x(1+y)} dy = e^{-x}$ for $0 \le x$; that of Y is

$\int_3^\infty xe^{-x(1+y)} dx = \frac{1}{(1+y)^2}$ for $0 \le y$. It is now clear that f(x,y) is not the product of the marginal pdf's, so the two r.v's are not independent.

c P(at least one exceeds 3) = 1 – P(X ≤ 3 and Y ≤ 3)

$= 1 - \int_0^3 \int_0^3 xe^{-x(1+y)} dydx = 1 - \int_0^3 \int_0^3 xe^{-x} e^{-xy} dy$

$= 1 - \int_0^3 e^{-x}(1-e^{-3x})dx = e^{-3} + .25 - .25e^{-12} = .300$

13.

a f(x,y) = $f_x(x) \cdot f_y(y) = \begin{cases} e^{-x-y} & x \ge 0, y \ge 0 \\ 0 & otherwise \end{cases}$

b P(X ≤ 1 and Y ≤ 1) = P(X ≤ 1) · P(Y ≤ 1) = $(1 - e^{-1})(1 - e^{-1}) = .400$

c P(X + Y ≤ 2) = $\int_0^2 \int_0^{2-x} e^{-x-y} dydx = \int_0^2 e^{-x}\left[1 - e^{-(2-x)}\right]dx$

$= \int_0^2 (e^{-x} - e^{-2})dx = 1 - e^{-2} - 2e^{-2} = .594$

d P(X + Y ≤ 1) = $\int_0^1 e^{-x}\left[1 - e^{-(1-x)}\right]dx = 1 - 2e^{-1} = .264$,

so P(1 ≤ X + Y ≤ 2) = P(X + Y ≤ 2) – P(X + Y ≤ 1) = .594 - .264 = .330

14.

a $P(X_1 < t, X_2 < t, \ldots, X_{10} < t) = P(X_1 < t) \ldots P(X_{10} < t) = (1-e^{-\lambda t})^{10}$

b If "success" = {fail before t}, then p = P(success) = $1-e^{-\lambda t}$,

and P(k successes among 10 trials) = $\binom{10}{k} 1 - e^{-\lambda t^k} (e^{-\lambda t})^{10-k}$

c P(exactly 5 fail) = P(5 of λ's fail and other 5 don't) + P(4 of λ's fail, μ fails, and other 5 don't) = $\binom{9}{5}\left(1-e^{-\lambda t}\right)^5 (e^{-\lambda t})^4 \left(e^{-\mu t}\right) + \binom{9}{4}\left(1-e^{-\lambda t}\right)^4 \left(1-e^{-\mu t}\right)\left(e^{-\lambda t}\right)^5$

15.

a F(y) = P(Y ≤ y) = P [(X_1 ≤y) ∪ ((X_2 ≤ y) ∩ (X_3 ≤ y))]
= P (X_1 ≤ y) + P[(X_2 ≤ y) ∩ (X_3 ≤ y)] - P[(X_1 ≤ y) ∩ (X_2 ≤ y) ∩ (X_3 ≤ y)]
= $(1-e^{-\lambda y}) + (1-e^{-\lambda y})^2 - (1-e^{-\lambda y})^3$ for y ≥ 0

f(y) = F′(y) = $\lambda e^{-\lambda y} + 2(1-e^{-\lambda y})\left(\lambda e^{-\lambda y}\right) - 3(1-e^{-\lambda y})^2\left(\lambda e^{-\lambda y}\right)$
= $4\lambda e^{-2\lambda y} - 3\lambda e^{-3\lambda y}$ for y ≥ 0

b E(Y) = $\int_0^\infty y \cdot \left(4\lambda e^{-2\lambda y} - 3\lambda e^{-3\lambda y}\right)dy = 2\left(\frac{1}{2\lambda}\right) - \frac{1}{3\lambda} = \frac{2}{3\lambda}$

16.

a f(x_1, x_3) = $\int_{-\infty}^{\infty} f(x_1, x_2, x_3)dx_2 = \int_0^{1-x_1-x_3} kx_1x_2\left(1-x_3\right)dx_2$

$72x_1\left(1-x_3\right)\left(1-x_1-x_3\right)^2$ 0 ≤ x_1, 0 ≤ x_3, x_1 + x_3 ≤ 1

b P(X_1 + X_3 ≤ .5) = $\int_0^{.5}\int_0^{.5-x_1} 72x_1(1-x_3)(1-x_1-x_3)^2\,dx_2dx_1$
= (after much algebra) .53125

c $f_{x_1}(x_1) = \int_{-\infty}^{\infty} f(x_1, x_3)dx_3 = \int 72x_1\left(1-x_3\right)\left(1-x_1-x_3\right)^2 dx_3$

$18x_1 - 48x_1^2 + 36x_1^3 - 6x_1^5$ 0 ≤ x_1 ≤ 1

17.

a $P\left((X,Y)\text{ within a circle of radius }\tfrac{R}{2}\right) = P(A) = \iint_A f(x,y)dxdy$

$= \frac{1}{\pi R^2}\iint_A dxdy = \frac{area.of.A}{\pi R^2} = \frac{1}{4} = .25$

b

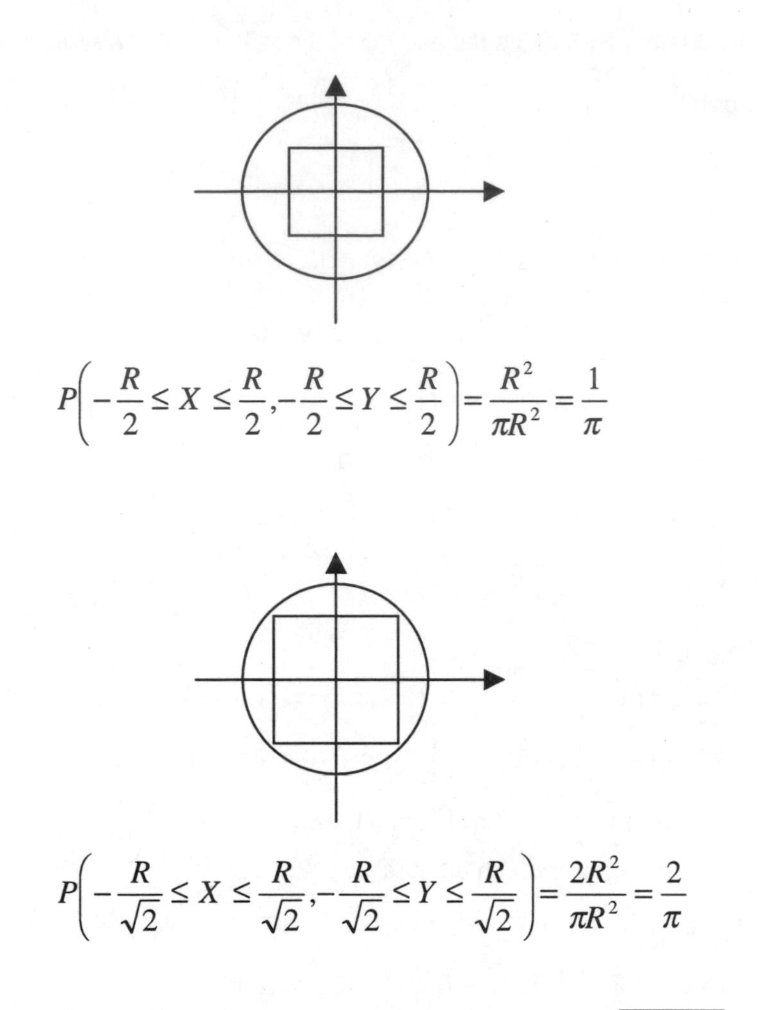

$$P\left(-\frac{R}{2} \le X \le \frac{R}{2}, -\frac{R}{2} \le Y \le \frac{R}{2}\right) = \frac{R^2}{\pi R^2} = \frac{1}{\pi}$$

c

$$P\left(-\frac{R}{\sqrt{2}} \le X \le \frac{R}{\sqrt{2}}, -\frac{R}{\sqrt{2}} \le Y \le \frac{R}{\sqrt{2}}\right) = \frac{2R^2}{\pi R^2} = \frac{2}{\pi}$$

d $f_x(x) = \int_{-\infty}^{\infty} f(x,y)dy = \int_{-\sqrt{R^2-x^2}}^{\sqrt{R^2-x^2}} \frac{1}{\pi R^2} dy = \frac{2\sqrt{R^2-x^2}}{\pi R^2}$ for $-R \le x \le R$ and similarly for $f_Y(y)$. X and Y are not independent since e.g. $f_x(.9R) = f_Y(.9R) > 0$, yet f(.9R, .9R) = 0 since (.9R, .9R) is outside the circle of radius R.

18.

a $P_{Y|X}(y|1)$ results from dividing each entry in x = 1 row of the joint probability table by $p_x(1) = .34$:

$$P_{y|x}(0|1) = \frac{.08}{.34} = .2353$$

$$P_{y|x}(1|1) = \frac{.20}{.34} = .5882$$

$$P_{y|x}(2|1) = \frac{.06}{.34} = .1765$$

b $P_{y|x}(x|2)$ is requested; to obtain this divide each entry in the y = 2 row by $p_x(2) = .50$:

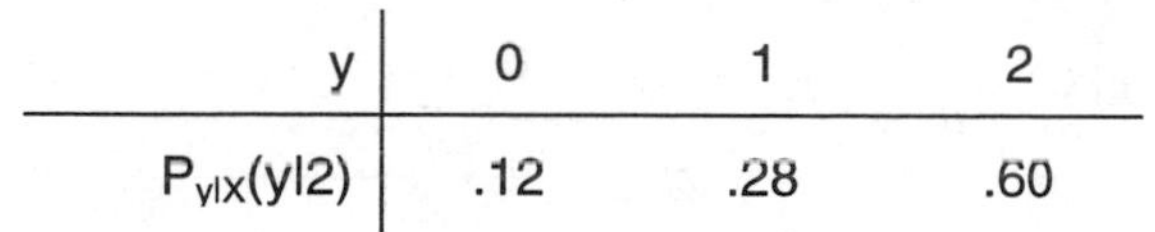

y	0	1	2
$P_{y|x}(y|2)$	.12	.28	.60

c $P(Y \le 1 \mid x = 2) = P_{y|x}(0|2) + P_{y|x}(1|2) = .12 + .28 = .40$

d $P_{X|Y}(x|2)$ results from dividing each entry in the y = 2 column by $p_y(2) = .38$:

x	0	1	2
$P_{x|y}(x|2)$	.0526	.1579	.7895

19.

a $$f_{Y|X}(y \mid x) = \frac{f(x,y)}{f_X(x)} = \frac{k(x^2+y^2)}{10kx^2+.05} \qquad 20 \le y \le 30$$

$$f_{X|Y}(x \mid y) = \frac{k(x^2+y^2)}{10ky^2+.05} \qquad 20 \le x \le 30 \qquad \left(k = \frac{3}{380{,}000}\right)$$

b $$P(Y \ge 25 \mid X = 22) = \int_{25}^{30} f_{Y|X}(y \mid 22)\,dy$$

$$= \int_{25}^{30} \frac{k((22)^2+y^2)}{10k(22)^2+.05}\,dy = .783$$

$$P(Y \ge 25) = \int_{25}^{30} f_Y(y)\,dy = \int_{25}^{30} (10ky^2+.05)\,dy = .75$$

c $$E(Y \mid X=22) = \int_{-\infty}^{\infty} y \cdot f_{Y|X}(y \mid 22)\,dy = \int_{20}^{30} y \cdot \frac{k((22)^2+y^2)}{10k(22)^2+.05}\,dy$$

$= 25.372912$

$$E(Y^2 \mid X=22) = \int_{20}^{30} y^2 \cdot \frac{k((22)^2+y^2)}{10k(22)^2+.05}\,dy = 652.028640$$

$V(Y \mid X = 22) = E(Y^2 \mid X=22) - [E(Y \mid X=22)]^2 = 8.243976$

20.

a $$f_{x_3|x_1,x_2}(x_3 \mid x_1, x_2) = \frac{f(x_1,x_2,x_3)}{f_{x_1,x_2}(x_1,x_2)}$$ where $f_{x_1,x_2}(x_1,x_2) =$ the marginal joint

pdf of $(X_1, X_2) = \int_{-\infty}^{\infty} f(x_1,x_2,x_3)\,dx_3$

b $$f_{X_2,X_3|X_1}(x_2,x_3 \mid x_1) = \frac{f(x_1,x_2,x_3)}{f_{X_1}(x_1)}$$ where

$$f_{X_1}(x_1) = \int_{-\infty}^{\infty}\int_{-\infty}^{\infty} f(x_1,x_2,x_3)dx_2dx_3$$

21. For every x and y, $f_{Y|X}(y|x) = f_y(y)$, since then $f(x,y) = f_{Y|X}(y|x) \cdot f_X(x) = f_Y(y) \cdot f_X(x)$, as required.

Section 5.2

22.

a $$E(X+Y) = \sum_x\sum_y (x+y)p(x,y) = (0+0)(.02)$$
$$+(0+5)(.06)+\ldots+(10+15)(.01) = 14.10$$

b $$E[\max(X,Y)] = \sum_x\sum_y \max(x+y)\cdot p(x,y)$$
$$= (0)(.02)+(5)(.06)+\ldots+(15)(.01) = 9.60$$

23. $$E(X_1 - X_2) = \sum_{x_1=0}^{4}\sum_{x_2=0}^{3}(x_1 - x_2)\cdot p(x_1,x_2) =$$
(0 – 0)(.08) + (0 – 1)(.07) + … + (4 – 3)(.06) = .15
(which also equals $E(X_1) - E(X_2) = 1.70 - 1.55$)

24. Let h(X,Y) = # of individuals who handle the message.

	h(x,y)	y 1	2	3	4	5	6
	1	-	2	3	4	3	2
	2	2	-	2	3	4	3
x	3	3	2	-	2	3	4
	4	4	3	2	-	2	3
	5	3	4	3	2	-	2
	6	2	3	4	3	2	-

Since $p(x,y) = \frac{1}{30}$ for each possible (x,y), $E[h(X,Y)] = \sum_x\sum_y h(x,y)\cdot\frac{1}{30} = \frac{84}{30} = 2.80$

25. E(XY) = E(X) · E(Y) = L · L = L²

26. Revenue = 3X + 10Y, so E (revenue) = E (3X + 10Y)

$$= \sum_{x=0}^{5}\sum_{y=0}^{2}(3x+10y)\cdot p(x,y) = 0\cdot p(0,0)+...+35\cdot p(5,2) = 15.4$$

27. E[h(X,Y)] = $\int_0^1\int_0^1 |x-y|\cdot 6x^2 y\,dx\,dy = 2\int_0^1\int_0^x (x-y)\cdot 6x^2 y\,dy\,dx$

$$12\int_0^1\int_0^x \left(x^3 y - x^2 y^2\right)dy\,dx = 12\int_0^1 \frac{x^5}{6}dx = \frac{1}{3}$$

28. E(XY) = $\sum_x\sum_y xy\cdot p(x,y) = \sum_x\sum_y xy\cdot p_x(x)\cdot p_y(y) = \sum_x xp_x(x)\cdot\sum_y yp_y(y)$

= E(X) · E(Y). (replace Σ with $\int$ in the continuous case)

29. Cov(X,Y) = $-\frac{2}{75}$ and $\mu_x = \mu_y = \frac{2}{5}$. E(X²) = $\int_0^1 x^2\cdot f_x(x)dx$

$= 12\int_0^1 x^3(1-x^2 dx) = \frac{12}{60} = \frac{1}{5}$, so Var (X) = $\frac{1}{5} - \frac{4}{25} = \frac{1}{25}$

Similarly, Var(Y) = $\frac{1}{25}$, so $\rho_{X,Y} = \frac{-2/75}{\sqrt{\frac{1}{25}}\cdot\sqrt{\frac{1}{25}}} = -\frac{50}{75} = -.667$

30.

a E(X) = 5.55, E(Y) = 8.55, E(XY) = (0)(.02) + (0)(.06) + ... + (150)(.01) = 44.25, so Cov(X,Y) = 44.25 – (5.55)(8.55) = -3.20

b $\sigma_X^2 = 12.45, \sigma_Y^2 = 19.15$, so $\rho_{X,Y} = \frac{-3.20}{\sqrt{(12.45)(19.15)}} = -.207$

31.

a E(X) = $\int_{20}^{30} xf_x(x)dx = \int_{20}^{30} x\left[10Kx^2 + .05\right]dx = 25.329 = E(Y)$

E(XY) = $\int_{20}^{30}\int_{20}^{30} xy\cdot K(x^2+y^2)dx\,dy = 641.447$

$\Rightarrow Cov(X,Y) = 641.447 - (25.329)^2 = -.111$

b E(X²) = $\int_{20}^{30} x^2\left[10Kx^2 + .05\right]dx = 649.8246 = E(Y^2)$,

so Var (X) = Var(Y) = 649.8246 – (25.329)² = 8.2664

$\Rightarrow \rho = \frac{-.111}{\sqrt{(8.2664)(8.2664)}} = -.0134$

32. There is a difficulty here. Existence of ρ requires that both X and Y have finite means and variances. Yet since the marginal pdf of Y is $\frac{1}{(1-y)^2}$ for $y \ge 0$,

$E(y) = \int_0^\infty \frac{y}{(1+y)^2}dy = \int_0^\infty \frac{(1+y-1)}{(1+y)^2}dy = \int_0^\infty \frac{1}{(1+y)}dy - \int_0^\infty \frac{1}{(1+y)^2}dy$, and the first integral is not finite. Thus ρ itself is undefined.

33. Since E(XY) = E(X) · E(Y), Cov(X,Y) = E(XY) – E(X) · E(Y) = E(X) · E(Y) - E(X) · E(Y) = 0, and since Corr(X,Y) = $\frac{Cov(X,Y)}{\sigma_x \sigma_y}$, then Corr(X,Y) = 0

34.

a In the discrete case, Var[h(X,Y)] = E{[h(X,Y) – E(h(X,Y))]^2} =

$$\sum_x \sum_y [h(x,y) - E(h(X,Y))]^2 p(x,y) = \sum_x \sum_y [h(x,y)^2 p(x,y)] - [E(h(X,Y))]^2$$

with $\iint$ replacing $\sum\sum$ in the continuous case.

b E[h(X,Y)] = E[max(X,Y)] = 9.60, and E[h^2(X,Y)] = E[(max(X,Y))^2] = $(0)^2(.02) + (5)^2(.06) + \ldots + (15)^2(.01) = 105.5$, so Var[max(X,Y)] = $105.5 - (9.60)^2 = 13.34$

35.

a Cov(aX + b, cY + d) = E[(aX + b)(cY + d)] – E(aX + b) · E(cY + d)
= E[acXY + adX + bcY + bd] – (aE(X) + b)(cE(Y) + d)
= acE(XY) – acE(X)E(Y) = acCov(X,Y)

b Corr(aX + b, cY + d) =

$$\frac{Cov(aX+b, cY+d)}{\sqrt{Var(aX+b)}\sqrt{Var(cY+d)}} = \frac{acCov(X,Y)}{|a| \cdot |c| \sqrt{Var(X) \cdot Var(Y)}}$$

= Corr(X,Y) when a and c have the same signs. When a and c differ in sign, Corr(aX + b, cY + d) = -Corr(X,Y).

36. Cov(X,Y) = Cov(X, aX+b) = E[X·(aX+b)] – E(X) ·E(aX+b) = a Var(X),

so Corr(X,Y) = $\frac{aVar(X)}{\sqrt{Var(X) \cdot Var(Y)}} = \frac{aVar(X)}{\sqrt{Var(X) \cdot a^2 Var(X)}}$ = 1 if a > 0, and –1 if a < 0

Chapter 5

Section 5.3

37.

$P(x_2)$	$P(x_1)$	.20	.50	.30
	$x_2 \mid x_1$	25	40	65
.20	25	.04	.10	.06
.50	40	.10	.25	.15
.30	65	.06	.15	.09

a

$\bar{x}$	25	32.5	40	45	52.5	65
$p(\bar{x})$	.04	.20	.25	.12	.30	.09

$$E(\bar{x}) = (25)(.04) + 32.5(.20) + ... + 65(.09) = 44.5 = \mu$$

b

s^2	0	112.5	312.5	800
$P(s^2)$	.38	.20	.30	.12

$E(s^2) = 212.25 = \sigma^2$

38.

a

T_0	0	1	2	3	4
$P(T_0)$	.04	.20	.37	.30	.09

b $\quad \mu_{T_0} = E(T_0) = 2.2 = 2 \cdot \mu$

c $\quad \sigma_{T_0}^2 = E(T_0^2) - E(T_0)^2 = 5.82 - (2.2)^2 = .98 = 2 \cdot \sigma^2$

39.

x	0	1	2	3	4	5	6	7	8	9	10
x/n	0	.1	.2	.3	.4	.5	.6	.7	.8	.9	1.0
p(x/n)	.000	.000	.000	.001	.005	.027	.088	.201	.302	.269	.107

X is a binomial random variable with p = .8.

40.

a Possible values of M are: 0, 5, 10. M = 0 when all 3 envelopes contain 0 money, hence $p(M = 0) = (.5)^3 = .125$. M = 10 when there is a single envelope with \$10, hence $p(M = 10) = 1 - p(\text{no envelopes with \$10}) = 1 - (.8)^3 = .488$.
$p(M = 5) = 1 - [.125 + .488] = .387$.

M	0	5	10
p(M)	.125	.387	.488

An alternative solution would be to list all 27 possible combinations using a tree diagram and computing probabilities directly from the tree.

b The statistic of interest is M, the maximum of x_1, x_2, or x_3, so that M = 0, 5, or 10. The population distribution is a s follows:

x	0	5	10
p(x)	1/2	3/10	1/5

Write a computer program to generate the digits 0 – 9 from a uniform distribution. Assign a value of 0 to the digits 0 – 4, a value of 5 to digits 5 – 7, and a value of 10 to digits 8 and 9. Generate samples of increasing sizes, keeping the number of replications constant and compute M from each sample. As n, the sample size, increases, p(M = 0) goes to zero, p(M = 10) goes to one. Furthermore, p(M = 5) goes to zero, but at a slower rate than p(M = 0).

41.

Outcome	1,1	1,2	1,3	1,4	2,1	2,2	2,3	2,4
Probability	.16	.12	.08	.04	.12	.09	.06	.03
$\overline{x}$	1	1.5	2	2.5	1.5	2	2.5	3
r	0	1	2	3	1	0	1	2

Outcome	3,1	3,2	3,3	3,4	4,1	4,2	4,3	4,4
Probability	.08	.06	.04	.02	.04	.03	.02	.01
$\overline{x}$	2	2.5	3	3.5	2.5	3	3.5	4
r	2	1	0	1	3	2	1	2

a

$\overline{x}$	1	1.5	2	2.5	3	3.5	4
$p(\overline{x})$	.16	.24	.25	.20	.10	.04	.01

b $P(\bar{x} \le 2.5) = .85$

c

r	0	1	2	3
p(r)	.30	.40	.22	.08

d $P(\bar{X} \le 1.5)$ = P(1,1,1,1) + P(2,1,1,1) + ... + P(1,1,1,2) + P(1,1,2,2) + ... + P(2,2,1,1) + P(3,1,1,1) + ... + P(1,1,1,3)
$= (.4)^4 + 4(.4)^3(.3) + 6(.4)^2(.3)^2 + 4(.4)^2(.2)^2 = .2400$

42.

a

$\bar{x}$	17.75	18.0	19.7	19.95	21.65	21.9	23.6
$p(\bar{x})$	$\frac{4}{30}$	$\frac{2}{30}$	$\frac{6}{30}$	$\frac{4}{30}$	$\frac{8}{30}$	$\frac{4}{30}$	$\frac{2}{30}$

b

$\bar{x}$	17.75	21.65	21.9
$p(\bar{x})$	$\frac{1}{3}$	$\frac{1}{3}$	$\frac{1}{3}$

c all three vales are the same: 20.4333

43. The statistic of interest is the fourth spread, or the difference between the medians of the upper and lower halves of the data. The population distribution is uniform with A = 8 and B = 10. Use a computer to generate samples of sizes n = 5, 10, 20, and 30 from a uniform distribution with A = 8 and B = 10. Keep the number of replications the same (say 500, for example). For each sample, compute the upper and lower fourth, then compute the difference. Plot the sampling distributions on separate histograms for n = 5, 10, 20, and 30.

44. Use a computer to generate samples of sizes n = 5, 10, 20, and 30 from a Weibull distribution with parameters as given, keeping the number of replications the same, as in problem 43 above. For each sample, calculate the mean. Below is a histogram, and a normal probability plot for the sampling distribution of $\bar{x}$ for n = 5, both generated by Minitab. This sampling distribution appears to be normal, so since larger sample sizes

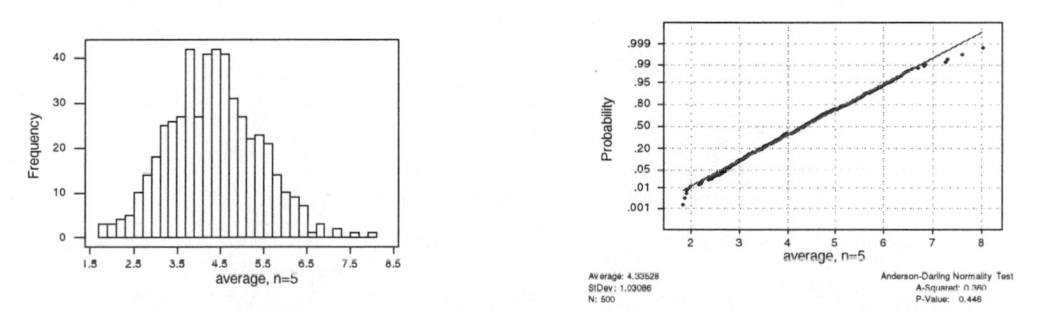

will produce distributions that are closer to normal, the others will also appear normal.

45. Using Minitab to generate the necessary sampling distribution, we can see that as n increases, the distribution slowly moves toward normality. However, even the sampling distribution for n = 50 is not yet approximately normal.

n = 10

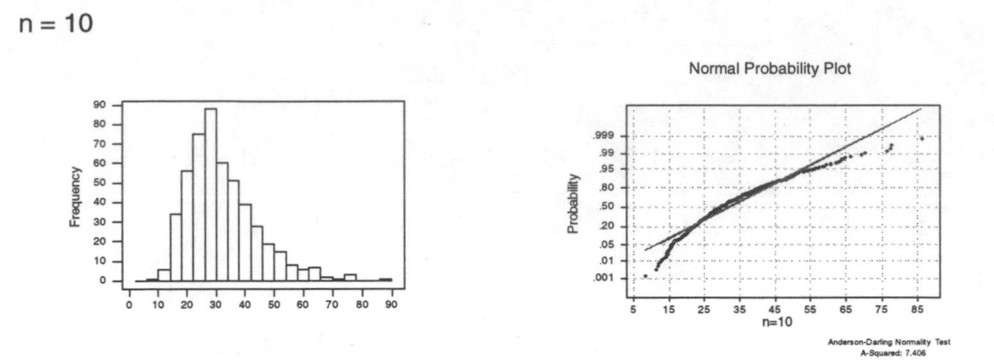

n = 50

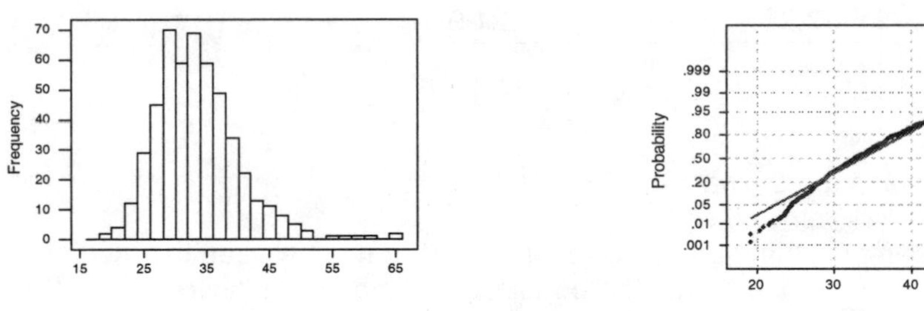

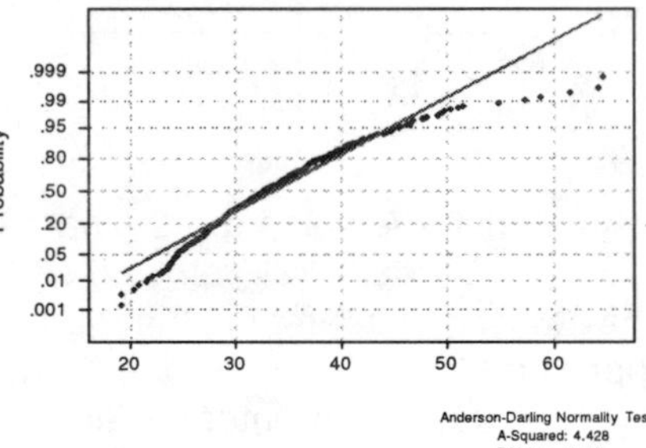

Section 5.4

46. μ = 12 cm σ = .04 cm

a n = 16 $E(\overline{X}) = \mu = 12cm$

$$\sigma_{\bar{x}} = \frac{\sigma_x}{\sqrt{n}} = \frac{.04}{4} = .01cm$$

b n = 64 $E(\overline{X}) = \mu = 12cm$

$$\sigma_{\bar{x}} = \frac{\sigma_x}{\sqrt{n}} = \frac{.04}{8} = .005cm$$

c $\overline{X}$ is more likely to be within .01 cm of the mean (12 cm) with the second, larger, sample. This is due to the decreased variability of $\overline{X}$ with a larger sample size.

47. $\mu = 12$ cm $\sigma = .04$ cm

a n = 16

$$P(11.99 \le \overline{X} \le 12.01) = P\left(\frac{11.99-12}{.01} \le Z \le \frac{12.01-12}{.01}\right)$$

$= P(-1 \le Z \le 1)$
$= \Phi(1) - \Phi(-1)$
$= .8413 - .1587$
$= .6826$

b n = 25

$$P(\overline{X} > 12.01) = P\left(Z > \frac{12.01-12}{.04/5}\right) = P(Z > 1.25)$$

$= 1 - \Phi(1.25)$
$= 1 - .8944$
$= .1056$

48.

a $\mu_{\overline{X}} = \mu = 50$, $\sigma_{\bar{x}} = \dfrac{\sigma_x}{\sqrt{n}} = \dfrac{1}{\sqrt{100}} = .10$

$$P(49.75 \le \overline{X} \le 50.25) = P\left(\frac{49.75-50}{.10} \le Z \le \frac{50.25-50}{.10}\right)$$

$= P(-2.5 \le Z \le 2.5) = .9876$

b $$P(49.75 \le \overline{X} \le 50.25) \approx P\left(\frac{49.75-49.8}{.10} \le Z \le \frac{50.25-49.8}{.10}\right)$$

$= P(-.5 \le Z \le 4.5) = .6915$

49.

a 11 P.M. – 6:50 P.M. = 250 minutes. With $T_0 = X_1 + \ldots + X_{40}$ = total grading time, $\mu_{T_0} = n\mu = (40)(6) = 240$ and $\sigma_{T_0} = \sigma\sqrt{n} = 37.95$, so $P(T_0 \le 250) \approx$

$$P\left(Z \le \frac{250-240}{37.95}\right) = P(Z \le .26) = .6026$$

b $$P(T_0 > 260) = P\left(Z > \frac{260-240}{37.95}\right) = P(Z > .53) = .2981$$

50. μ = 10,000 psi σ = 500 psi

a n = 40

P(9,900 ≤ $\bar{X}$ ≤ 10,200) ≈ $P\left(\frac{9{,}900-10{,}000}{500/\sqrt{40}} \le Z \le \frac{10{,}200-10{,}000}{500/\sqrt{40}}\right)$

= P(-1.26 ≤ Z ≤ 2.53)
= Φ(2.53) - Φ(-1.26)
= .9943 - .1038
= .8905

b According to the Rule of Thumb given in Section 5.4, n should be greater than 30 in order to apply the C.L.T., thus using the same procedure for n = 15 as was used for n = 40 would not be appropriate.

51. X ~ N(10,4). For day 1, n = 5

P($\bar{X}$ ≤ 11)= $P\left(Z \le \frac{11-10}{2/\sqrt{5}}\right) = P(Z \le 1.12) = .8686$

For day 2, n = 6

P($\bar{X}$ ≤ 11)= $P\left(Z \le \frac{11-10}{2/\sqrt{6}}\right) = P(Z \le 1.22) = .8888$

For both days,

P($\bar{X}$ ≤ 11)= (.8686)(.8888) = .7720

52. X ~ N(10), n =4

$\mu_{T_0} = n\mu = (4)(10) = 40$ and $\sigma_{T_0} = \sigma\sqrt{n} = (2)(1) = 2,$

We desire the 95th percentile: 40 + (1.645)(2) = 43.29

53. μ = 50, σ = 1.2

a n = 9

P($\bar{X}$ ≥ 51) = $P\left(Z \ge \frac{51-50}{1.2/\sqrt{9}}\right) = P(Z \ge 2.5) = 1 - .9938 = .0062$

b n = 40

P($\bar{X}$ ≥ 51) = $P\left(Z \ge \frac{51-50}{1.2/\sqrt{40}}\right) = P(Z \ge 5.27) \approx 0$

54.

a $\mu_{\bar{X}} = \mu = 2.65$, $\sigma_{\bar{x}} = \frac{\sigma_x}{\sqrt{n}} = \frac{.85}{5} = .17$

P($\bar{X}$ ≤ 3.00)= $P\left(Z \le \frac{3.00-2.65}{.17}\right) = P(Z \le 2.06) = .9803$

P(2.65 ≤ $\bar{X}$ ≤ 3.00)= $= P(\bar{X} \le 3.00) - P(\bar{X} \le 2.65) = .4803$

b $P(\overline{X} \le 3.00) = P\left(Z \le \frac{3.00-2.65}{.85/\sqrt{n}}\right) = .99$ implies that $\frac{.35}{85/\sqrt{n}} = 2.33$, from which n = 32.02. Thus n = 33 will suffice.

55. $\mu = np = 20 \quad \sigma = \sqrt{npq} = 3.464$

a $P(25 \le X) \approx P\left(\frac{24.5-20}{3.464} \le Z\right) = P(1.30 \le Z) = .0968$

b $P(15 \le X \le 25) \approx P\left(\frac{14.5-20}{3.464} \le Z \le \frac{25.5-20}{3.464}\right)$
$= P(-1.59 \le Z \le 1.59) = .8882$

56.

a With Y = # of tickets, Y has approximately a normal distribution with $\mu = \lambda = 50$, $\sigma = \sqrt{\lambda} = 7.071$, so $P(35 \le Y \le 70) \approx P\left(\frac{34.5-50}{7.071} \le Z \le \frac{70.5-50}{7.071}\right)$
$= P(-2.19 \le Z \le 2.90) = .9838$

b Here $\mu = 250$, $\sigma^2 = 250, \sigma = 15.811$, so $P(225 \le Y \le 275) \approx$
$P\left(\frac{224.5-250}{15.811} \le Z \le \frac{275.5-250}{15.811}\right) = P(-1.61 \le Z \le 1.61) = .8926$

57. E(X) = 100, Var(X) = 200, $\sigma_x = 14.14$, so $P(X \le 125) \approx P\left(Z \le \frac{125-100}{14.14}\right)$
$= P(Z \le 1.77) = .9616$

Section 5.5

58.

a $E(27X_1 + 125X_2 + 512X_3) = 27\,E(X_1) + 125\,E(X_2) + 512\,E(X_3)$
$= 27(200) + 125(250) + 512(100) = 87{,}850$
$V(27X_1 + 125X_2 + 512X_3) = 27^2\,V(X_1) + 125^2\,V(X_2) + 512^2\,V(X_3)$
$= 27^2\,(10)^2 + 125^2\,(12)^2 + 512^2\,(8)^2 = 19{,}100{,}116$

b The expected value is still correct, but the variance is not because the covariances now also contribute to the variance.

59.

a $E(X_1 + X_2 + X_3) = 180$, $V(X_1 + X_2 + X_3) = 45$, $\sigma_{x_1+x_2+x_3} = 6.708$

$P(X_1 + X_2 + X_3 \le 200) = P\left(Z \le \frac{200-180}{6.708}\right) = P(Z \le 2.98) = .9986$

$P(150 \le X_1 + X_2 + X_3 \le 200) = P(-4.47 \le Z \le 2.98) \approx .9986$

b $\mu_{\overline{X}} = \mu = 60$, $\sigma_{\overline{x}} = \frac{\sigma_x}{\sqrt{n}} = \frac{\sqrt{15}}{\sqrt{3}} = 2.236$

$P(\overline{X} \ge 55) = P\left(Z \ge \frac{55-60}{2.236}\right) = P(Z \ge -2.236) = .9875$

$P(58 \le \overline{X} \le 62) = P(-.89 \le Z \le .89) = .6266$

c $E(X_1 - .5X_2 - .5X_3) = 0$;

$V(X_1 - .5X_2 - .5X_3) = \sigma_1^2 + .25\sigma_2^2 + .25\sigma_3^2 = 22.5$, sd = 4.7434

$P(-10 \le X_1 - .5X_2 - .5X_3 \le 5) = P\left(\frac{-10-0}{4.7434} \le Z \le \frac{5-0}{4.7434}\right)$

$= P(-2.11 \le Z \le 1.05)$ = .8531 - .0174 = .8357

d $E(X_1 + X_2 + X_3) = 150$, $V(X_1 + X_2 + X_3) = 36$, $\sigma_{x_1+x_2+x_3} = 6$

$P(X_1 + X_2 + X_3 \le 200) = P\left(Z \le \frac{160-150}{6}\right) = P(Z \le 1.67) = .9525$

We want $P(X_1 + X_2 \ge 2X_3)$, or written another way, $P(X_1 + X_2 - 2X_3 \ge 0)$.
$E(X_1 + X_2 - 2X_3) = 40 + 50 - 2(60) = -30$,
$V(X_1 + X_2 - 2X_3) = \sigma_1^2 + \sigma_2^2 + 4\sigma_3^2 = 78,36$, sd = 8.832, so

$P(X_1 + X_2 - 2X_3 \ge 0) = P\left(Z \ge \frac{0-(-30)}{8.832}\right) = P(Z \ge 3.40) = .0003$

60. Y is normally distributed with $\mu_Y = \frac{1}{2}(\mu_1 + \mu_2) - \frac{1}{3}(\mu_3 + \mu_4 + \mu_5) = -1$, and

$\sigma_Y^2 = \frac{1}{4}\sigma_1^2 + \frac{1}{4}\sigma_2^2 + \frac{1}{9}\sigma_3^2 + \frac{1}{9}\sigma_4^2 + \frac{1}{9}\sigma_5^2 = 3.167, \sigma_Y = 1.7795$.

Thus, $P(0 \le Y) = P\left(\frac{0-(-1)}{1.7795} \le Z\right) = P(.56 \le Z) = .2877$ and

$P(-1 \le Y \le 1) = P\left(0 \le Z \le \frac{2}{1.7795}\right) = P(0 \le Z \le 1.12) = .3686$

61.

a The marginal pmf's of X and Y are given in the solution to Exercise 7, from which $E(X) = 2.8$, $E(Y) = .7$, $V(X) = 1.66$, $V(Y) = .61$. Thus $E(X+Y) = E(X) + E(Y) = 3.5$, $V(X+Y) = V(X) + V(Y) = 2.27$, and the standard deviation of $X + Y$ is 1.51

b $E(3X+10Y) = 3E(X) + 10E(Y) = 15.4$, $V(3X+10Y) = 9V(X) + 100V(Y) = 75.94$, and the standard deviation of revenue is 8.71

62. $E(X_1 + X_2 + X_3) = E(X_1) + E(X_2) + E(X_3) = 15 + 30 + 20 = 65$ min.,
$V(X_1 + X_2 + X_3) = 1^2 + 2^2 + 1.5^2 = 7.25$, $\sigma_{x_1+x_2+x_3} = \sqrt{7.25} = 2.6926$

Thus, $P(X_1 + X_2 + X_3 \le 60) = P\left(Z \le \frac{60-65}{2.6926}\right) = P(Z \le -1.86) = .0314$

63.

a $E(X_1) = 1.70$, $E(X_2) = 1.55$, $E(X_1X_2) = \sum_{x_1}\sum_{x_2} x_1x_2 p(x_1, x_2) = 3.33$, so
$Cov(X_1,X_2) = E(X_1X_2) - E(X_1)E(X_2) = 3.33 - 2.635 = .695$

b $V(X_1 + X_2) = V(X_1) + V(X_2) + 2\,Cov(X_1,X_2) = 1.59 + 1.0875 + 2(.695) = 4.0675$

64. Let $X_1, \ldots, X_5$ denote morning times and $X_6, \ldots, X_{10}$ denote evening times.

a $E(X_1 + \ldots + X_{10}) = E(X_1) + \ldots + E(X_{10}) = 5\,E(X_1) + 5\,E(X_6) = 5(2.5) + 5(5) = 37.5$

b $Var(X_1 + \ldots + X_{10}) = Var(X_1) + \ldots + Var(X_{10}) = 5\,Var(X_1) + 5Var(X_6)$

$$= 5\left[\frac{25}{12} + \frac{100}{12}\right] = \frac{625}{12} = 52.083$$

c $E(X_1 - X_6) = E(X_1) - E(X_6) = 2.5 - 5 = -2.5$

$Var(X_1 - X_6) = Var(X_1) + Var(X_6) = \frac{25}{12} + \frac{100}{12} = \frac{125}{12} = 10.417$

d $E[(X_1 + \ldots + X_5) - (X_6 + \ldots + X_{10})] = 5(2.5) - 5(5) = -12.5$
$Var[(X_1 + \ldots + X_5) - (X_6 + \ldots + X_{10})]$
$= Var(X_1 + \ldots + X_5) + Var(X_6 + \ldots + X_{10})] = 52.083$

65. $\mu = 5.00$, $\sigma = .2$

a $E(\bar{X} - \bar{Y}) = 0;$ $V(\bar{X} - \bar{Y}) = \frac{\sigma^2}{25} + \frac{\sigma^2}{25} = .0032$, $\sigma_{\bar{X}-\bar{Y}} = .0566$

$\Rightarrow P(-.1 \le \bar{X} - \bar{Y} \le .1) \approx P(-1.77 \le Z \le 1.77) = .9232$ (by the CLT)

b $V(\bar{X} - \bar{Y}) = \frac{\sigma^2}{36} + \frac{\sigma^2}{36} = .0022222$, $\sigma_{\bar{X}-\bar{Y}} = .0471$
$\Rightarrow P(-.1 \le \bar{X} - \bar{Y} \le .1) \approx P(-2.12 \le Z \le 2.12) = .9660$

66.

a. With $M = 5X_1 + 10X_2$, $E(M) = 5(2) + 10(4) = 50$,
$Var(M) = 5^2 (.5)^2 + 10^2 (1)^2 = 106.25$, $\sigma_M = 10.308$.

b. $P(75 < M) = P\left(\frac{75-50}{10.308} < Z\right) = P(2.43 < Z) = .0075$

c. $M = A_1X_1 + A_2X_2$ with the A_i's and X_i's all independent, so
$E(M) = E(A_1X_1) + E(A_2X_2) = E(A_1)E(X_1) + E(A_2)E(X_2) = 50$

d. $Var(M) = E(M^2) - [E(M)]^2$. Recall that for any r.v. Y,
$E(Y^2) = Var(Y) + [E(Y)]^2$. Thus, $E(M^2) = E\left(A_1^2X_1^2 + 2A_1X_1A_2X_2 + A_2^2X_2^2\right)$
$= E\left(A_1^2\right)E\left(X_1^2\right) + 2E(A_1)E(X_1)E(A_2)E(X_2) + E\left(A_2^2\right)E\left(X_2^2\right)$
(by independence)
$= (.25 + 25)(.25 + 4) + 2(5)(2)(10)(4) + (.25 + 100)(1 + 16) = 2611.5625$, so
$Var(M) = 2611.5625 - (50)^2 = 111.5625$

e. $E(M) = 50$ still, but now
$$Var(M) = a_1^2Var(X_1) + 2a_1a_2Cov(X_1, X_2) + a_2^2Var(X_2)$$
$= 6.25 + 2(5)(10)(-.25) + 100 = 81.25$

67. Letting X_1, X_2, and X_3 denote the lengths of the three pieces, the total length is $X_1 + X_2 - X_3$. This has a normal distribution with mean value $20 + 15 - 1 = 34$, variance $.25+.16+.01 = .42$, and standard deviation .6481. Standardizing gives
$P(34.5 \le X_1 + X_2 - X_3 \le 35) = P(.77 \le Z \le 1.54) = .1588$

68. Let X_1 and X_2 denote the (constant) speeds of the two planes.

a. After two hours, the planes have traveled $2X_1$ km. and $2X_2$ km., respectively, so the second will not have caught the first if $2X_1 + 10 > 2X_2$, i.e. if $X_2 - X_1 < 5$. $X_2 - X_1$ has a mean $500 - 520 = -20$, variance $100 + 100 = 200$, and standard deviation 14.14. Thus,
$$P(X_2 - X_1 < 5) = P\left(Z < \frac{5-(-20)}{14.14}\right) = P(Z < 1.77) = .9616.$$

b. After two hours, #1 will be $10 + 2X_1$ km from where #2 started, whereas #2 will be $2X_2$ from where it started. Thus the separation distance will be al most 10 if $|2X_2 - 10 - 2X_1| \le 10$, i.e. $-10 \le 2X_2 - 10 - 2X_1 \le 10$, i.e. $0 \le X_2 - X_1 \le 10$. The corresponding probability is $P(0 \le X_2 - X_1 \le 10) = P(1.41 \le Z \le 2.12)$
$= .9830 - .9207 = .0623$.

69.

a. $E(X_1 + X_2 + X_3) = 800 + 1000 + 600 = 2400$.

b. Assuming independence of X_1, X_2, X_3, $Var(X_1 + X_2 + X_3)$
$= (16)^2 + (25)^2 + (18)^2 = 12.05$

c. $E(X_1 + X_2 + X_3) = 2400$ as before, but now $Var(X_1 + X_2 + X_3)$
$= Var(X_1) + Var(X_2) + Var(X_3) + 2Cov(X_1,X_2) + 2Cov(X_1, X_3) + 2Cov(X_2, X_3)$
$= 1745$, with sd $= 41.77$

70.

a $$E(Y_i) = .5, \text{ so } E(W) = \sum_{i=1}^{n} i \cdot E(Y_i) = .5\sum_{i=1}^{n} i = \frac{n(n+1)}{4}$$

b $$Var(Y_i) = .25, \text{ so } Var(W) = \sum_{i=1}^{n} i^2 \cdot Var(Y_i) = .25\sum_{i=1}^{n} i^2 = \frac{n(n+1)(2n+1)}{24}$$

71.

a $M = a_1X_1 + a_2X_2 + W\int_0^{12} x\,dx = a_1X_1 + a_2X_2 + 72W$, so

E(M) = (5)(2) + (10)(4) + (72)(1.5) = 158m

$\sigma_M^2 = (5)^2(.5)^2 + (10)^2(1)^2 + (72)^2(.25)^2 = 430.25$, $\sigma_M = 20.74$

b $$P(M \le 200) = P\left(Z \le \frac{200-158}{20.74}\right) = P(Z \le 2.03) = .9788$$

72. The total elapsed time between leaving and returning is $T_o = X_1 + X_2 + X_3 + X_4$, with $E(T_o) = 40$, $\sigma_{T_o}^2 = 40$, $\sigma_{T_o} = 5.477$. T_o is normally distributed, and the desired value t is the 99th percentile of the lapsed time distribution added to 10 A.M.:
10:00 + [40+(5.477)(2.33)] = 10:52.76

73.

a Both approximately normal by the C.L.T.

b The difference of two r.v.'s is just a special linear combination, and a linear combination of normal r.v's has a normal distribution, so $\overline{X} - \overline{Y}$ has approximately a normal distribution with $\mu_{\overline{X}-\overline{Y}} = 5$ and

$$\sigma_{\overline{X}-\overline{Y}}^2 = \frac{8^2}{40} + \frac{6^2}{35} = 2.629, \sigma_{\overline{X}-\overline{Y}} = 1.621$$

c $$P(-1 \le \overline{X} - \overline{Y} \le 1) \doteq P\left(\frac{-1-5}{1.6213} \le Z \le \frac{1-5}{1.6213}\right) = P(-3.70 \le Z \le -2.47) \approx .0068$$

d $P(\overline{X} - \overline{Y} \ge 10) \doteq P\left(Z \ge \frac{10-5}{1.6213}\right) = P(Z \ge 3.08) = .0010.$ This probability is quite small, so such an occurrence is unlikely if $\mu_1 - \mu_2 = 5$, and we would thus doubt this claim.

74. X is approximately normal with $\mu_1 = (50)(.7) = 35$ and $\sigma_1^2 = (50)(.7)(.3) = 10.5$, as is Y with $\mu_2 = 30$ and $\sigma_2^2 = 12$. Thus $\mu_{X-Y} = 5$ and $\sigma_{X-Y}^2 = 22.5$, so

$$p(-5 \le X - Y \le 5) \approx P\left(\frac{-10}{4.74} \le Z \le \frac{0}{4.74}\right) = P(-2.11 \le Z \le 0) = .4826$$

Supplementary Exercises

75.

a $p_X(x)$ is obtained by adding joint probabilities across the row labeled x, resulting in $p_X(x)$ = .2, .5, .3 for x = 7, 9, 10 respectively. Similarly, from column sums $p_y(y)$ = .1, .35, .55 for y = 7, 9, 10 respectively.

b $P(X \le 9 \text{ and } Y \le 9) = p(7,7) + p(7,9) + p(9,7) + p(9,9) = .25$

c $p_x(7) \cdot p_y(7) = (.2)(.1) \ne .05 = p(7,7)$, so X and Y are not independent. (Almost any other (x,y) pair yields the same conclusion).

d $E(X+Y) = \sum\sum (x+y)p(x,y) = 18.25$ (or = E(X) + E(Y) = 18.25)

e $E(|X-Y|) = \sum\sum |x+y| p(x,y) = 1.05$

76. The roll-up procedure is not valid for the 75th percentile unless $\sigma_1 = 0$ or $\sigma_2 = 0$ or both σ_1 and $\sigma_2 = 0$, as described below.

Sum of percentiles: $\mu_1 + (Z)\sigma_1 + \mu_2 + (Z)\sigma_2 = \mu_1 + \mu_2 + (Z)(\sigma_1 + \sigma_2)$

Percentile of sums: $\mu_1 + \mu_2 + (Z)\sqrt{\sigma_1^2 + \sigma_2^2}$

These are equal when Z = 0 (i.e. for the median) or in the unusual case when $\sigma_1 + \sigma_2 = \sqrt{\sigma_1^2 + \sigma_2^2}$, which happens when $\sigma_1 = 0$ or $\sigma_2 = 0$ or both σ_1 and $\sigma_2 = 0$.

77.

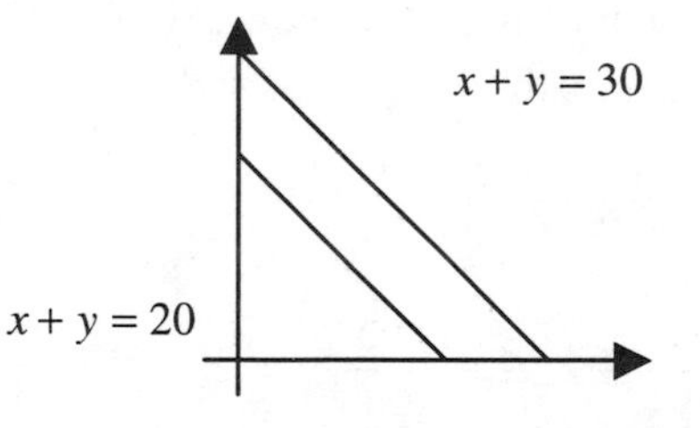

a $1 = \int_{-\infty}^{\infty}\int_{-\infty}^{\infty} f(x,y)\,dxdy = \int_0^{20}\int_{20-x}^{30-x} kxy\,dydx + \int_{20}^{30}\int_0^{30-x} kxy\,dydx$

$$= \frac{81{,}250}{3} \cdot k \Rightarrow k = \frac{3}{81{,}250}$$

b $$f_X(x) = \begin{cases} \int_{20-x}^{30-x} kxy\,dy = k(250x - 10x^2) & 0 \le x \le 20 \\ \int_0^{30-x} kxy\,dy = k(450x - 30x^2 + \frac{1}{2}x^3) & 20 \le x \le 30 \end{cases}$$

and by symmetry $f_Y(y)$ is obtained by substituting y for x in $f_X(x)$. Since $f_X(25) > 0$, and $f_Y(25) > 0$, but $f(25, 25) = 0$, $f_X(x) \cdot f_Y(y) \ne f(x,y)$ for all x,y so X and Y are not independent.

c $P(X+Y \le 25) = \int_0^{20}\int_{20-x}^{25-x} kxydydx + \int_{20}^{25}\int_0^{25-x} kxydydx$

$= \frac{3}{81{,}250}\cdot\frac{230{,}625}{24} = .355$

d $E(X+Y) = E(X) + E(Y) = 2\left\{\int_0^{20} x\cdot k\left(250x - 10x^2\right)dx\right.$

$\left. + \int_{20}^{30} x\cdot k\left(450x - 30x^2 + \tfrac{1}{2}x^3\right)dx\right\}$

$= 2k(351{,}666.67) = 25.969$

e $E(XY) = \int_{-\infty}^{\infty}\int_{-\infty}^{\infty} xy\cdot f(x,y)dxdy = \int_0^{20}\int_{20-x}^{30-x} kx^2y^2dydx$

$+ \int_{20}^{30}\int_0^{30-x} kx^2y^2dydx = \frac{k}{3}\cdot\frac{33{,}250{,}000}{3} = 136.4103$, so

Cov(X,Y) = 136.4103 – $(12.9845)^2$ = -32.19, and $E(X^2) = E(Y^2)$ = 204.6154, so

$\sigma_x^2 = \sigma_y^2 = 204.6154 - (12.9845)^2 = 36.0182$ and $\rho = \frac{-32.19}{36.0182} = -.894$

f Var (X + Y) = Var(X) + Var(Y) + 2Cov(X,Y) = 7.66

78. $F_Y(y) = P(\max(X_1, \ldots, X_n) \le y) = P(X_1 \le y, \ldots, X_n \le y) = [P(X_1 \le y)]^n = \left(\frac{y-100}{100}\right)^n$ for

$100 \le y \le 200$. Thus $f_Y(y) = \frac{n}{100^n}(y-100)^{n-1}$ for $100 \le y \le 200$.

$$E(Y) = \int_{100}^{200} y\cdot\frac{n}{100^n}(y-100)^{n-1}dy = \frac{n}{100^n}\int_0^{100}(u+100)u^{n-1}du$$

$$= 100 + \frac{n}{100^n}\int_0^{100} u^n du = 100 + 100\frac{n}{n+1} = \frac{2n+1}{n+1}\cdot 100$$

79. $E(\overline{X}+\overline{Y}+\overline{Z}) = 500 + 900 + 2000 = 3400$

$Var(\overline{X}+\overline{Y}+\overline{Z}) = \frac{50^2}{365} + \frac{100^2}{365} + \frac{180^2}{365} = 123.014$, and the std dev = 11.09.

$P(\overline{X}+\overline{Y}+\overline{Z} \le 3500) = P(Z \le 9.0) \approx 1$

80.

a Let $X_1, \ldots, X_{12}$ denote the weights for the business-class passengers and $Y_1, \ldots, Y_{50}$ denote the tourist-class weights. Then T = total weight
$= X_1 + \ldots + X_{12} + Y_1 + \ldots + Y_{50} = X + Y$
$E(X) = 12E(X_1) = 12(30) = 360$; $V(X) = 12V(X_1) = 12(36) = 432$.
$E(Y) = 50E(Y_1) = 50(40) = 2000$; $V(Y) = 50V(Y_1) = 50(100) = 5000$.
Thus E(T) = E(X) + E(Y) = 360 + 2000 = 2360
And V(T) = V(X) + V(Y) = 432 + 5000 = 5432, std dev = 73.7021

b $$P(T \le 2500) = P\left(Z \le \frac{2500 - 2360}{73.7021} \right) = P(Z \le 1.90) = .9713$$

81.

a E(N) · μ = (10)(40) = 400 minutes

b We expect 20 components to come in for repair during a 4 hour period, so E(N) · μ = (20)(3.5) = 70

82. X ~ Bin (200, .45) and Y ~ Bin (300, .6). Because both n's are large, both X and Y are approximately normal, so X + Y is approximately normal with mean (200)(.45) + (300)(.6) = 270, variance 200(.45)(.55) + 300(.6)(.4) = 121.40, and standard deviation 11.02.

Thus, P(X + Y ≥ 250) $= P\left(Z \ge \frac{249.5 - 270}{11.02} \right) = P(Z \ge -1.86) = .9686$

83. $0.95 = P(\mu - .02 \le \overline{X} \le \mu + .02) \doteq P\left(\frac{-.02}{.01/\sqrt{n}} \le Z \le \frac{.02}{.01/\sqrt{n}} \right)$

$= P\left(-.2\sqrt{n} \le Z \le .2\sqrt{n}\right)$, but $P(-1.96 \le Z \le 1.96) = .95$ so

$.2\sqrt{n} = 1.96 \Rightarrow n = 97$. The C.L.T.

84. I have 192 oz. The amount which I would consume if there were no limit is $T_o = X_1 + \ldots + X_{14}$ where each X_i is normally distributed with μ = 13 and σ = 2. Thus T_o is normal with $\mu_{T_o} = 182$ and $\sigma_{T_o} = 7.483$, so $P(T_o < 192) = P(Z < 1.34) = .9099$.

85. The expected value and standard deviation of volume are 87,850 and 4370.37, respectively, so

$$P(volume \le 100{,}000) = P\left(Z \le \frac{100{,}000 - 87{,}850}{4370.37} \right) = P(Z \le 2.78) = .9973$$

86. The student will not be late if $X_1 + X_3 \le X_2$, i.e. if $X_1 - X_2 + X_3 \le 0$. This linear combination has mean –2, variance 4.25, and standard deviation 2.06, so

$$P(X_1 - X_2 + X_3 \le 0) = P\left(Z \le \frac{0 - (-2)}{2.06} \right) = P(Z \le .97) = .8340$$

87.

a $$Var(aX + Y) = a^2\sigma_x^2 + 2aCov(X,Y) + \sigma_y^2 = a^2\sigma_x^2 + 2a\sigma_X\sigma_Y\rho + \sigma_y^2.$$

Substituting $a = \frac{\sigma_Y}{\sigma_X}$ yields $\sigma_Y^2 + 2\sigma_Y^2\rho + \sigma_Y^2 = 2\sigma_Y^2(1-\rho) \ge 0$, so $\rho \ge -1$

b Same argument as in **a**

c Suppose $\rho = 1$. Then $Var(aX - Y) = 2\sigma_Y^2(1-\rho) = 0$, which implies that $aX - Y = k$ (a constant), so $aX - Y = aX - k$, which is of the form $aX + b$.

88. $E(X+Y-t)^2 = \int_0^1\int_0^1 (x+y-t)^2 \cdot f(x,y)dxdy$. To find the minimizing value of t, take the derivative with respect to t and equate it to 0:

$0 = \int_0^1\int_0^1 2(x+y-t)(-1)f(x,y) = 0 \Rightarrow \int_0^1\int_0^1 tf(x,y)dxdy = t$

$= \int_0^1\int_0^1 (x+y)\cdot f(x,y)dxdy = E(X+Y)$, so the best prediction is the individual's expected score (= 1.167).

89.

a With Y = $X_1 + X_2$,

$$F_Y(y) = \int_0^y \left\{ \int_0^{y-x_1} \frac{1}{2^{\nu_1/2}\Gamma(\nu_1/2)} \cdot \frac{1}{2^{\nu_{21}/2}\Gamma(\nu_2/2)} \cdot x_1^{\frac{\nu_1}{2}-1} x_2^{\frac{\nu_2}{2}-1} e^{-\frac{x_1+x_2}{2}} dx_2 \right\} dx_1 .$$

But the inner integral can be shown to be equal to

$\frac{1}{2^{(\nu_1+\nu_2)/2}\Gamma((\nu_1+\nu_2)/2)} y^{[(\nu_1+\nu_2)/2]-1} e^{-y/2}$, from which the result follows.

b By **a**, $Z_1^2 + Z_2^2$ is chi-squared with $\nu = 2$, so $(Z_1^2 + Z_2^2) + Z_3^2$ is chi-squared with $\nu = 3$, etc, until $Z_1^2 + ... + Z_n^2$ 9s chi-squared with $\nu = n$

c $\frac{X_i - \mu}{\sigma}$ is standard normal, so $\left[\frac{X_i - \mu}{\sigma}\right]^2$ is chi-squared with $\nu = 1$, so the sum is chi-squared with $\nu = n$.

90.

a Cov(X, Y + Z) = E[X(Y + Z)] – E(X) · E(Y + Z) = E(XY) + E(XZ) – E(X) · E(Y) – E(X) · E(Z) = E(XY) – E(X) · E(Y) + E(XZ) – E(X) · E(Z) = Cov(X,Y) + Cov(X,Z).

b Cov(X_1 + X_2 , Y_1 + Y_2) = Cov(X_1 , Y_1) + Cov(X_1 ,Y_2) + Cov(X_2 , Y_1) + Cov(X_2 ,Y_2) (apply **a** twice) = 16.

c By repeated application of **a**, $Cov\left(\sum_i a_i X_i, \sum_j b_j Y_j\right) = \sum_i \sum_j a_i b_j Cov(X_i, Y_j)$

91.

a $V(X_1) = V(W + E_1) = \sigma_W^2 + \sigma_E^2 = V(W + E_2) = V(X_2)$ and $Cov(X_1, X_2) = Cov(W + E_1, W + E_2) = Cov(W,W) + Cov(W,E_2) + Cov(E_1,W) + Cov(E_1,E_2) = Cov(W,W) = V(W) = \sigma_w^2$.

Thus, $\rho = \dfrac{\sigma_W^2}{\sqrt{\sigma_W^2+\sigma_E^2}\cdot\sqrt{\sigma_W^2+\sigma_E^2}} = \dfrac{\sigma_W^2}{\sigma_W^2+\sigma_E^2}$

b $\rho = \dfrac{1}{1+.0001} = .9999$

92.

a Cov(X,Y) = Cov(A+D, B+E)
= Cov(A,B) + Cov(D,B) + Cov(A,E) + Cov(D,E)
= Cov(A,B). Thus

$$Corr(X,Y) = \frac{Cov(A,B)}{\sqrt{\sigma_A^2+\sigma_D^2}\cdot\sqrt{\sigma_B^2+\sigma_E^2}} = \frac{Cov(A,B)}{\sigma_A\sigma_B}\cdot\frac{\sigma_A}{\sqrt{\sigma_A^2+\sigma_D^2}}\cdot\frac{\sigma_B}{\sqrt{\sigma_B^2+\sigma_E^2}}$$

The first factor in this expression is Corr(A,B), and (by the result of exercise 70**a**) the second and third factors are the square roots of Corr(X_1, X_2) and Corr(Y_1, Y_2), respectively.
Clearly, measurement error reduces the correlation, since both square-root factors are between 0 and 1.

b $\sqrt{.8100}\cdot\sqrt{.9025} = .855$. This is disturbing, because measurement error substantially reduces the correlation.

93. $E(Y) \doteq h(\mu_1,\mu_2,\mu_3,\mu_4) = 120\left[\frac{1}{10}+\frac{1}{15}+\frac{1}{20}\right] = 26$

The partial derivatives of $h(\mu_1,\mu_2,\mu_3,\mu_4)$ with respect to x_1, x_2, x_3, and x_4 are $-\dfrac{x_4}{x_1^2}$, $-\dfrac{x_4}{x_2^2}$, $-\dfrac{x_4}{x_3^2}$, and $\dfrac{1}{x_1}+\dfrac{1}{x_2}+\dfrac{1}{x_3}$, respectively. Substituting $x_1 = 10$, $x_2 = 15$, $x_3 = 20$, and $x_4 = 120$ gives –1.2, -.5333, -.3000, and .2167, respectively, so V(Y) = $(1)(-1.2)^2 + (1)(-.5333)^2 + (1.5)(-.3000)^2 + (4.0)(.2167)^2 = 2.6783$, and the approximate sd of y is 1.64.

94. The four second order partials are $\dfrac{2x_4}{x_1^3}$, $\dfrac{2x_4}{x_2^3}$, $\dfrac{2x_4}{x_3^3}$, and 0 respectively. Substitution gives E(Y) = 26 + .1200 + .0356 + .0338 = 26.1894.

Chapter 6

Section 6.1

1.

a We use the sample mean, $\bar{x}$ to estimate the population mean μ.

$$\hat{\mu} = \bar{x} = \frac{\Sigma x_i}{n} = \frac{219.80}{27} = 8.1407$$

b We use the sample median, $\tilde{x} = 7.7$ (the middle observation when arranged in ascending order).

c We use the sample standard deviation, $s = \sqrt{s^2} = \sqrt{\dfrac{1860.94 - \frac{(219.8)^2}{27}}{26}} = 1.660$

d With "success" = observation greater than 10, x = # of successes = 4, and $\hat{p} = \frac{x}{n} = \frac{4}{27} = .1481$

e We use the sample (std dev)/(mean), or $\dfrac{s}{\bar{x}} = \dfrac{1.660}{8.1407} = .2039$

2.

a With X = # of T's in the sample, the estimator is $\hat{p} = \frac{X}{n}; x = 10,$ so

$$\hat{p} = \frac{10}{20}, = .50.$$

b Here, X = # in sample which are not reverse polish, and x = 3, so $\hat{p} = \dfrac{3}{20}, = .85$

3.

a We use the sample mean, $\bar{x} = 1.3481$

b Because we assume normality, the mean = median, so we also use the sample mean $\bar{x} = 1.3481$. We could also easily use the sample median.

c We use the 90th percentile of the sample:

$$\hat{\mu} + (1.28)\hat{\sigma} = \bar{x} + 1.28s = 1.3481 + (1.28)(.3385) = 1.7814.$$

d Since we can assume normality,

$$P(X < 1.5) \approx P\left(Z < \frac{1.5 - \bar{x}}{s}\right) = P\left(Z < \frac{1.5 - 1.3481}{.3385}\right) = P(Z < .45) = .6736$$

e The estimated standard error of $\bar{x} = \dfrac{\hat{\sigma}}{\sqrt{n}} = \dfrac{s}{\sqrt{n}} = \dfrac{.3385}{\sqrt{16}} = .0846$

4.

a $E(\overline{X}-\overline{Y})=E(\overline{X})-E(\overline{Y})=\mu_1-\mu_2$; $\overline{x}-\overline{y}=8.141-8.575=.434$

b $V(\overline{X}-\overline{Y})=V(\overline{X})+V(\overline{Y})=\sigma_{\overline{X}}^2+\sigma_{\overline{Y}}^2=\dfrac{\sigma_1^2}{n_1}+\dfrac{\sigma_2^2}{n_2}$

$\sigma_{\overline{X}-\overline{Y}}=\sqrt{V(\overline{X}-\overline{Y})}=\sqrt{\dfrac{\sigma_1^2}{n_1}+\dfrac{\sigma_2^2}{n_2}}$; The estimate would be

$$s_{\overline{X}-\overline{Y}}=\sqrt{\frac{s_1^2}{n_1}+\frac{s_2^2}{n_2}}=\sqrt{\frac{1.66^2}{27}+\frac{2.104^2}{20}}=.5687.$$

c $\dfrac{s_1}{s_2}=\dfrac{1.660}{2.104}=.7890$

d $V(X-Y)=V(X)+V(Y)=\sigma_1^2+\sigma_2^2=1.66^2+2.104^2=7.1824$

5. N = 5,000 T = 1,761,300

$\overline{y}=374.6$ $\overline{x}=340.6$ $\overline{d}=34.0$

$\hat{\theta}_1=N\overline{x}=(5{,}000)(340.6)=1{,}703{,}000$

$\hat{\theta}_2=T-N\overline{d}=1{,}761{,}300-(5{,}000)(34.0)=1{,}591{,}300$

$\hat{\theta}_3=T\left(\dfrac{\overline{x}}{\overline{y}}\right)=1{,}761{,}300\left(\dfrac{340.6}{374.6}\right)=1{,}601{,}438.281$

6.

a Let $y_i=\ln(x_i)$ for I = 1, .., 31. It is easily verified that the sample mean and sample sd of the y_i's are $\overline{y}=5.102$ and $s_y=.4961$. Using the sample mean and sample sd to estimate μ and σ, respectively, gives $\hat{\mu}=5.102$ and $\hat{\sigma}=.4961$ (whence $\hat{\sigma}^2=s_y^2=.2461$).

b $E(X)\equiv\exp\left[\mu+\dfrac{\sigma^2}{2}\right]$. It is natural to estimate E(X) by using $\hat{\mu}$ and $\hat{\sigma}^2$ in place of μ and σ^2 in this expression:

$$E(\hat{X})=\exp\left[5.102+\frac{.2461}{2}\right]=\exp(5.225)=185.87$$

Chapter 6

7.

a $\hat{\mu} = \bar{x} = \frac{\sum x_i}{n} = \frac{1206}{10} = 120.6$

b $\hat{\tau} = 10{,}000 \qquad \hat{\mu} = 1{,}206{,}000$

c 8 of 10 houses in the sample used at least 100 therms (the "successes"), so $\hat{p} = \frac{8}{10} = .80.$

d The ordered sample values are 89, 99, 103, 109, 118, 122, 125, 138, 147, 156, from which the two middle values are 118 and 122, so

$$\hat{\tilde{\mu}} = \tilde{x} = \frac{118+122}{2} = 120.0$$

8.

a With q denoting the true proportion of defective components,

$$\hat{q} = \frac{(\#defective.in.sample)}{sample.size} = \frac{12}{80} = .150$$

b P(system works) = p^2, so an estimate of this probability is $\hat{p}^2 = \left(\frac{68}{80}\right)^2 = .723$

9.

a $E(\overline{X}) = \mu = E(X) = \lambda$, so $\overline{X}$ is an unbiased estimator for the Poisson parameter λ; $\sum x_i = (0)(18) + (1)(37) + ... + (7)(1) = 317$, since n = 150,

$$\hat{\lambda} = \bar{x} = \frac{317}{150} = 2.11.$$

b $\sigma_{\bar{x}} = \frac{\sigma}{\sqrt{n}} = \frac{\sqrt{\lambda}}{\sqrt{n}}$, so the estimated standard error is $\sqrt{\frac{\hat{\lambda}}{n}} = \frac{\sqrt{2.11}}{\sqrt{150}} = .119$

10.

a $E(\overline{X}^2) = Var(\overline{X}) + [E(\overline{X})]^2 = \frac{\sigma^2}{n} + \mu^2$, so the bias of the estimator $\overline{X}^2$ is $\frac{\sigma^2}{n}$; thus $\overline{X}^2$ tends to overestimate μ^2.

b $E(\overline{X}^2 - kS^2) = E(\overline{X}^2) - kE(S^2) = \mu^2 + \frac{\sigma^2}{n} - k\sigma^2$, so with $k = \frac{1}{n}$,

$E(\overline{X}^2 - kS^2) = \mu^2$.

11.

a $E\left(\frac{X_1}{n_1}-\frac{X_2}{n_2}\right)=\frac{1}{n_1}E(X_1)-\frac{1}{n_2}E(X_2)=\frac{1}{n_1}(n_1p_1)-\frac{1}{n_2}(n_2p_2)=p_1-p_2.$

b $Var\left(\frac{X_1}{n_1}-\frac{X_2}{n_2}\right)=Var\left(\frac{X_1}{n_1}\right)+Var\left(\frac{X_2}{n_2}\right)=\left(\frac{1}{n_1}\right)^2Var(X_1)+\left(\frac{1}{n_2}\right)^2Var(X_2)$

$\frac{1}{n_1^2}(n_1p_1q_1)+\frac{1}{n_2^2}(n_2p_2q_2)=\frac{p_1q_1}{n_1}+\frac{p_2q_2}{n_2}$, and the standard error is the square root of this quantity.

c With $\hat{p}_1=\frac{x_1}{n_1}$, $\hat{q}_1=1-\hat{p}_1$, $\hat{p}_2=\frac{x_2}{n_2}$, $\hat{q}_2=1-\hat{p}_2$, the estimated standard error is $\sqrt{\frac{\hat{p}_1\hat{q}_1}{n_1}+\frac{\hat{p}_2\hat{q}_2}{n_2}}$.

d $(\hat{p}_1-\hat{p}_2)=\frac{127}{200}-\frac{176}{200}=.635-.880=-.245$

e $\sqrt{\frac{(.635)(.365)}{200}+\frac{(.880)(.120)}{200}}=.041$

12. $E\left[\frac{(n_1-1)S_1^2+(n_2-1)S_2^2}{n_1+n_2-2}\right]=\frac{(n_1-1)}{n_1+n_2-2}E(S_1^2)+\frac{(n_2-1)}{n_1+n_2-2}E(S_2^2)$

$=\frac{(n_1-1)}{n_1+n_2-2}\sigma^2+\frac{(n_2-1)}{n_1+n_2-2}\sigma^2=\sigma^2.$

13. $E(X)=\int_{-1}^{1}x\cdot\tfrac{1}{2}(1+\theta x)dx=\frac{x^2}{4}+\frac{\theta x^3}{6}\Big|_{-1}^{1}=\frac{1}{3}\theta \qquad E(X)=\frac{1}{3}\theta \quad E(\overline{X})=\frac{1}{3}\theta$

$\hat{\theta}=3\overline{X}\Rightarrow E(\hat{\theta})=E(3\overline{X})=3E(\overline{X})=3\left(\frac{1}{3}\right)\theta=\theta$

14.

a min(x_i) = 202 and max(x_i) = 525, so the estimate of the number of planes manufactured is max(x_i) - min(x_i) + 1 = 525 – 202 + 1 = 324.

b The estimate will equal the true number of planes manufactured iff min(x_i) = α and max(x_i) = β, i.e., iff the smallest serial number in the population and the largest serial number in the population both appear in the sample. The estimator is not unbiased. This is because max(x_i) never overestimates β and will usually underestimate it (unless max(x_i) = β) , so that E[max(x_i)] < β. Similarly, E[min(x_i)] > α ,so E[max(x_i) - min(x_i)] < β - α + 1; The estimate will usually be smaller than β - α + 1, and can never exceed it.

15.

a $E(X^2)=2\theta$ implies that $E\left(\frac{X^2}{2}\right)=\theta$. Consider $\hat{\theta}=\frac{\sum X_i^2}{2n}$. Then

$$E(\hat{\theta})=E\left(\frac{\sum X_i^2}{2n}\right)=\frac{\sum E(X_i^2)}{2n}=\frac{\sum 2\theta}{2n}=\frac{2n\theta}{2n}=\theta$$, implying that $\hat{\theta}$ is an unbiased estimator for θ.

b $\sum x_i^2=1490.1058$, so $\hat{\theta}=\frac{1490.1058}{20}=74.505$

16.

a $E\left[\delta\overline{X}+(1-\delta)\overline{Y}\right]=\delta E(\overline{X})+(1-\delta)E(\overline{Y})=\delta\mu+(1-\delta)\mu=\mu$

b $Var\left[\delta\overline{X}+(1-\delta)\overline{Y}\right]=\delta^2 Var(\overline{X})+(1-\delta)^2 Var(\overline{Y})=\frac{\delta^2\sigma^2}{m}+\frac{4(1-\delta)^2\sigma^2}{n}$.

Setting the derivative with respect to δ equal to 0 yields

$\frac{2\delta\sigma^2}{m}+\frac{8(1-\delta)\sigma^2}{n}=0$, from which $\delta=\frac{4m}{4m+n}$.

17.

a $$E(\hat{p})=\sum_{x=0}^{\infty}\frac{r-1}{x+r-1}\cdot\binom{x+r-1}{x}\cdot p^r\cdot(1-p)^x$$

$$=p\sum_{x=0}^{\infty}\frac{(x+r-2)!}{x!(r-2)!}\cdot p^{r-1}\cdot(1-p)^x=p\sum_{x=0}^{\infty}\binom{x+r-2}{x}p^{r-1}(1-p)^x$$

$$=p\sum_{x=0}^{\infty}nb(x;r-1,p)=p.$$

b For the given sequence, x = 5, so $\hat{p}=\frac{5-1}{5+5-1}=\frac{4}{9}=.444$

18.

a $f(x;\mu,\sigma^2)=\frac{1}{\sqrt{2\pi}\sigma}e^{-\left((x-\mu)^2/2\sigma^2\right)}$, so $f(\mu;\mu,\sigma^2)=\frac{1}{\sqrt{2\pi}\sigma}$ and

$\frac{1}{4n[[f(\mu)]^2}=\frac{2\pi\sigma^2}{4n}=\frac{\pi}{2}\cdot\frac{\sigma^2}{n}$; since $\frac{\pi}{2}>1$, $Var(\tilde{X})>Var(\overline{X})$.

b $f(\mu)=\frac{1}{\pi}$, so $Var(\tilde{X})\approx\frac{\pi^2}{4n}=\frac{2.467}{n}$.

19.

a $\lambda = .5p + .15 \Rightarrow 2\lambda = p + .3$, so $p = 2\lambda - .3$ and $\hat{p} = 2\hat{\lambda} - .3 = 2\left(\frac{Y}{n}\right) - .3$; the estimate is $2\left(\frac{20}{80}\right) - .3 = .2$.

b $E(\hat{p}) = E\left(2\hat{\lambda} - .3\right) = 2E\left(\hat{\lambda}\right) - .3 = 2\lambda - .3 = p$, as desired.

c Here $\lambda = .7p + (.3)(.3)$, so $p = \frac{10}{7}\lambda - \frac{9}{70}$ and $\hat{p} = \frac{10}{7}\left(\frac{Y}{n}\right) - \frac{9}{70}$.

Section 6.2

20.

a We wish to take the derivative of $\ln\left[\binom{n}{x}p^x(1-p)^{n-x}\right]$, set it equal to zero and solve for p. $\frac{d}{dp}\left[\ln\binom{n}{x} + x\ln(p) + (n-x)\ln(1-p)\right] = \frac{x}{p} - \frac{n-x}{1-p}$; setting this equal to zero and solving for p yields $\hat{p} = \frac{x}{n}$. For n = 20 and x = 3,

$\hat{p} = \frac{3}{20} = .15$

b $E(\hat{p}) = E\left(\frac{X}{n}\right) = \frac{1}{n}E(X) = \frac{1}{n}(np) = p$; thus $\hat{p}$ is an unbiased estimator of p.

c $(1 - .15)^5 = .4437$

21.

a $E(X) = \beta \cdot \Gamma\left(1+\frac{1}{\alpha}\right)$ and $E(X^2) = Var(X) + [E(X)]^2 = \beta^2\Gamma\left(1+\frac{2}{\alpha}\right)$, so the moment estimators $\hat{\alpha}$ and $\hat{\beta}$ are the solution to $\overline{X} = \hat{\beta} \cdot \Gamma\left(1+\frac{1}{\hat{\alpha}}\right)$, $\frac{1}{n}\sum X_i^2 = \hat{\beta}^2\Gamma\left(1+\frac{2}{\hat{\alpha}}\right)$. Thus $\hat{\beta} = \frac{\overline{X}}{\Gamma\left(1+\frac{1}{\hat{\alpha}}\right)}$, so once $\hat{\alpha}$ has been determined $\Gamma\left(1+\frac{1}{\hat{\alpha}}\right)$ is evaluated and $\hat{\beta}$ then computed. Since

$\overline{X}^2 = \hat{\beta}^2 \cdot \Gamma^2\left(1+\frac{1}{\hat{\alpha}}\right)$, $\frac{1}{n}\sum \frac{X_i^2}{\overline{X}^2} = \frac{\Gamma\left(1+\frac{2}{\hat{\alpha}}\right)}{\Gamma^2\left(1+\frac{1}{\hat{\alpha}}\right)}$, so this equation must be solved to obtain $\hat{\alpha}$.

b From **a**, $\frac{1}{20}\left(\frac{16{,}500}{28.0^2}\right) = 1.05 = \frac{\Gamma\left(1+\frac{2}{\hat{\alpha}}\right)}{\Gamma^2\left(1+\frac{1}{\hat{\alpha}}\right)}$, so $\frac{1}{1.05} = .95 = \frac{\Gamma^2\left(1+\frac{1}{\hat{\alpha}}\right)}{\Gamma\left(1+\frac{2}{\hat{\alpha}}\right)}$, and from the hint, $\frac{1}{\hat{\alpha}} = .2 \Rightarrow \hat{\alpha} = 5$. Then $\hat{\beta} = \frac{\overline{x}}{\Gamma(1.2)} = \frac{28.0}{\Gamma(1.2)}$.

22.

a $E(X) = \int_0^1 x(\theta+1)x^\theta dx = \frac{\theta+1}{\theta+2} = 1 - \frac{1}{\theta+2}$, so the moment estimator $\hat{\theta}$ is the solution to $\overline{X} = 1 - \frac{1}{\hat{\theta}+2}$, yielding $\hat{\theta} = \frac{1}{1-\overline{X}} - 2$. Since $\overline{x} = .80, \hat{\theta} = 5 - 2 = 3.$

b $f(x_1,\ldots,x_n;\theta) = (\theta+1)^n (x_1 x_2 \ldots x_n)^\theta$, so the log likelihood is $n\ln(\theta+1) + \theta\sum\ln(x_i)$. Taking $\frac{d}{d\theta}$ and equating to 0 yields $\frac{n}{\theta+1} = -\sum\ln(x_i)$, so $\hat{\theta} = -\frac{n}{\sum\ln(X_i)} - 1$. Taking $\ln(x_i)$ for each given x_i yields ultimately $\hat{\theta} = 3.12$.

23. For a single sample from a Poisson distribution,

$$f(x_1,\dots,x_n;\lambda)=\frac{e^{-\lambda}\lambda^{x_1}}{x_1!}\cdots\frac{e^{-\lambda}\lambda^{x_n}}{x_n!}=\frac{e^{-n\lambda}\lambda^{\sum x_1}}{x_1!\dots x_n!}$$, so

$\ln[f(x_1,\dots,x_n;\lambda)]=-n\lambda+\sum x_i\ln(\lambda)-\sum\ln(x_i!)$. Thus

$\frac{d}{d\lambda}[\ln[f(x_1,\dots,x_n;\lambda)]]=-n+\frac{\sum x_i}{\lambda}=0\Rightarrow\hat{\lambda}=\frac{\sum x_i}{n}=\bar{x}$. For our problem, $f(x_1,\dots,x_n,y_1\dots y_n;\lambda_1,\lambda_2)$ is a product of the x sample likelihood and the y sample likelihood, implying that $\hat{\lambda}_1=\bar{x},\hat{\lambda}_2=\bar{y}$, and (by the invariance principle) $(\lambda_1\hat{-}\lambda_2)=\bar{x}-\bar{y}$.

24. We wish to take the derivative of $\ln\left[\binom{x+r-1}{x}p^r(1-p)^x\right]$ with respect to p, set it equal to zero, and solve for p: $\frac{d}{dp}\left[\ln\binom{x+r-1}{x}+r\ln(p)+x\ln(1-p)\right]=\frac{r}{p}-\frac{x}{1-p}$.

Setting this equal to zero and solving for p yields $\hat{p}=\frac{r}{r+x}$. This is the number of successes over the total number of trials, which is the same estimator for the binomial in exercise 6.20. The unbiased estimator from exercise 6.17 is $\hat{p}=\frac{r-1}{r+x-1}$, which is not the same as the maximum likelihood estimator.

25.

a $\hat{\mu}=\bar{x}=384.4; s^2=395.16$, so $\frac{1}{n}\sum(x_i-\bar{x})^2=\hat{\sigma}^2=\frac{9}{10}(395.16)=355.64$ and $\hat{\sigma}=\sqrt{355.64}=18.86$ (this is not s).

b The 95th percentile is $\mu+1.645\sigma$, so the mle of this is (by the invariance principle) $\hat{\mu}+1.645\hat{\sigma}=415.42$.

26. The mle of $P(X\le 400)$ is (by the invariance principle)

$$\Phi\left(\frac{400-\hat{\mu}}{\hat{\sigma}}\right)=\Phi\left(\frac{400-384.4}{18.86}\right)=\Phi(.80)=.7881$$

27.

a $f(x_1,\ldots,x_n;\alpha,\beta)=\frac{(x_1x_2\ldots x_n)^{\alpha-1}e^{-\Sigma x_i/\beta}}{\beta^{n\alpha}\Gamma^n(\alpha)}$, so the log likelihood is $(\alpha-1)\sum\ln(x_i)-\frac{\sum x_i}{\beta}-n\alpha\ln(\beta)-n\ln\Gamma(\alpha)$. Equating both $\frac{d}{d\alpha}$ and $\frac{d}{d\beta}$ to 0 yields $\sum\ln(x_i)-n\ln(\beta)-n\frac{d}{d\alpha}\Gamma(\alpha)=0$ and $\frac{\sum x_i}{\beta^2}=\frac{n\alpha}{\beta}=0$, a very difficult system of equations to solve.

b From the second equation in **a**, $\frac{\sum x_i}{\beta}=n\alpha\Rightarrow\overline{x}=\alpha\beta=\mu$, so the mle of μ is $\hat{\mu}=\overline{X}$.

28.

a $\left(\frac{x_1}{\theta}\exp\left[-x_1^2/2\theta\right]\right)\ldots\left(\frac{x_n}{\theta}\exp\left[-x_n^2/2\theta\right]\right)=(x_1\ldots x_n)\frac{\exp\left[-\Sigma x_i^2/2\theta\right]}{\theta^n}$. The natural log of the likelihood function is $\ln(x_i\ldots x_n)-n\ln(\theta)-\frac{\Sigma x_i^2}{2\theta}$. Taking the derivative wrt θ and equating to 0 gives $-\frac{n}{\theta}+\frac{\Sigma x_i^2}{2\theta^2}=0$, so $n\theta=\frac{\Sigma x_i^2}{2}$ and $\theta=\frac{\Sigma x_i^2}{2n}$. The mle is therefore $\hat{\theta}=\frac{\Sigma X_i^2}{2n}$, which is identical to the unbiased estimator suggested in Exercise 15.

b For x > 0 the cdf of X if $F(x;\theta)=P(X\le x)$ is equal to $1-\exp\left[\frac{-x^2}{2\theta}\right]$.

Equating this to .5 and solving for x gives the median in terms of θ:

$.5=\exp\left[\frac{-x^2}{2\theta}\right]$ implies that $\ln(.5)=\frac{-x^2}{2\theta}$, so $x=\tilde{\mu}=\sqrt{1.38630}$. The mle of $\tilde{\mu}$ is therefore $\left(1.38630\hat{\theta}\right)^{\frac{1}{2}}$.

29.

a The joint pdf (likelihood function) is

$$f(x_1,...,x_n;\lambda,\theta)=\begin{cases}\lambda^n e^{-\lambda\Sigma(x_i-\theta)} & x_1\ge\theta,...,x_n\ge\theta\\ 0 & otherwise\end{cases}$$

Notice that $x_1\ge\theta,...,x_n\ge\theta$ iff $\min(x_i)\ge\theta$,
and that $-\lambda\Sigma(x_i-\theta)=-\lambda\Sigma x_i+n\lambda\theta$.

Thus likelihood = $\begin{cases}\lambda^n\exp(-\lambda\Sigma x_i)\exp(n\lambda\theta) & \min(x_i)\ge\theta\\ 0 & \min(x_i)<\theta\end{cases}$

Consider maximization wrt θ. Because the exponent $n\lambda\theta$ is positive, increasing θ will increase the likelihood provided that $\min(x_i)\ge\theta$; if we make θ larger than $\min(x_i)$, the likelihood drops to 0. This implies that the mle of θ is $\hat\theta=\min(x_i)$. The log likelihood is now $n\ln(\lambda)-\lambda\Sigma(x_i-\hat\theta)$. Equating the derivative wrt λ to 0 and solving yields $\hat\lambda=\dfrac{n}{\Sigma(x_i-\hat\theta)}=\dfrac{n}{\Sigma x_i-n\hat\theta}$.

b $\hat\theta=\min(x_i)=.64$, and $\Sigma x_i=55.80$, so $\hat\lambda=\dfrac{10}{55.80-6.4}=.202$

30. The likelihood is $f(y;n,p)=\binom{n}{y}p^y(1-p)^{n-y}$ where

$p=P(X\ge 24)=1-\int_0^{24}\lambda e^{-\lambda x}dx=e^{-24\lambda}$. We know $\hat p=\dfrac{y}{n}$, so by the invariance principle $e^{-24\lambda}=\dfrac{y}{n}\Rightarrow\hat\lambda=-\dfrac{\left[\ln\left(\frac{y}{n}\right)\right]}{24}=.0120$ for n = 20, y = 15.

Supplementary Exercises

31. $P(|\bar X-\mu|>\varepsilon)=P(\bar X-\mu>\varepsilon)+P(\bar X-\mu<-\varepsilon)=P\left(\dfrac{\bar X-\mu}{\sigma/\sqrt n}>\dfrac{\varepsilon}{\sigma/\sqrt n}\right)+P\left(\dfrac{\bar X-\mu}{\sigma/\sqrt n}<\dfrac{-\varepsilon}{\sigma/\sqrt n}\right)$

$$=P\left(Z>\frac{\sqrt n\varepsilon}{\sigma}\right)+P\left(Z<\frac{-\sqrt n\varepsilon}{\sigma}\right)=\int_{\sqrt n\varepsilon/\sigma}^{\infty}\frac{1}{\sqrt{2\pi}}e^{-z^2/2}dz+\int_{-\infty}^{-\sqrt n\varepsilon/\sigma}\frac{1}{\sqrt{2\pi}}e^{-z^2/2}dz.$$

As $n\to\infty$, both integrals $\to 0$ since $\lim_{c\to\infty}\int_c^{\infty}\frac{1}{\sqrt{2\pi}}e^{-z^2/2}dz=0$.

32. $F_Y(y) = P(Y \le y) = P(X_1 \le y, ..., X_n \le y) = P(X_1 \le y)...P(X_n \le y) = \left(\frac{y}{\theta}\right)^n$ for $0 \le y \le \theta$, so $f_Y(y) = \frac{ny^{n-1}}{\theta^n}$ and $E(Y) = \int_0^\theta y \cdot \frac{ny^{n-1}}{n} dy = \frac{n}{n+1}\theta$. While $\hat{\theta} = Y$ is not unbiased, $\frac{n+1}{n}Y$ is, since $E\left[\frac{n+1}{n}Y\right] = \frac{n+1}{n}E(Y) = \frac{n+1}{n} \cdot \frac{n}{n+1}\theta = \theta$, so $K = \frac{n+1}{n}$ does the trick.

33. Let x_1 = the time until the first birth, x_2 = the elapsed time between the first and second births, and so on. Then $f(x_1, ..., x_n; \lambda) = \lambda e^{-\lambda x_1} \cdot (2\lambda)e^{-2\lambda x_2} ...(n\lambda)e^{-n\lambda x_n} = n!\lambda^n e^{-\lambda \Sigma k x_k}$. Thus the log likelihood is $\ln(n!) + n\ln(\lambda) - \lambda\Sigma k x_k$. Taking $\frac{d}{d\lambda}$ and equating to 0 yields $\hat{\lambda} = \frac{n}{\sum_{k=1}^{n} k x_k}$. For the given sample, n = 6, x_1 = 25.2, x_2 = 41.7 – 25.2 = 16.5, x_3 = 9.5, x_4 = 4.3, x_5 = 4.0, x_6 = 2.3; so $\sum_{k=1}^{6} k x_k = (1)(25.2) + (2)(16.5) + ... + (6)(2.3) = 137.7$ and $\hat{\lambda} = \frac{6}{137.7} = .0436$.

34. $MSE(KS^2) = Var(KS^2) + Bias(KS^2)$.
$Bias(KS^2) = E(KS^2) - \sigma^2 = K\sigma^2 - \sigma^2 = \sigma^2(K-1)$, and
$Var(KS^2) = K^2 Var(S^2) = K^2\left(E[(S^2)^2] - [E(S^2)]^2\right) = K^2\left(\frac{(n+1)\sigma^4}{n-1} - (\sigma^2)^2\right)$
$= \left[\frac{2K^2}{n-1} + (k-1)^2\right]\sigma^4$. To find the minimizing value of K, take $\frac{d}{dK}$ and equate to 0; the result is $K = \frac{n-1}{n+1}$; thus the estimator which minimizes MSE is neither the unbiased estimator (K = 1) nor the mle $K = \frac{n-1}{n}$.

35.

$x_i + x_j$	23.5	26.3	28.0	28.2	29.4	29.5	30.6	31.6	33.9	49.3
23.5	23.5	24.9	25.75	25.85	26.45	26.5	27.05	27.55	28.7	36.4
26.3		26.3	27.15	27.25	27.85	27.9	28.45	28.95	30.1	37.8
28.0			28.0	28.1	28.7	28.75	29.3	29.8	30.95	38.65
28.2				28.2	28.8	28.85	29.4	29.9	31.05	38.75
29.4					29.4	29.45	30.0	30.5	30.65	39.35
29.5						29.5	30.05	30.55	31.7	39.4
30.6							30.6	31.1	32.25	39.95
31.6								31.6	32.75	40.45
33.9									33.9	41.6
49.3										49.3

There are 55 averages, so the median is the 28th in order of increasing magnitude. Therefore, $\hat{\mu} = 29.5$

36. With $\sum x = 555.86$ and $\sum x^2 = 15{,}490$, $s = \sqrt{s^2} = \sqrt{2.1570} = 1.4687$. The $|x_i - \tilde{x}|'s$ are, in increasing order, .02, .02, .08, .22, .32, .42, .53, .54, .65, .81, .91, 1.15, 1.17, 1.30, 1.54, 1.54, 1.71, 2.35, 2.92, 3.50. The median of these values is $\frac{(.81 + .91)}{2} = .86$. The estimate based on the resistant estimator is then $\frac{.86}{.6745} = 1.275$. This estimate is in reasonably close agreement with s.

37. Let $c = \frac{\Gamma\left(\frac{n-1}{2}\right)}{\Gamma\left(\frac{n}{2}\right)\cdot\sqrt{\frac{2}{n-1}}}$. Then E(cS) = cE(S), and c cancels with the two Γ factors and the square root in E(S), leaving just σ. When n = 20, $c = \frac{\Gamma(9.5)}{\Gamma(10)\cdot\sqrt{\frac{2}{19}}}$. $\Gamma(10) = 9!$ and $\Gamma(9.5) = (8.5)(7.5)...(1.5)(.5)\Gamma(.5)$, but $\Gamma(.5) = \sqrt{\pi}$. Straightforward calculation gives c = 1.0132.

38.

a The likelihood is

$$\prod_{i=1}^{n} \frac{1}{\sqrt{2\pi\sigma^2}} e^{-\frac{(x_i-\mu_i)}{2\sigma^2}} \cdot \frac{1}{\sqrt{2\pi\sigma^2}} e^{-\frac{(y_i-\mu_i)}{2\sigma^2}} = \frac{1}{(2\pi\sigma^2)^n} e^{-\frac{\left(\Sigma(x_i-\mu_i)^2+\Sigma(y_i-\mu_i)^2\right)}{2\sigma^2}}$$

. The log likelihood is thus $-n\ln(2\pi\sigma^2) - \frac{\left(\Sigma(x_i-\mu_i)^2+\Sigma(y_i-\mu_i)^2\right)}{2\sigma^2}$. Taking $\frac{d}{d\mu_i}$ and equating to zero gives $\hat{\mu}_i = \frac{x_i+y_i}{2}$. Substituting these estimates of the $\hat{\mu}_i$'s into the log likelihood gives

$$-n\ln(2\pi\sigma^2) - \frac{1}{2\sigma^2}\left(\sum\left(x_i - \frac{x_i+y_i}{2}\right)^2 + \sum\left(y_i - \frac{x_i+y_i}{2}\right)^2\right)$$

$= -n\ln(2\pi\sigma^2) - \frac{1}{2\sigma^2}\left(\frac{1}{2}\Sigma(x_i - y_i)^2\right)$. Now taking $\frac{d}{d\sigma^2}$, equating to zero, and solving for σ^2 gives the desired result.

b $E(\hat{\sigma}) = \frac{1}{4n}E\left(\Sigma(X_i - Y_i)^2\right) = \frac{1}{4n} \cdot \Sigma E(X_i - Y)^2$, but

$E(X_i - Y)^2 = V(X_i - Y) + [E(X_i - Y)]^2 = 2\sigma^2 + 0 = 2\sigma^2$. Thus

$E(\hat{\sigma}^2) = \frac{1}{4n}\Sigma(2\sigma^2) = \frac{1}{4n}2n\sigma^2 = \frac{\sigma^2}{2}$, so the mle is definitely not unbiased; the expected value of the estimator is only half the value of what is being estimated!

Chapter 7

Section 7.1

1.

a $z_{\alpha/2} = 2.81$ implies that $\alpha/2 = 1 - \Phi(2.81) = .0025$, so $\alpha = .005$ and the confidence level is $100(1-\alpha)\% = 99.5\%$.

b $z_{\alpha/2} = 1.44$ for $\alpha = 2[1 - \Phi(1.44)] = .15$, and $100(1-\alpha)\% = 85\%$.

c 99.7% implies that $\alpha = .003$, $\alpha/2 = .0015$, and $z_{.0015} = 2.96$. (Look for cumulative area .9985 in the main body of table A.3, the Z table.)

d 75% implies $\alpha = .25$, $\alpha/2 = .125$, and $z_{.125} = 1.15$.

2.

a The sample mean is the center of the interval, so $\bar{x} = \dfrac{114.4 + 115.6}{2} = 115$.

b The interval (114.4, 115.6) has the 90% confidence level. The higher confidence level will produce a wider interval.

3.

a A 90% confidence interval will be narrower (See 2b, above) Also, the z critical value for a 90% confidence level is 1.645, smaller than the z of 1.96 for the 95% confidence level, thus producing a narrower interval.

b Not a correct statement. Once and interval has been created from a sample, the mean μ is either enclosed by it, or not. The 95% confidence is in the general procedure, for repeated sampling.

c Not a correct statement. The interval is an estimate for the population mean, not a boundary for population values.

d Not a correct statement. In theory, if the process were repeated an infinite number of times, 95% of the intervals would contain the population mean μ.

4.

a $58.3 \pm \dfrac{1.96(3)}{\sqrt{25}} = 58.3 \pm 1.18 = (57.1, 59.5)$

b $58.3 \pm \dfrac{1.96(3)}{\sqrt{100}} = 58.3 \pm .59 = (57.7, 58.9)$

c $58.3 \pm \dfrac{2.58(3)}{\sqrt{100}} = 58.3 \pm .77 = (57.5, 59.1)$

(continued)

d 82% confidence $\Rightarrow 1-\alpha = .82 \Rightarrow \alpha = .18 \Rightarrow \alpha/2 = .09$, so $z_{\alpha/2} = z_{.09} = 1.34$

and the interval is $58.3 \pm \dfrac{1.34(3)}{\sqrt{100}} = (57.9, 58.7)$.

e $n = \left[\dfrac{2(2.58)3}{1}\right]^2 = 239.62$ so n = 240.

5.

a $4.85 \pm \dfrac{(1.96)(.75)}{\sqrt{20}} = 4.85 \pm .33 =$ (4.52, 5.18).

b $z_{\alpha/2} = z_{.02/2} = z_{.01} = 2.33$, so the interval is $4.56 \pm \dfrac{(2.33)(.75)}{\sqrt{16}} =$ (4.12, 5.00).

c $n = \left[\dfrac{2(1.96)(.75)}{.40}\right]^2 = 54.02$, so n = 55.

d $n = \left[\dfrac{2(2.58)(.75)}{.2}\right]^2 = 93.61$, so n = 94.

6.

a $8439 \pm \dfrac{(1.645)(100)}{\sqrt{25}} = 8439 \pm 32.9 =$ (8406.1, 8471.9).

b $1-\alpha = .92 \Rightarrow \alpha = .08 \Rightarrow \alpha/2 = .04$ so $z_{\alpha/2} = z_{.04} = 1.75$

7. If $L = 2z_{\alpha/2}\dfrac{\sigma}{\sqrt{n}}$ and we increase the sample size by a factor of 4, the new length is

$L' = 2z_{\alpha/2}\dfrac{\sigma}{\sqrt{4n}} = \left[2z_{\alpha/2}\dfrac{\sigma}{\sqrt{n}}\right]\left(\dfrac{1}{2}\right) = \dfrac{L}{2}$. Thus halving the length requires n to be

increased fourfold. If $n' = 25n$, then $L' = \dfrac{L}{5}$, so the length is decreased by a factor of

5.

8.

a With probability $1-\alpha$, $z_{\alpha_1} \le (\overline{X}-\mu)\left(\frac{\sigma}{\sqrt{n}}\right) \le z_{\alpha_2}$. These inequalities can be manipulated exactly as was done in the text to isolate μ; the result is $\overline{X} - z_{\alpha_2}\frac{\sigma}{\sqrt{n}} \le \mu \le \overline{X} + z_{\alpha_1}\frac{\sigma}{\sqrt{n}}$, so a $100(1-\alpha)\%$ interval is

$$\left(\overline{X} - z_{\alpha_2}\frac{\sigma}{\sqrt{n}}, \overline{X} + z_{\alpha_1}\frac{\sigma}{\sqrt{n}}\right)$$

b The usual 95% interval has length $3.92\frac{\sigma}{\sqrt{n}}$, while this interval will have length $\left(z_{\alpha_1} + z_{\alpha_2}\right)\frac{\sigma}{\sqrt{n}}$. With $z_{\alpha_1} = z_{.0125} = 2.24$ and $z_{\alpha_2} = z_{.0375} = 1.78$, the length is $(2.24+1.78)\frac{\sigma}{\sqrt{n}} = 4.02\frac{\sigma}{\sqrt{n}}$, which is longer.

9.

a $\left(\overline{x} - 1.645\frac{\sigma}{\sqrt{n}}, \infty\right)$. From 5**a**, $\overline{x} = 4.85$, $\sigma = .75$, n = 20;

$4.85 - 1.645\frac{.75}{\sqrt{20}} = 4.5741$, so the interval is $(4.5741, \infty)$.

b $\left(\overline{x} - z_\alpha\frac{\sigma}{\sqrt{n}}, \infty\right)$

c $\left(-\infty, \overline{x} + z_\alpha\frac{\sigma}{\sqrt{n}}\right)$; From 4**a**, $\overline{x} = 58.3$, $\sigma = 3.0$, n = 25;

$58.3 + 2.33\frac{3}{\sqrt{25}} = (-\infty, 59.70)$

10.

a When n = 15, $2\lambda\sum X_i$ has a chi-squared distribution with 30 d.f. From the 30 d.f. row of Table A.6, the critical values that capture lower and upper tail areas of .025 (and thus a central area of .95) are 16.791 and 46.979. An argument parallel to that given in Example 7.5 gives $\left(\frac{2\sum x_i}{46.979}, \frac{2\sum x_i}{16.791}\right)$ as a 95% C. I. for $\mu = \frac{1}{\lambda}$. Since $\sum x_i = 63.2$ the interval is (2.69, 7.53).

b A 99% confidence level requires using critical values that capture area .005 in each tail of the chi-squared curve with 30 d.f.; these are 13.787 and 53.672, which replace 16.791 and 46.979 in **a**.

c $V(X)=\frac{1}{\lambda^2}$ when X has an exponential distribution, so the standard deviation is $\frac{1}{\lambda}$, the same as the mean. Thus the interval of **a** is also a 95% C.I. for the standard deviation of the lifetime distribution.

11. Y is a binomial r.v. with n = 1000 and p = .95, so E(Y) = np = 950, the expected number of intervals that capture μ, and $\sigma_Y = \sqrt{npq} = 6.892$. Using the normal approximation to the binomial distribution, $P(940 \le Y \le 960) = P(939.5 \le Y_{normal} \le 960.5)$
$= P(-1.52 \le Z \le 1.52) = .9357 - .0643 = .8714$.

Section 7.2

12. $\bar{x} \pm 2.58\frac{s}{\sqrt{n}} = .81 \pm 2.58\frac{.34}{\sqrt{110}} = .81 \pm .08 = (.73,.89)$

13.

a $\bar{x} \pm z_{.025}\frac{s}{\sqrt{n}} = 1.028 \pm 1.96\frac{.163}{\sqrt{69}} = 1.028 \pm .038 = (.990,1.066)$

b $w = .05 = \frac{2(1.96)(.16)}{\sqrt{n}} \Rightarrow \sqrt{n} = \frac{2(1.96)(.16)}{.05} = 12.544 \Rightarrow n = (12.544)^2 \approx 158$

14.

a $89.10 \pm 1.96\frac{3.73}{\sqrt{169}} = 89.10 \pm .56 = (88.54,.89.66)$. Yes, this is a very narrow interval. It appears quite precise.

b $n = \left[\frac{(1.96)(.16)}{.5}\right]^2 = 245.86 \Rightarrow n = 246$.

15.

a $z_\alpha = .84$, and $\Phi(.84) = .7995 \approx .80$, so the confidence level is 80%.

b $z_\alpha = 2.05$, and $\Phi(2.05) = .9798 \approx .98$, so the confidence level is 98%.

c $z_\alpha = .67$, and $\Phi(.67) = .7486 \approx .75$, so the confidence level is 75%.

16. n = 46, $\bar{x} = 382.1$, s = 31.5; The 95% upper confidence bound =

$$\bar{x} + z_\alpha \frac{s}{\sqrt{n}} = 382.1 + 1.645\frac{31.5}{\sqrt{46}} = 382.1 + 7.64 = 389.74$$

17. 99% lower bound: $\bar{x} - z_{.01}\frac{s}{\sqrt{n}} = 64.3 - 2.33\frac{6.0}{\sqrt{50}} = 64.3 - 2.0 = 62.3$

18. 90% lower bound: $\bar{x} - z_{.10}\frac{s}{\sqrt{n}} = 4.25 - 1.28\frac{1.30}{\sqrt{75}} = 4.06$

19. $\hat{p} = \frac{201}{356} = .5646$; We calculate a 95% confidence interval for the proportion of all dies that pass the probe:

$$\frac{.5646 + \frac{(1.96)^2}{2(356)} \pm 1.96\sqrt{\frac{(.5646)(.4354)}{356} + \frac{(1.96)^2}{4(356)^2}}}{1 + \frac{(1.96)^2}{356}} = \frac{.5700 \pm .0518}{1.01079} = (.513, .615)$$

20. n = 507, x = # of successes = 142 $\Rightarrow \hat{p} = \frac{142}{507} = .28$; the 99% two-sided interval is:

$$\frac{.28 + \frac{(2.58)^2}{2(507)} \pm 2.58\sqrt{\frac{(.28)(.72)}{507} + \frac{(2.58)^2}{4(507)^2}}}{1 + \frac{(2.58)^2}{507}} = \frac{.2866 \pm .0519}{1.0131} = (.2317, .3341)$$

21. $\hat{p} = \frac{133}{539} = .2468$; the 95% lower confidence bound is:

$$\frac{.2468 + \frac{(1.645)^2}{2(539)} - 1.645\sqrt{\frac{(.2468)(.7532)}{539} + \frac{(1.645)^2}{4(539)^2}}}{1 + \frac{(1.645)^2}{539}} = \frac{.2493 - .0307}{1.005} = .218$$

22. $\hat{p} = .072$; the 99% upper confidence bound is:

$$\frac{.072 + \frac{(2.33)^2}{2(487)} + 2.33\sqrt{\frac{(.072)(.928)}{487} + \frac{(2.33)^2}{4(487)^2}}}{1 + \frac{(2.33)^2}{487}} = \frac{.0776 + .0279}{1.0111} = .1043$$

23.

a $\hat{p} = \frac{24}{37} = .6486$; The 99% confidence interval for p is

$$\frac{.6486 + \frac{(2.58)^2}{2(37)} \pm 2.58\sqrt{\frac{(.6486)(.3514)}{37} + \frac{(2.58)^2}{4(37)^2}}}{1 + \frac{(2.58)^2}{37}} = \frac{.7386 \pm .2216}{1.1799} = (.438, .814)$$

b $$n = \frac{2(2.58)^2(.25) - (2.58)^2(.01) \pm \sqrt{4(2.58)^4(.25)(.25 - .01) + .01(2.58)^4}}{.01}$$

$$= \frac{3.261636 \pm 3.3282}{.01} \approx 659$$

24. n = 56, $\overline{x} = 8.17$, s = 1.42; For a 95% C.I., $z_{\alpha/2} = 1.96$. The interval is $8.17 \pm 1.96\left(\frac{1.42}{\sqrt{56}}\right) = (7.798, 8.542)$. We make no assumptions about the distribution if percentage elongation.

25.

a $$n = \frac{2(1.96)^2(.25) - (1.96)^2(.01) \pm \sqrt{4(1.96)^4(.25)(.25 - .01) + .01(1.96)^4}}{.01} \approx 381$$

b $$n = \frac{2(1.96)^2\left(\frac{1}{3} \cdot \frac{2}{3}\right) - (1.96)^2(.01) \pm \sqrt{4(1.96)^4\left(\frac{1}{3} \cdot \frac{2}{3}\right)\left(\frac{1}{3} \cdot \frac{2}{3} - .01\right) + .01(1.96)^4}}{.01} \approx 253$$

26. With $\theta = \lambda$, $\hat{\theta} = \overline{X}$ and $\sigma_{\hat{\theta}} = \sqrt{\frac{\lambda}{n}}$ so $\hat{\sigma}_{\hat{\theta}} = \sqrt{\frac{\overline{X}}{n}}$. The large sample C.I. is then $\overline{x} \pm z_{\alpha/2}\sqrt{\frac{\overline{x}}{n}}$. We calculate $\sum x_i = 203$, so $\overline{x} = 4.06$, and a 95% interval for λ is

$$4.06 \pm 1.96\sqrt{\frac{4.06}{50}} = 4.06 \pm .56 = (3.50, 4.62)$$

27. Note that the midpoint of the new interval is $\frac{x+\frac{z^2}{2}}{n+z^2}$, which is roughly $\frac{x+2}{n+4}$ with a confidence level of 95% and approximating $1.96 \approx 2$. The variance of this quantity is $\frac{np(1-p)}{(n+z^2)^2}$, or roughly $\frac{p(1-p)}{n+4}$. Now replacing p with $\frac{x+2}{n+4}$, we have $\left(\frac{x+2}{n+4}\right) \pm z_{\alpha/2}\sqrt{\frac{\left(\frac{x+2}{n+4}\right)\left(1-\frac{x+2}{n+4}\right)}{n+4}}$; For clarity, let $x^* = x+2$ and $n^* = n+4$, then $\hat{p}^* = \frac{x^*}{n^*}$ and the formula reduces to $\hat{p}^* \pm z_{\alpha/2}\sqrt{\frac{\hat{p}^*\hat{q}^*}{n^*}}$, the desired conclusion. For further discussion, see the Agresti article.

Section 7.3

28.

- a 1.341
- b 1.753
- c 1.708
- d 1.684
- e 2.704

29.

- a $t_{.025,10} = 2.228$
- b $t_{.025,20} = 2.086$
- c $t_{.005,20} = 2.845$
- d $t_{.005,50} = 2.678$
- e $t_{.01,25} = 2.485$
- f $-t_{.025,5} = -2.571$

30.

- a $t_{.025,10} = 2.228$
- b $t_{.025,15} = 2.131$
- c $t_{.005,15} = 2.947$
- d $t_{.005,4} = 4.604$
- e $t_{.01,24} = 2.492$
- f $t_{.005,37} \approx 2.712$

31.

a $t_{.05,10} = 1.812$

b $t_{.05,15} = 1.753$

c $t_{.01,15} = 2.602$

d $t_{.01,4} = 3.747$

e $\approx t_{.025,24} = 2.064$

f $t_{.01,37} \approx 2.429$

32. d.f. = n – 1 = 7, so the critical value for a 95% C.I. is $t_{.025,7} = 2.365$. The interval is

$$30.2 \pm (2.365)\left(\frac{3.1}{\sqrt{8}}\right) = 30.2 \pm 2.6 = (27.6, 32.8).$$

33.

a The boxplot indicates a very slight positive skew, with no outliers. The data appears to center near 438.

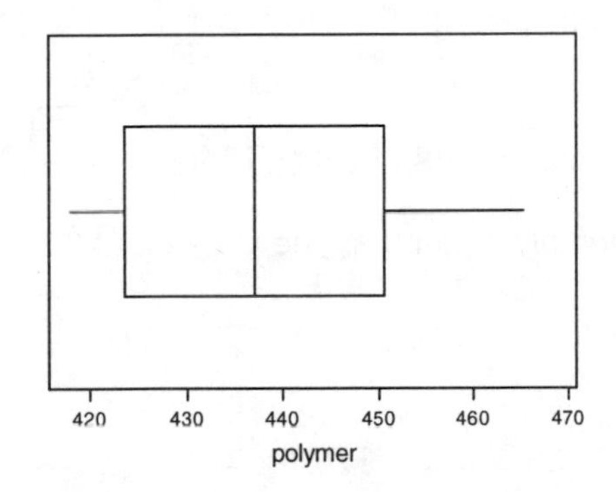

b Based on a normal probability plot, it is reasonable to assume the sample observations came from a normal distribution.

c With d.f. = n – 1 = 16, the critical value for a 95% C.I. is $t_{.025,16} = 2.120$, and the interval is $438.29 \pm (2.120)\left(\frac{15.14}{\sqrt{17}}\right) = 438.29 \pm 7.785 = (430.51, 446.08)$.

Since 440 is within the interval, 440 is a plausible value for the true mean. 450, however, is not, since it lies outside the interval.

34. n = 14, $\bar{x} = 8.48$, s = .79; $t_{.05,13} = 1.771$

a A 95% lower confidence bound: $8.48 - 1.771\left(\frac{.79}{\sqrt{14}}\right) = 8.48 - .37 = 8.11$. With 95% confidence, the value of the true mean proportional limit stress of all such joints lies in the interval $(8.11, \infty)$. If this interval is calculated for sample after sample, in the long run 95% of these intervals will include the true mean proportional limit stress of all such joints. We must assume that the sample observations were taken from a normally distributed population.

b A 95% lower prediction bound: $8.48 - 1.771(.79)\sqrt{1 + \frac{1}{14}} = 8.48 - 1.45 = 7.03$.

If this bound is calculated for sample after sample, in the long run 95% of these bounds will provide a lower bound for the corresponding future values of the proportional limit stress of a single joint of this type.

35. n = 5, $\bar{x} = 2887.6$, s = .84.0; $t_{.025,4} = 2.776$

a A 95% C.I. for the mean: $2887.6 \pm (2.776)\left(\frac{84}{\sqrt{5}}\right) \Rightarrow (2783.3, 2991.9)$

b A 95% Prediction Interval: $2887.6 \pm 2.776(84)\sqrt{1 + \frac{1}{5}} \Rightarrow (2632.1, 3143.1)$.

The P.I. is considerably larger than the C.I., about 2.5 times larger.

36. n = 26, $\bar{x} = 370.69$, s = 24.36; $t_{.05,25} = 1.708$

a A 95% upper confidence bound:

$$370.69 + (1.708)\left(\frac{24.36}{\sqrt{26}}\right) = 370.69 + 8.16 = 378.85$$

b A 95% upper prediction bound:

$$370.69 + 1.708(24.36)\sqrt{1 + \frac{1}{26}} = 370.69 + 42.45 = 413.14$$

(continued)

c Following a similar argument as that on p. 300 of the text, we need to find the variance of $\overline{X} - \overline{X}_{new}$:

$$V(\overline{X} - \overline{X}_{new}) = V(\overline{X}) + V(\overline{X}_{new}) = V(\overline{X}) + V(\tfrac{1}{2}(X_{27} + X_{28}))$$
$$= V(\overline{X}) + V(\tfrac{1}{2}X_{27}) + V(\tfrac{1}{2}X_{28}) = V(\overline{X}) + \tfrac{1}{4}V(X_{27}) + \tfrac{1}{4}V(X_{28})$$
$$= \frac{\sigma^2}{n} + \frac{1}{4}\sigma^2 + \frac{1}{4}\sigma^2 = \sigma^2\left(\frac{1}{2} + \frac{1}{n}\right)$$. We eventually arrive at

$T = \dfrac{\overline{X} - \overline{X}_{new}}{s\sqrt{\frac{1}{2} + \frac{1}{n}}}$ ~ t distribution with n – 1 d.f., so the new prediction interval is

$\overline{x} \pm t_{\alpha/2,n-1} \cdot s\sqrt{\frac{1}{2} + \frac{1}{n}}$. For this situation, we have

$$370.69 \pm 1.708(24.36)\sqrt{\frac{1}{2} + \frac{1}{26}} = 370.69 \pm 30.53 = (39.47, 400.53)$$

37.

a A 95% C.I. : $.9255 \pm 2.093(.0181) = .9255 \pm .0379 \Rightarrow (.8876, .9634)$

b A 95% P.I. : $.9255 \pm 2.093(.0809)\sqrt{1 + \frac{1}{20}} = .9255 \pm .1735 \Rightarrow (.7520, 1.0990)$

c A tolerance interval is requested, with k = 99, confidence level 95%, and n = 20. The tolerance critical value, from Table A.6, is 3.615. The interval is $.9255 \pm 3.615(.0809) \Rightarrow (.6330, 1.2180)$.

38. N = 25, $\overline{x} = .0635$, s = .0065

a 95% P.I. : $.0635 \pm 2.064(.0065)\sqrt{1 + \frac{1}{25}} = .0635 \pm .0137 \Rightarrow (.0498, .0772)$.

b 99% Tolerance Interval, with k = 95, critical value 2.972 (table A.6): $.0635 \pm 2.972(.0065) \Rightarrow (.0442, .0828)$.

39. The 20 d.f. row of Table A.5 shows that 1.725 captures upper tail area .05 and 1.325 captures uppertail area .10 The confidence level for each interval is 100(central area)%. For the first interval, central area = 1 – sum of tail areas = 1 – (.25 + .05) = .70, and for the second and third intervals the central areas are 1 – (.20 + .10) = .70 and 1 – (.15 + .15) = 70. Thus each interval has confidence level 70%. The width of the first interval is $\dfrac{s(.687 + 1.725)}{\sqrt{n}} = \dfrac{.2412s}{\sqrt{n}}$, whereas the widths of the second and third intervals are 2.185 and 2.128 respectively. The third interval, with symmetrically placed critical values, is the shortest, so it should be used. This will always be true for a t interval.

Section 7.4

40.

a $\chi^2_{.1,15} = 22.307$ (.1 column, 15 d.f. row)

b $\chi^2_{.1,25} = 34.381$

c $\chi^2_{.01,25} = 44.313$

d $\chi^2_{.005,25} = 46.925$

e $\chi^2_{.99,25} = 11.523$ (from .99 column, 25 d.f. row)

f $\chi^2_{.995,25} = 10.519$

41.

a $\chi^2_{.05,10} = 18.307$

b $\chi^2_{.95,10} = 3.940$

c Since $10.987 = \chi^2_{.975,22}$ and $36.78 = \chi^2_{.025,22}$, $P(\chi^2_{.975,22} \le \chi^2 \le \chi^2_{.025,22}) = .95$.

d Since $14.61 = \chi^2_{.95,25}$ and $37.65 = \chi^2_{.05,25}$, $P(\chi^2_{.95,25} \le \chi^2 \le \chi^2_{.05,25}) = .90$.

42. n – 1 = 8, $\chi^2_{.025,8} = 17.543$, $\chi^2_{.975,8} = 2.180$, so the 95% interval for σ^2 is $\left(\frac{8(7.90)}{17.543}, \frac{8(7.90)}{2.180}\right) = (3.60, 28.98)$. The 95% interval for σ is $\left(\sqrt{3.60}, \sqrt{28.98}\right) = (1.90, 5.38)$.

43. n = 22 implies that d.f. = n – 1 = 21, so the .995 and .005 columns of Table A.7 give the necessary chi-squared critical values as 8.033 and 41.399. $\Sigma x_i = 1701.3$ and $\Sigma x_i^2 = 132{,}097.35$, so $s^2 = 25.368$. The interval for σ^2 is $\left(\frac{21(25.368)}{41.399}, \frac{21(25.368)}{8.033}\right) = (12.868, 66.317)$ and that for σ is $(3.6, 8.1)$ Validity of this interval requires that fracture toughness be (at least approximately) normally distributed.

44.

a Using a normal probability plot, we ascertain that it is plausible that this sample was taken from a normal population distribution.

b With s = 1.579 , n = 15, and $\chi^2_{.05,14} = 23.685$ the 95% upper confidence bound for σ is $\sqrt{\frac{14(1.579)^2}{23.685}} = 1.214$

Chapter 7

Supplementary Exercises

45.

a n = 48, $\bar{x} = 8.079$, $s^2 = 23.7017$, and s = 4.868.
A 95% C.I. for μ = the true average strength is

$$\bar{x} \pm 1.96\frac{s}{\sqrt{n}} = 8.079 \pm 1.96\frac{4.868}{\sqrt{48}} = 8.079 \pm 1.377 = (6.702, 9.456)$$

b $\hat{p} = \frac{13}{48} = .2708$. A 95% C.I. is

$$\frac{.2708 + \frac{1.96^2}{2(48)} \pm 1.96\sqrt{\frac{(.2708)(.7292)}{48} + \frac{1.96^2}{4(48)^2}}}{1 + \frac{1.96^2}{48}} = \frac{.3108 \pm .1319}{1.0800} = (.166, .410)$$

46. A 98% t C.I. requires $t_{\alpha/2,n-1} = t_{.01,8} = 2.896$. The interval is

$$188.0 \pm 2.896\frac{7.2}{\sqrt{9}} = 188.0 \pm 7.0 = (181.0, 195.0).$$

47.

a There appears to be a slight positive skew in the middle half of the sample, but the lower whisker is much longer than the upper whisker. The extent of variability is rather substantial, although there are no outliers.

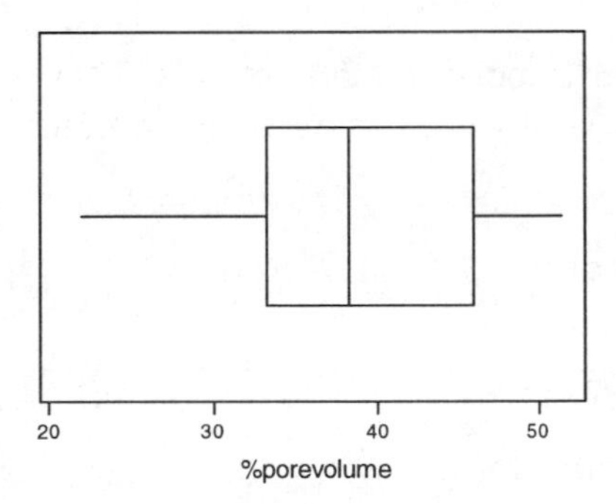

b The pattern of points in a normal probability plot is reasonably linear, so, yes, normality is plausible.

c n = 18, $\bar{x} = 38.66$, s = 8.473, and $t_{.01,17} = 2.586$. The 98% confidence interval is $38.66 \pm 2.586\frac{8.473}{\sqrt{18}} = 38.66 \pm 5.13 = (33.53, 43.79)$.

48. $\bar{x}$ = the middle of the interval = $\frac{229.764+233.502}{2} = 231.633$. To find *s* we use $width = 2\left(t_{.025,4}\right)\left(\frac{s}{\sqrt{n}}\right)$, and solve for *s*. Here, n = 5, $t_{.025,4} = 2.776$, and width = upper limit – lower limit = 3.738. $3.738 = 2(2776)\frac{s}{\sqrt{5}} \Rightarrow s = \frac{\sqrt{5}(3.738)}{2(2.776)} = 1.5055$. So for a 99% C.I., $t_{.005,4} = 4.604$, and the interval is

$$231.633 \pm 4.604\frac{1.5055}{\sqrt{5}} = 213.633 \pm 3.100 = (228.533, 234.733).$$

49.

a $\hat{p} = \frac{136}{200} = .680 \Rightarrow$ a 90% C.I. is

$$\frac{.680 + \frac{1.645^2}{2(200)} \pm 1.645\sqrt{\frac{(.680)(.320)}{200} + \frac{1.645^2}{4(200)^2}}}{1 + \frac{1.645^2}{200}} = \frac{.6868 \pm .0547}{1.01353} = (.624, .732).$$

b $$n = \frac{2(1.645)^2(.25) - (1.645)^2(.05)^2 \pm \sqrt{4(1.645)^4(.25)(.25 - .0025) + .05^2(1.645)^4}}{.0025}$$

$$= \frac{1.3462 \pm 1.3530}{.0025} = 1079.7 \Rightarrow \text{ use n = 1080}$$

c No, it gives a 95% upper bound.

50.

a Assuming normality, $t_{.05,15} = 1.753$, do s 95% C.I. for μ is

$$.214 \pm 1.753\frac{.036}{\sqrt{16}} = .214 \pm .016 = (.198, .230)$$

b A 90% upper bound for σ, with $\chi^2_{.10,15} = 1.341$, is

$$\sqrt{\frac{15(.036)^2}{1.341}} = \sqrt{.0145} = .120$$

c A 95% prediction interval, with $t_{.025,15} = 2.131$, is

$$.214 \pm 2.131(.036)\sqrt{1 + \tfrac{1}{16}} = .214 \pm .0791 = (.1349, .2931).$$

51. With $\hat{\theta} = \frac{1}{3}(\overline{X}_1 + \overline{X}_2 + \overline{X}_3) - \overline{X}_4$, $\sigma^2_{\hat{\theta}} = \frac{1}{9}Var(\overline{X}_1 + \overline{X}_2 + \overline{X}_3) + Var(\overline{X}_4) =$ $\frac{1}{9}\left(\frac{\sigma_1^2}{n_1} + \frac{\sigma_2^2}{n_2} + \frac{\sigma_3^2}{n_3}\right) + \frac{\sigma_4^2}{n_4}$; $\hat{\sigma}_{\hat{\theta}}$ is obtained by replacing each $\hat{\sigma}_i^2$ by s_i^2 and taking the square root. The large-sample interval for θ is then $\frac{1}{3}(\overline{x}_1 + \overline{x}_2 + \overline{x}_3) - \overline{x}_4 \pm z_{\alpha/2}\sqrt{\frac{1}{9}\left(\frac{s_1^2}{n_1} + \frac{s_2^2}{n_2} + \frac{s_3^2}{n_3}\right) + \frac{s_4^2}{n_4}}$. For the given data, $\hat{\theta} = -.50$, $\hat{\sigma}_{\hat{\theta}} = .1718$, so the interval is $-.50 \pm 1.96(.1718) = (-.84, -.16)$.

52. $\hat{p} = \frac{11}{55} = .2 \Rightarrow$ a 90% C.I. is

$$\frac{.2 + \frac{1.645^2}{2(55)} \pm 1.645\sqrt{\frac{(.2)(.8)}{55} + \frac{1.645^2}{4(55)^2}}}{1 + \frac{1.645^2}{55}} = \frac{.2246 \pm .0887}{1.0492} = (.1295, .2986).$$

53. The specified condition is that the interval be length .2, so $n = \left[\frac{2(1.96)(.8)}{.2}\right]^2 = 245.86$, so n = 246 should be used.

54.

a A normal probability plot lends support to the assumption that pulmonary compliance is normally distributed. Note also that the lower and upper fourths are 192.3 and 228,1, so the fourth spread is 35.8, and the sample contains no outliers.

b $t_{.025,15} = 2.131$, so the C.I. is

$$209.75 \pm 2.131\frac{24.156}{\sqrt{16}} = 209.75 \pm 12.87 = (196.88, 222.62).$$

c K = 95, n = 16, and the tolerance critical value is 2.903, so the 95% tolerance interval is $209.75 \pm 2.903(24.156) = 209.75 \pm 70.125 = (139.625, 279.875)$.

55. Proceeding as in Example 7.5 with T_r replacing ΣX_i, the C.I. for $\frac{1}{\lambda}$ is $\left(\frac{2t_r}{\chi^2_{1-\alpha/2,2r}}, \frac{2t_r}{\chi^2_{\alpha/2,2r}}\right)$ where $t_r = y_1 + ... + y_r + (n-r)y_r$. In Example 6.7, n = 20, r = 10, and t_r = 1115. With d.f. = 20, the necessary critical values are 9.591 and 34.170, giving the interval (65.3, 232.5). This is obviously an extremely wide interval. The censored experiment provides less information about $\frac{1}{\lambda}$ than would an uncensored experiment with n = 20.

56.

a $P(\min(X_i) \le \tilde{\mu} \le \max(X_i)) = 1 - P(\tilde{\mu}, < \min(X_i) or \max(X_i) < \tilde{\mu})$
$= 1 - P(\tilde{\mu}, < \min(X_i)) - P(\max(X_i) < \tilde{\mu})$
$= 1 - P(\tilde{\mu} < X_1, ..., \tilde{\mu} < X_n) - P(X_1 < \tilde{\mu}, ..., X_n < \tilde{\mu})$
$= 1 - (.5)^n - (.5)^n = 1 - 2(.5)^{n-1}$, from which the confidence interval follows.

b Since $\min(x_i) = 1.44$ and $\max(x_i) = 3.54$, the C.I. is (1.44, 3.54).

c $P(X_{(2)} \le \tilde{\mu} \le X_{(n-1)}) = 1 - P(\tilde{\mu}, < X_{(2)}) - P(X_{(n-1)} < \tilde{\mu})$
= 1 – P(at most one X_I is below $\tilde{\mu}$) – P(at most one X_I exceeds $\tilde{\mu}$)

$$1 - (.5)^n - \binom{n}{1}(.5)^1(.5)^{n-1} - (.5)^n - \binom{n}{1}(.5)^{n-1}(.5).$$

$$= 1 - 2(n+1)(.5)^n = 1 - (n+1)(.5)^{n-1}$$

Thus the confidence coefficient is $1 - (n+1)(.5)^{n-1}$, or in another way, a $100(1 - (n+1)(.5)^{n-1})\%$ confidence interval.

57.

a $\int_{(\alpha/2)^{1/n}}^{(1-\alpha/2)^{1/n}} nu^{n-1} du = u^n \Big]_{(\alpha/2)^{1/n}}^{(1-\alpha/2)^{1/n}} = 1 - \frac{\alpha}{2} - \frac{\alpha}{2} = 1 - \alpha$. From the probability statement, $\frac{(\alpha/2)^{1/n}}{\max(X_i)} \le \frac{1}{\theta} \le \frac{(1-\alpha/2)^{1/n}}{\max(X_i)}$ with probability $1-\alpha$, so taking the reciprocal of each endpoint and interchanging gives the C.I.

$\left(\frac{\max(X_i)}{(1-\alpha/2)^{1/n}}, \frac{\max(X_i)}{(\alpha/2)^{1/n}}\right)$ for θ.

b $\alpha^{1/n} \le \frac{\max(X_i)}{\theta} \le 1$ with probability $1-\alpha$, so $1 \le \frac{\theta}{\max(X_i)} \le \frac{1}{\alpha^{1/n}}$ with probability $1-\alpha$, which yields the interval $\left(\max(X_i), \frac{\max(X_i)}{\alpha^{1/n}}\right)$.

c It is easily verified that the interval of **b** is shorter – draw a graph of $f_U(u)$ and verify that the shortest interval which captures area $1-\alpha$ under the curve is the rightmost such interval, which leads to the C.I. of **b**. With $\alpha = .05$, n = 5, max(x_I)=4.2; this yields (4.2, 7.65).

58. The length of the interval is $\left(z_\gamma + z_{\alpha-\gamma}\right)\frac{s}{\sqrt{n}}$, which is minimized when $z_\gamma + z_{\alpha-\gamma}$ is minimized, i.e. when $\Phi^{-1}(1-\gamma)+\Phi^{-1}(1-\alpha+\gamma)$ is minimized. Taking $\frac{d}{d\gamma}$ and equating to 0 yields $\frac{1}{\Phi(1-\gamma)}=\frac{1}{\Phi(1-\alpha+\gamma)}$ where $\Phi(\bullet)$ is the standard normal p.d.f., whence $\gamma=\frac{\alpha}{2}$.

59. $\tilde{x}=76.2$, the lower and upper fourths are 73.5 and 79.7, respectively, and $f_s=6.2$. The robust interval is $76.2\pm(1.93)\left(\frac{6.2}{\sqrt{22}}\right)=76.2\pm 2.6=(73.6,78.8)$.

$\bar{x}=77.33$, s = 5.037, and $t_{.025,21}=2.080$, so the t interval is $77.33\pm(2.080)\left(\frac{5.037}{\sqrt{22}}\right)=77.33\pm 2.23=(75.1,79.6)$. The t interval is centered at $\bar{x}$, which is pulled out to the right of $\tilde{x}$ by the single mild outlier 93.7; the interval widths are comparable.

60.

a Since $2\lambda\Sigma X_i$ has a chi-squared distribution with 2n d.f. and the area under this chi-squared curve to the right of $\chi^2_{.95,2n}$ is .95, $P\left(\chi^2_{.95,2n}<2\lambda\Sigma X_i\right)=.95$. This implies that $\frac{\chi^2_{.95,2n}}{2\Sigma X_i}$ is a lower confidence bound for λ with confidence coefficient 95%. Table A.7 gives the chi-squared critical value for 20 d.f. as 10.851, so the bound is $\frac{10.851}{2(550.87)}=.0098$. We can be 95% confident that λ exceeds .0098.

b Arguing as in **a**, $P\left(2\lambda\Sigma X_i<\chi^2_{.05,2n}\right)=.95$. The following inequalities are equivalent to the one in parentheses:

$$\lambda<\frac{\chi^2_{.05,2n}}{2\Sigma X_i}\quad\Rightarrow -\lambda t<\frac{-t\chi^2_{.05,2n}}{2\Sigma X_i}\quad\Rightarrow e^{-\lambda t}<\exp\left[\frac{-t\chi^2_{.05,2n}}{2\Sigma X_i}\right].$$

Replacing the ΣX_i by Σx_i in the expression on the right hand side of the last inequality gives a 95% lower confidence bound for $e^{-\lambda t}$. Substituting t = 100, $\chi^2_{.05,20}=31.410$ and $\Sigma x_i=550.87$ gives .058 as the lower bound for the probability that time until breakdown exceeds 100 minutes.

Chapter 8

Section 8.1

1.

- **a** Yes. It is an assertion about the value of a parameter.
- **b** No. The sample mean $\overline{X}$ is not a parameter.
- **c** No. The sample standard deviation s is not a parameter.
- **d** Yes. The assertion is that the standard deviation of population #2 exceeds that of population #1.
- **e** No. $\overline{X}$ and $\overline{Y}$ are statistics rather than parameters, so cannot appear in a hypothesis.
- **f** Yes. H is an assertion about the value of a parameter.

2.

- **a** These hypotheses comply with our rules.
- **b** H_o is not an equality claim (e.g. $\sigma = 20$), so these hypotheses are not in compliance.
- **c** H_o should contain the equality claim, whereas H_a does here, so these are not legitimate.
- **d** The asserted value of $\mu_1 - \mu_2$ in H_o should also appear in H_a. It does not here, so our conditions are not met.
- **e** Each S^2 is a statistic, so does not belong in a hypothesis.
- **f** We are not allowing both H_o and H_a to be equality claims (though this is allowed in more comprehensive treatments of hypothesis testing).
- **g** These hypotheses comply with our rules.
- **h** These hypotheses are in compliance.

3. In this formulation, H_o states the welds do not conform to specification. This assertion will not be rejected unless there is strong evidence to the contrary. Thus the burden of proof is on those who wish to assert that the specification is satisfied. Using H_a: $\mu < 100$ results in the welds being believed in conformance unless provided otherwise, so the burden of proof is on the non-conformance claim.

4. When the alternative is H_a: $\mu < 5$, the formulation is such that the water is believed unsafe until proved otherwise. A type I error involved deciding that the water is safe (rejecting H_o) when it isn't (H_o is true). This is a very serious error, so a test which ensures that this error is highly unlikely is desirable. A type II error involves judging the water unsafe when it is actually safe. Though a serious error, this is less so than the type I error. It is generally desirable to formulate so that the type 1 error is more serious, so that the probability of this error can be explicitly controlled. Using H_a: $\mu > 5$, the type II error (now stating that the water is safe when it isn't) is the more serious of the two errors.

5. Let σ denote the population standard deviation. The appropriate hypotheses are $H_o: \sigma = .05$ vs $H_a: \sigma < .05$. With this formulation, the burden of proof is on the data to show that the requirement has been met (the sheaths will not be used unless H_o can be rejected in favor of H_a. Type I error: Conclude that the standard deviation is < .05 mm when it is really equal to .05 mm. Type II error: Conclude that the standard deviation is .05 mm when it is really < .05.

6. $H_o: \mu = 40$ vs $H_a: \mu \neq 40$, where μ is the true average burn-out amperage for this type of fuse. The alternative reflects the fact that a departure from $\mu = 40$ in either direction is of concern. Notice that in this formulation, it is initially believed that the value of μ is the design value of 40.

7. A type I error here involves saying that the plant is not in compliance when in fact it is. A type II error occurs when we conclude that the plant is in compliance when in fact it isn't. Reasonable people may disagree as to which of the two errors is more serious. If in your judgement it is the type II error, then the reformulation $H_o: \mu = 150$ vs $H_a: \mu < 150$ makes the type I error more serious.

8. Let μ_1 = the average amount of warpage for the regular laminate, and μ_2 = the analogous value for the special laminate. Then the hypotheses are $H_o: \mu_1 = \mu_2$ vs $H_o: \mu_1 > \mu_2$. Type I error: Conclude that the special laminate produces less warpage than the regular, when it really does not. Type II error: Conclude that there is no difference in the two laminates when in reality, the special one produces less warpage.

9.

a R_1 is most appropriate, because x either too large or too small contradicts p = .5 and supports p $\neq$ .5.

b A type I error consists of judging one of the tow candidates favored over the other when in fact there is a 50-50 split in the population. A type II error involves judging the split to be 50-50 when it is not.

c X has a binomial distribution with n = 25 and p = 0.5. α =P(type I error) = $P(X \leq 7 or X \geq 18$ when X ~ Bin(25, .5)) = B(7; 25,.5) + 1 – B(17; 25,.5) = .044

d $\beta(.4) = P(8 \leq X \leq 17$ when p = .4) = B(17; 25,.5) – B(7, 25,.4) = 0.845, and $\beta(.6) = 0.845$ also. $\beta(.3) = B(17;25,.3) - B(7;25,.3) = .488 = \beta(.7)$

e x = 6 is in the rejection region R_1 , so H_o is rejected in favor of H_a.

10.

a $H_o: \mu = 1300$ vs $H_a: \mu > 1300$

b $\bar{x}$ is normally distributed with mean $E(\bar{x}) = \mu$ and standard deviation $\frac{\sigma}{\sqrt{n}} = \frac{60}{\sqrt{20}} = 13.416$. When H_o is true, $E(\bar{x}) = 1300$. Thus $\alpha = P(\bar{x} \ge 1331.26$ when H_o is true) = $P\left(z \ge \frac{1331.26-1300}{13.416}\right) = P(z \ge 2.33) = .01$

c When $\mu = 1350$, $\bar{x}$ has a normal distribution with mean 1350 and standard deviation 13.416, so $\beta(1350) = P(\bar{x} < 1331.26$ when $\mu = 1350)$ = $P\left(z \le \frac{1331.26-1350}{13.416}\right) = P(z \le -1.40) = .0808$

d Replace 1331.26 by c, where c satisfies $\frac{c-1300}{13.416} = 1.645$ (since $P(z \ge 1.645) = .05$). Thus c = 1322.07. Increasing α gives a decrease in β; now $\beta(1350) = P(z \le -2.08) = .0188$.

e $\bar{x} \ge 1331.26$ iff $z \ge \frac{1331.26-1300}{13.416}$ i.e. iff $z \ge 2.33$.

11.

a $H_o: \mu = 10$ vs $H_a: \mu \ne 10$

b $\alpha = P($ rejecting H_o when H_o is true) = $P(\bar{x} \ge 10.1032$ or $\le 9.8968 when \mu = 10)$. Since $\bar{x}$ is normally distributed with standard deviation $\frac{\sigma}{\sqrt{n}} = \frac{.2}{5} = .04$, $\alpha = P(z \ge 2.58 or \le -2.58) = .005 + .005 = .01$

c When $\mu = 10.1$, $E(\bar{x}) = 10.1$, so $\beta(10.1) = P(9.8968 < \bar{x} < 10.1032$ when $\mu = 10.1) = P(-5.08 < z < .08) = .5319$. Similarly, $\beta(9.8) = P(2.42 < z < 7.58) = .0078$

d $c = \pm 2.58$

e Now $\frac{\sigma}{\sqrt{n}} = \frac{.2}{3.162} = .0632$. Thus 10.1032 is replaced by c, where $\frac{c-10}{.0632} = 2.58$ and so c = 10.163. Similarly, 9.8968 is replaced by 9.837.

f $\bar{x} = 10.020$. Since $\bar{x}$ is neither ≥ 10.163 nor ≤ 9.837, it is not in the rejection region. H_o is not rejected; it is still plausible that $\mu = 10$.

g $\bar{x} \ge 10.1032$ or ≤ 9.8968 iff $z \ge 2.58$ or ≤ -2.58.

12.

a Let μ = true average braking distance for the new design at 40 mph. The hypotheses are $H_o : \mu = 120$ vs $H_a : \mu < 120$.

b R_2 should be used, since support for H_a is provided only by an $\bar{x}$ value substantially smaller than 120. ($E(\bar{x}) = 120$ when H_o is true and , 120 when H_a is true).

c $\sigma_{\bar{x}} = \frac{\sigma}{\sqrt{n}} = \frac{10}{6} = 1.6667$, so $\alpha = P(\bar{x} \ge 115.20$ when $\mu = 120) =$

$P\left(z \le \frac{115.20 - 120}{1.6667}\right) = P(z \le -2.88) = .002$. To obtain $\alpha = .001$, replace 115.20 by $c = 120 - 3.08(1.6667) = 114.87$, so that $P(\bar{x} \le 114.87$ when $\mu = 120) = P(z \le -3.08) = .001$.

d $\beta(115) = P(\bar{x} > 115.2$ when $\mu = 115) = P(z > .12) = .4522$

e $\alpha = P(z \le -2.33) = .01$, because when H_o is true Z has a standard normal distribution ($\bar{X}$ has been standardized using 120). Similarly $P(z \le -2.88) = .002$, so this second rejection region is equivalent to R_2.

13.

a $P(\bar{x} \ge \mu_o + 2.33\frac{\sigma}{\sqrt{n}} \, when \mu = \mu_o) = P\left(z \ge \frac{\left(\mu_o + 2.33\frac{\sigma}{\sqrt{n}}\right)}{\frac{\sigma}{\sqrt{n}}}\right)$

$= P(z \ge 2.33) = .01$, where Z is a standard normal r.v.

b P(rejecting H_o when $\mu = 99) = P(\bar{x} \ge 102.33$ when $\mu = 99) = P\left(z \ge \frac{102 - 99}{1}\right)$

$= P(z \ge 3.33) = .0004$. Similarly, $\alpha(98) = P(\bar{x} \ge 102.33$ when $\mu = 98) = P(z \ge 4.33) = 0$. In general, we have P(type I error) < .01 when this probability is calculated for a value of μ less than 100. The boundary value $\mu = 100$ yields the largest α.

14.

a $\sigma_{\bar{x}} = .04$, so $P(\bar{x} \ge 10.1004 or \le 9.8940$ when $\mu = 10) = P(z \ge 2.51 or \le -2.65) = .006 + .004 = .01$

b $\beta(10.1) = P(9.8940 < \bar{x} < 10.1004$ when $\mu = 10.1) = P(-5.15 < z < .01) = .5040$, whereas $\beta(9.9) = P(-.15 < z < 5.01) = .5596$. Since $\mu = 9.9$ and $\mu = 10.1$ represent equally serious departures from H_o, one would probably want to use a test procedure for which $\beta(9.9) = \beta(10.1)$. A similar result and comment apply to any other pair of alternative values symmetrically placed about 10.

Chapter 8

Section 8.2

15.

a $\alpha = P(z \geq 1.88$ when z has a standard normal distribution) $= 1 - \Phi(1.88) = .0301$

b $\alpha = P(z \leq -2.75$ when z ~ N(0, 1) $= \Phi(-2.75) = .003$

c $\alpha = \Phi(-2.88) + (1 - \Phi(2.88)) = .004$

16.

a $\alpha = P(t \geq 3.733$ when t has a t distribution with 15 d.f.) =.001, because the 15 d.f. row of Table A.5 shows that $t_{.001,15} = .3733$

b d.f. = n – 1 = 23, so $\alpha = P(t \leq -2.500) = .01$

c d.f. = 30, and $\alpha = P(t \geq 1.697) + P(t \leq -1.697) = .05 + .05 = .10$

17.

a $z = \dfrac{20{,}960 - 20{,}000}{1500/\sqrt{16}} = 2.56 > 2.33$ so reject H_o.

b $\beta(20{,}500) : \Phi\left(2.33 + \dfrac{20{,}000 - 20{,}500}{1500/\sqrt{16}}\right) = \Phi(1.00) = .8413$

c $\beta(20{,}500) = .05 : n = \left[\dfrac{1500(2.33 + 1.645)}{20{,}000 - 20{,}500}\right]^2 = 142.2$, so use n = 143

d $\alpha = 1 - \Phi(2.56) = .0052$

18.

a $\dfrac{72.3 - 75}{1.8} = -1.5$ so 72.3 is 1.5 SD's (of $\bar{x}$) below 75.

b H_o is rejected if $z \leq -2.33$; since $z = -2.87 \leq -2.33$, reject H_o.

c $\alpha =$ area under standard normal curve below –2.88 $= \Phi(-2.88) = .0020$

d $\Phi\left(-2.88 + \dfrac{75 - 70}{9/5}\right) = \Phi(-.1) = .4602$ so $\beta(70) = .5398$

e $n = \left[\dfrac{9(2.88 + 2.33)}{75 - 70}\right]^2 = 87.95$, so use n = 88

f $\alpha(76) = P(Z < -2.33$ when $\mu = 76) = P(\bar{X} < 72.9$ when $\mu = 76)$

$= \Phi\left(\dfrac{72.9 - 76}{.9}\right) = \Phi(-3.44) = .0003$

19.

a Reject H_o if either $z \geq 2.58$ or $z \leq -2.58$; $\frac{\sigma}{\sqrt{n}} = 0.3$, so $z = \frac{94.32 - 95}{0.3} = -2.27$.

Since –2.27 is not < -2.58, don't reject H_o.

b $\beta(94) = \Phi\left(\frac{2.58 - 1}{0.3}\right) - \Phi\left(\frac{-2.58 - 1}{0.3}\right) = \Phi(-.75) - \Phi(-5.91) = .2266$

c $n = \left[\frac{1.20(2.58 + 1.28)}{95 - 94}\right]^2 = 21.46$, so use n = 22.

20. With H_o: $\mu = 750$, and H_a: $\mu < 750$ and a significance level of .05, we reject H_o if z < -1.645; z = -2.14 < -1.645, so we reject the null hypothesis and do not continue with the purchase. At a significance level of .01, we reject H_o if z < -2.33; z = -2.14 > -2.33, so we don't reject the null hypothesis and thus continue with the purchase.

21. With H_o: $\mu = .5$, and H_a: $\mu \neq .5$ we reject H_o if $t > t_{\alpha/2,n-1}$ or $t < -t_{\alpha/2,n-1}$

a 1.6 < $t_{.025,12}$ = 2.179, so don't reject H_o
b -1.6 > -$t_{.025,12}$ = -2.179, so don't reject H_o
c – 2.6 > -$t_{.005,24}$ = -2.797, so don't reject H_o
d –3.9 < the negative of all t values in the df = 24 row, so we reject H_o in favor of H_a.

22.

a It appears that the true average weight could be more than the production specification of 200 lb per pipe.

b H_o: $\mu = 200$, and H_a: $\mu > 200$ we reject H_o if $t > t_{.05,29} = 1.699$.

$t = \frac{206.73 - 200}{6.35/\sqrt{30}} = \frac{6.73}{1.16} = 5.80 > 1.699$, so reject H_o. The test appears to substantiate the statement in part **a**.

23. H_o: $\mu = 360$ vs. H_a: $\mu > 360$; $t = \frac{\bar{x} - 360}{s/\sqrt{n}}$; reject H_o if $t > t_{.05,25} = 1.708$;

$t = \frac{370.69 - 360}{24.36/\sqrt{26}} = 2.24 > 1.708$. Thus H_o should be rejected. There appears to be a contradiction of the prior belief.

24. H_o: $\mu = 3000$ vs. H_a: $\mu \neq 3000$; $t = \frac{\bar{x} - 3000}{s/\sqrt{n}}$; reject H_o if $|t| > t_{.025,4} = 2.776$;

$t = \frac{2887.6 - 3000}{84/\sqrt{5}} = -2.99 < -2.776$, so we reject H_o. This requirement is not satisfied.

25.

a H_o: $\mu = 5.5$ vs. H_a: $\mu \neq 5.5$; for a level .01 test, (not specified in the problem description), reject H_o if either $z \geq 2.58$ or $z \leq -2.58$. Since $z = \frac{5.25 - 5.5}{.075} = -3.33 \leq -2.58$, reject H_o.

b $$1 - \beta(5.6) = 1 - \Phi\left(2.58 + \frac{(-.1)}{.075}\right) + \Phi\left(-2.58 - \frac{(-.1)}{.075}\right)$$
$$= 1 - \Phi(1.25) + \Phi(-3.91) = .105$$

c $n = \left[\frac{.3(2.58 + 2.33)}{-.1}\right]^2 = 216.97$, so use n = 217.

26. Reject H_o if $z \geq 1.645$; $\frac{s}{\sqrt{n}} = .811$, so $z = \frac{52.7 - 50}{.811} = 3.33$. Since 3.33 is ≥ 1.645, reject H_o at level .05 and conclude that true average penetration exceeds 50 mils.

27. We wish to test H_o: $\mu = 75$ vs. H_a: $\mu < 75$; Using $\alpha = .01$, H_o is rejected if $t \leq -t_{.01,31} \approx -2.457$ (from the df 30 row of the t-table). Since $t = \frac{73.1 - 75}{5.9/\sqrt{32}} = -1.82$, which is not ≤ -2.457, H_o is not rejected. The alloy is not suitable.

28. With $\mu =$ true average recumbency time, the hypotheses are H_o: $\mu = 20$ vs H_a: $\mu < 20$. The test statistic value is $z = \frac{\bar{x} - 20}{s/\sqrt{n}}$, and H_o should be rejected if $z \leq -z_{.10} = -1.28$ Since $z = \frac{18.86 - 20}{8.6/\sqrt{73}} = -1.13$, which is not ≤ -1.28, H_o is not rejected. The sample data does not strongly suggest that true average time is less than 20.

29.

a For n = 8, n – 1 = 7, and $t_{.05,7} = 1.895$, so H_o is rejected at level .05 if $t \geq 1.895$. Since $\frac{s}{\sqrt{n}} = \frac{1.25}{\sqrt{8}} = .442$, $t = \frac{3.72 - 3.50}{.442} = .498$; this does not exceed 1.895, so H_o is not rejected.

b $d = \frac{|\mu_o - \mu|}{\sigma} = \frac{|3.50 - 4.00|}{1.25} = .40$, and n = 8, so from table A.17, $\beta(4.0) \approx .72$

30. n = 115, $\bar{x} = 11.3$, $s = 6.43$

1 Parameter of Interest: $\mu =$ true average dietary intake of zinc among males aged 65 – 74 years.

2 Null Hypothesis: H_o: $\mu = 15$

3 Alternative Hypothesis: H_a: $\mu < 15$

4 $z = \dfrac{\bar{x} - \mu_o}{s/\sqrt{n}} = \dfrac{\bar{x} - 15}{s/\sqrt{n}}$

5 Rejection Region: No value of α was given, so select a reasonable level of significance, such as α= .05. $z \le z_\alpha$ or $z \le -1.645$

6 $z = \dfrac{11.3 - \mu_o}{6.43/\sqrt{115}} = -6.17$

7 –6.17 < -1.645, so reject H_o. The data does support the claim that average daily intake of zinc for males aged 65 - 74 years falls below the recommended daily allowance of 15 mg/day.

31. The hypotheses of interest are H_o: $\mu = 7$ vs H_a: $\mu < 7$, so a lower-tailed test is appropriate; H_o should be rejected if $t \le -t_{.1,8} = -1.397$. $t = \dfrac{6.32 - 7}{1.65/\sqrt{9}} = -1.24$.

Because -1.24 is not ≤ -1.397, H_o (prior belief) is not rejected (contradicted) at level .01.

32. n = 12, $\bar{x} = 98.375$, $s = 6.1095$

a

1 Parameter of Interest: $\mu =$ true average reading of this type of radon detector when exposed to 100 pCi/L of radon.

2 Null Hypothesis: H_o: $\mu = 100$

3 Alternative Hypothesis: H_a: $\mu \ne 100$

4 $t = \dfrac{\bar{x} - \mu_o}{s/\sqrt{n}} = \dfrac{\bar{x} - 100}{s/\sqrt{n}}$

5 $t \le -2.201$ or $t \ge 2.201$

6 $t = \dfrac{98.375 - 100}{6.1095/\sqrt{12}} = -.9213$

7 Fail to reject H_o. The data does not indicate that these readings differ significantly from 100.

b σ = 7.5, β = 0.10. From table A.17, df ≈ 29, thus n ≈30.

33. $\beta(\mu_o - \Delta) = \Phi\left(z_{\alpha/2} + \Delta\sqrt{n}/\sigma\right) - \Phi\left(-z_{\alpha/2} - \Delta\sqrt{n}/\sigma\right)$

$= 1 - \left[\Phi\left(-z_{\alpha/2} - \Delta\sqrt{n}/\sigma\right) + \Phi\left(z_{\alpha/2} - \Delta\sqrt{n}/\sigma\right)\right] = \beta(\mu_o + \Delta)$ (since 1 - Φ(c) = Φ(-c)).

34. For an upper-tailed test, $= \beta(\mu) = \Phi\left(z_\alpha + \sqrt{n}(\mu_o - \mu)/\sigma\right)$. Since in this case we are considering $\mu > \mu_o$, $\mu_o - \mu$ is negative so $\sqrt{n}(\mu_o - \mu)/\sigma \to -\infty$ as n $\to \infty$. The desired conclusion follows since $\Phi(-\infty) = 0$. The arguments for a lower-tailed and tow-tailed test are similar.

Section 8.3

35.

1. Parameter of interest: p = true proportion of cars in this particular county passing emissions testing on the first try.
2. H_o: p = .70
3. H_a: p $\neq$ .70
4. $$z = \frac{\hat{p} - p_o}{\sqrt{p_o(1-p_o)/n}} = \frac{\hat{p} - .70}{\sqrt{.70(.30)/n}}$$
5. either z $\geq$ 1.96 or z $\leq$ -1.96
6. $$z = \frac{156/200 - .70}{\sqrt{.70(.30)/200}} = 2.469$$
7. Reject H_o. The data indicates that the proportion of cars passing the first time on emission testing or this county differs from the proportion of cars passing statewide.

36.

a

1. p = true proportion of all nickel plates that blister under the given circumstances.
2. H_o: p = .10
3. H_a: p > .10
4. $$z = \frac{\hat{p} - p_o}{\sqrt{p_o(1-p_o)/n}} = \frac{\hat{p} - .10}{\sqrt{.10(.90)/n}}$$
5. Reject H_o if $z \geq 1.645$
6. $$z = \frac{14/100 - .10}{\sqrt{.10(.90)/100}} = 1.33$$
7. Fail to Reject H_o. The data does not give compelling evidence for concluding that more than 10% of all plates blister under the circumstances.

The possible error we could have made is a Type II error: Failing to reject the null hypothesis when it is actually true.

(continued)

b $\beta(.15)=\Phi\left[\dfrac{.10-.15+1.645\sqrt{.10(.90)/100}}{\sqrt{.15(.85)/100}}\right]=\Phi(-.02)=.4920$. When n = 200, $\beta(.15)=\Phi\left[\dfrac{.10-.15+1.645\sqrt{.10(.90)/200}}{\sqrt{.15(.85)/200}}\right]=\Phi(-.60)=.2743$

c $n=\left[\dfrac{1.645\sqrt{.10(.90)}+1.28\sqrt{.15(.85)}}{.15-.10}\right]=19.01$, so use n = 20

37.

1 p = true proportion of the population with type A blood

2 H_o: p = .40

3 H_a: p ≠ .40

4 $z=\dfrac{\hat{p}-p_o}{\sqrt{p_o(1-p_o)/n}}=\dfrac{\hat{p}-.40}{\sqrt{.40(.60)/n}}$

5 Reject H_o if $z \ge 2.58$ or $z \le -2.58$

6 $z=\dfrac{92/150-.40}{\sqrt{.40(.60)/150}}=\dfrac{.213}{.04}=5.3$

7 Reject H_o. The data does suggest that the percentage of the population with type A blood differs from 40%. (at the .01 significance level). Since the z critical value for a significance level of .05 is less than that of .01, the conclusion would not change.

38.

a We wish to test H_o: p = .02 vs H_a: p < .02; only if H_o can be rejected will the inventory be postponed. The lower-tailed test rejects H_o if $z \le -1.645$. With $\hat{p}=\dfrac{15}{1000}=.015$, z = -1.01, which is not ≤ -1.645. Thus, H_o cannot be rejected, so the inventory should be carried out.

b $\beta(.01)=\Phi\left[\dfrac{.02-.01+1.645\sqrt{.02(.98)/1000}}{\sqrt{.01(.99)/1000}}\right]=\Phi(5.49)\approx 1$

c $\beta(.05)=\Phi\left[\dfrac{.02-.05+1.645\sqrt{.02(.98)/1000}}{\sqrt{.05(.95)/1000}}\right]=\Phi(-3.30)=.0005$, so is p = .05 it is highly unlikely that H_o will be rejected and the inventory will almost surely be carried out.

39. Let p denote the true proportion of those called to appear for service who are black. We wish to test H_o: p = .25 vs H_a: p < .25. We use $z = \dfrac{\hat{p} - .25}{\sqrt{.25(.75)/n}}$, with the rejection region $z \le -z_{.01} = -2.33$. We calculate $\hat{p} = \dfrac{177}{1050} = .1686$, and $z = \dfrac{.1686 - .25}{.0134} = -6.1$. Because –6.1 < -2.33, H_o is rejected. A conclusion that discrimination exists is very compelling.

40.

a P = true proportion of current customers who qualify. H_o: p = .05 vs H_a: p ≠ .05, $z = \dfrac{\hat{p} - .05}{\sqrt{.05(.95)/n}}$, reject H_o if $z \ge 2.58$ or $z \le -2.58$. $\hat{p} = .08$, so

$z = \dfrac{.03}{.00975} = 3.07 \ge 2.58$, so H_o is rejected. The company's premise is not correct.

b $\beta(.10) = \Phi\left[\dfrac{.05 - .10 + 2.58\sqrt{.05(.95)/500}}{\sqrt{.10(.90)/500}}\right] = \Phi(-1.85) = .0332$

41.

a The alternative of interest here is H_a: p > .50 (which states that more than 50% of all enthusiasts prefer gut), so the rejection region should consist of large values of X (an upper-tailed test). Thus (15, 16, 17, 18, 19, 20) is the appropriate region.

b $\alpha = P(15 \le X$ when $p = .5)$ = 1 – B(14; 20, .05) = .021, so this is a level .05 test. For R = {14, 15, …, 20}, α = .058, so this R does not specify a level .05 test and the region of **a** is the best level .05 test. (α ≤ .05 along with smallest possible β).

c β(.6) = B(14; 20, .6) = .874, and β(.8) = B(14; 20, .8) = .196.

d The best level .10 test is specified by R = (14, …, 20} (with α = .052) Since 13 is not in R, H_o is not rejected at this level.

42. The hypotheses are H_o: p = .10 vs. H_a: p > .10, so R has the form {c, …, n}. For n = 10, c = 3 (i.e. R = {3, 4, …, 10}) yields α = 1 – B(2; 10, .1) = .07 while no larger R has α ≤ .10; however β(.3) = B(2; 10, .3) = .383. For n = 20, c = 5 yields α = 1 – B(4; 20, .1) = .043, but again β(.3) = B(4; 20, .3) = .238. For n = 25, c = 5 yields α = 1 – B(4; 25, .1) = .098 while β(.7) = B(4; 25, .3) = .090 ≤ .10, so n = 25 should be used.

43. H_o: p = .03 vs H_a: p < .03. We use $z = \dfrac{\hat{p} - .03}{\sqrt{.03(.97)/n}}$, with the rejection region $z \le -z_{.01}$ = -2.33. With $\hat{p} = .028$, $z = \dfrac{-.002}{.00763} = -.26$. Because -.26 isn't ≤ -2.33, H_o is not rejected. Robots have not demonstrated their superiority.

Chapter 8

Section 8.4

44. Using $\alpha = .05$, H_o should be rejected whenever p-value < .05.

- **a** P-value = .001 < .05, so reject H_o
- **b** .021 < .05, so reject H_o.
- **c** .078 is not < .05, so don't reject H_o.
- **d** .047 < .05, so reject H_o (a close call).
- **e** .148 > .05, so H_o can't be rejected at level .05.

45.

- **a** p-value = .084 > .05 = α, so don't reject H_o.
- **b** p-value = .003 < .001 = α, so reject H_o.
- **c** .498 >> .05, so H_o can't be rejected at level .05
- **d** .084 < .10, so reject H_o at level .10
- **e** .039 is not < .01, so don't reject H_o.
- **f** p-value = .218 > .10, so H_o cannot be rejected.

46. In each case the p-value = $1-\Phi(z)$

- **a** .0778
- **b** .1841
- **c** .0250
- **d** .0066
- **e** .4562

47.

- **a** .0358
- **b** .0802
- **c** .5824
- **d** .1586
- **e** 0

48.

- **a** In the df = 8 row of table A.5, t = 2.0 is between 1.860 and 2.306, so the p-value is between .025 and .05: .025 < p-value < .05.
- **b** 2.201 < | -2.4 | < 2.718, so .01 < p-value < .025.
- **c** 1.341 < | -1.6 | < 1.753, so .05 < P(t < -1.6) < .10. Thus a two-tailed p-value: 2(.05 < P(t < -1.6) < .10), or .10 < p-value < .20
- **d** With an upper-tailed test and t = -.4, the p-value = P(t > -.4) > .50.
- **e** 4.032 < t=5 < 5.893, so .001 < p-value < .005
- **f** 3.551 < | -4.8 |, so P(t < -4.8) < .0005. A two-tailed p-value = 2[P(t < -4.8)] < 2(.0005), or p-value < .001.

49. An upper-tailed test

- **a** Df = 14, .001 < p-value < .005, which is less than .05, so reject H_o.
- **b** Df = 8, .05 < p-value < .10, which is greater than .01, so don't reject H_o.
- **c** Df = 23, p-value > .50, so fail to reject H_o at any significance level.

50. The p-value is greater than the level of significance $\alpha = .01$, therefore fail to reject H_o that $\mu = 5.63$. The data does not indicate a difference in average serum receptor concentration between pregnant women and all other women.

51. Here we might be concerned with departures above as well as below the specified weight of 5.0, so the relevant hypotheses are H$_o$: $\mu = 5.0$ vs H$_a$: $\mu \neq 5.0$. At level .01, reject H$_o$ if either $z \geq 2.58$ or $z \leq -2.58$. Since $\frac{s}{\sqrt{n}} = .035$, $z = \frac{-.13}{.035} = -3.71$, which is ≤ -2.58, so H$_o$ should be rejected. Because 3.71 is "off" the z-table, p-value < 2(.0002) = .0004 (.0002 corresponds to z = -3.49).

52.

a For testing H$_o$: p = .2 vs H$_a$: p > .2, an upper-tailed test is appropriate. The computed Z is z = .97, so p-value = $1 - \Phi(.97) = .166$. Because the p-value is rather large, H$_o$ would not be rejected at any reasonable α (it can't be rejected for any α < .166), so no modification appears necessary.

b With p = .5, $1 - \beta(.5) = 1 - \Phi[(-.3 + 2.33(.0516))/.0645] = 1 - \Phi(-2.79) = .9974$

53. The hypotheses to be tested are H$_o$: $\mu = 25$ vs H$_a$: $\mu > 25$, and H$_o$ should be rejected if $t \geq t_{.05,12} = 1.782$. The computed summary statistics are $\bar{x} = 27.923$, $s = 5.619$, so $\frac{s}{\sqrt{n}} = 1.559$ and $t = \frac{2.923}{1.559} = 1.88$. Because $t_{.025,12} = 2.179 > 1.88 > t_{.05,12} = 1.782$, the p-value is between .025 and .05, so H$_o$ is rejected at level .05.

54.

a The appropriate hypotheses are H$_o$: $\mu = 10$ vs H$_a$: $\mu < 10$

b $-2.3 < -t_{.05,17} = -1.740$, so we would reject H$_o$. The data indicates that the pens do not meet the design specifications.

c -1.8 is not $\leq -t_{.01,17} = -2.567$ so we would not reject H$_o$. There is not enough evidence to say that the pens don't satisfy the design specifications.

d Since $t_{.001,17} = 3.646 > 3.6 > t_{.005,17} = 2.898$, the p-value is between .001 and .005, which gives strong evidence to support the alternative hypothesis.

55. $\mu =$ true average reading, H$_o$: $\mu = 70$ vs H$_a$: $\mu \neq 70$, and

$$t = \frac{\bar{x} - 70}{s/\sqrt{n}} = \frac{75.5 - 70}{7/\sqrt{6}} = \frac{5.5}{2.86} = 1.92$$

From the 5 d.f. row of the t table, $t_{.10,5} < 1.92 < t_{.05,5}$, so .10 < p-value < .20.

56. With H$_o$: $\mu = .60$ vs H$_a$: $\mu \neq .60$, and a two-tailed p-value of .0711, we fail to reject H$_o$ at levels .01 and .05 (thus concluding that the amount of impurities need not be adjusted) , but we would reject H$_o$ at level .10 (and conclude that the amount of impurities does need adjusting).

Section 8.5

57.

a The formula for β is $1-\Phi\left(-2.33+\frac{\sqrt{n}}{9.4}\right)$, which gives .8980 for n = 100, .1049 for n = 900, and .0014 for n = 2500.

b Z = -5.3, which is "off the z table," so p-value < .0002; this value of z is quite statistically significant.

c No. Eve when the departure from H_o is insignificant from a practical point of view, a statistically significant result is highly likely to appear; the test is too likely to detect small departures from H_o.

58.

a Here $\beta=\Phi\left(\frac{-.01+.9320/\sqrt{n}}{.4073/\sqrt{n}}\right)=\Phi\left(\frac{\left(-.01\sqrt{n}+.9320\right)}{.4073}\right)$ = .9793, .8554, .4325, .0944, and 0 for n = 100, 2500, 10,000, 40,000, and 90,000, respectively.

b Here $z=.025\sqrt{n}$ which equals .25, 1.25, 2.5, and 5 for the four n's, whence p-value = .4213, .1056, .0062, .0000, respectively.

c No; the reasoning is the same as in 54 (c).

Supplementary Exercises

59. Because n = 50 is large, we use a z test here, rejecting H_o: $\mu=3.2$ in favor of H_a: $\mu\neq 3.2$ if either $z>z_{.025}=1.96$ or $z\leq -1.96$. The computed z value is $z=\frac{3.05-3.20}{.34/\sqrt{50}}=-3.12$. Since –3.12 is ≤ -1.96, H_o should be rejected in favor of H_a.

60. Here we assume that thickness is normally distributed, so that for any *n* a t test is appropriate, and use Table A.17 to determine n. We wish $\pi(3)=.95$ when $d=\frac{|3.2-3|}{.3}=.667$. By inspection, n = 20 satisfies this requirement, so n = 50 is too large.

61.

a H_o: $\mu=3.2$ vs H_a: $\mu\neq 3.2$ (Because H_a: $\mu>3.2$ gives a p-value of roughly .15)

b With a p-value of .30, we would reject the null hypothesis at any reasonable significance level, which includes both .05 and .10.

62.

a H_o: $\mu = 2150$ vs H_a: $\mu > 2150$

b $t = \dfrac{\bar{x} - 2150}{s / \sqrt{n}}$

c $t = \dfrac{2160 - 2150}{30 / \sqrt{16}} = \dfrac{10}{7.5} = 1.33$

d Since $t_{.10,15} = 1.341$, p-value > .10 (actually $\approx .10$)

e From **d**, p-value > .05, so H_o cannot be rejected at this significance level.

63.

a The relevant hypotheses are H_o: $\mu = 548$ vs H_a: $\mu \neq 548$. At level .05, H_o will be rejected if either $t \geq t_{.025,10} = 2.228$ or $t \leq -t_{.025,10} = -2.228$. The test statistic value is $t = \dfrac{587 - 548}{10 / \sqrt{11}} = \dfrac{39}{3.02} = 12.9$. This clearly falls into the upper tail of the two-tailed rejection region, so H_o should be rejected at level .05, or any other reasonable level).

b The population sampled was normal or approximately normal.

64. $n = 8, \bar{x} = 30.7875, s = 6.5300$

1. Parameter of interest: $\mu =$ true average heat-flux of plots covered with coal dust
2. H_o: $\mu = 29.0$
3. H_a: $\mu > 29.0$
4. $t = \dfrac{\bar{x} - 29.0}{s / \sqrt{n}}$
5. RR: $t \geq t_{\alpha,n-1}$ or $t \geq 1.895$
6. $t = \dfrac{30.7875 - 29.0}{6.53 / \sqrt{8}} = .7742$
7. Fail to reject H_o. The data does not indicate the mean heat-flux for pots covered with coal dust is greater than for plots covered with grass.

65. N = 47, $\bar{x} = 215$ mg, s = 235 mg. Range 5 mg to 1,176 mg.

a No, the distribution does not appear to be normal, it appears to be skewed to the right. It is not necessary to assume normality if the sample size is large enough due to the central limit theorem. This sample size is large enough so we can conduct a hypothesis test about the mean.

(continued)

b

1. Parameter of interest: $\mu =$ true daily caffeine consumption of adult women.
2. H_o: $\mu = 200$
3. H_a: $\mu > 200$
4. $z = \dfrac{\bar{x} - 200}{s / \sqrt{n}}$
5. RR: $z \geq 1.282$ or if p-value $\leq .10$
6. $z = \dfrac{215 - 200}{235 / \sqrt{47}} = .44$; p-value $= 1 - \Phi(.44) = .33$
7. Fail to reject H_o. because .33 > .10. The data does not indicate that daily consumption of all adult women exceeds 200 mg.

66. At the .05 significance level, reject H_o because .043 < .05. At the level .01, fail to reject H_o because .043 > .01. Thus the data contradicts the design specification that sprinkler activation is less than 25 seconds at the level .05, but not at the .01 level.

67.

a From table A.17, when $\mu = 9.5$, d = .625, df = 9, and $\beta \approx .60$, when $\mu = 9.0$, d = 1.25, df = 9, and $\beta \approx .20$.

b From Table A.17, $\beta = .25$, d = .625, n ≈ 28

68. A normality plot reveals that these observations could have come from a normally distributed population, therefore a t-test is appropriate. The relevant hypotheses are H_o: $\mu = 9.75$ vs H_a: $\mu > 9.75$. Summary statistics are n = 20, $\bar{x} = 9.8525$, and s = .0965, which leads to a test statistic $t = \dfrac{9.8525 - 9.75}{.0965 / \sqrt{20}} = 4.75$, from which the p-value = .0001. (From MINITAB output). With such a small p-value, the data strongly supports the alternative hypothesis. The condition is not met.

69.

a With H_o: p = $1/75$ vs H_o: p $\neq 1/75$, we reject H_o if either $z \geq 1.96$ or $z \leq -1.96$. With $\hat{p} = \dfrac{16}{800} = .02$, $z \dfrac{.02 - .01333}{\sqrt{\dfrac{.01333(.98667)}{800}}} = 1.645$, which is not in either rejection region. Thus, we fail to reject the null hypothesis. There is not evidence that the incidence rate among prisoners differs from that of the adult population. The possible error we could have made is a type II.

b P-value $= 2[1 - \Phi(1.645)] = 2[.05] = .10$. Yes, since .10 < .20, we could reject H_o.

70.

a Assuming normality, a t test is appropriate; H_o: $\mu = 1.75$ is rejected in favor of H_a: $\mu \neq 1.75$ if either $t \geq t_{.025,25} = 2.060$ or $t \leq -2.060$. The computed t is $t = \dfrac{1.89 - 1.75}{.42/\sqrt{26}} = 1.70$. Since 1.70 is neither ≥ 2.060 nor ≤ -2.060, do not reject H_o; the data does not contradict prior research.

b $1.70 \doteq 1.708 = t_{.025,25}$, so $P \doteq 2(.05) = .10$ (since for a two-tailed test, $.05 = \alpha / 2$.)

71. Even though the underlying distribution may not be normal, a z test can be used because n is large. H_o: $\mu = 3200$ should be rejected in favor of H_a: $\mu < 3200$ if $z \leq -z_{.001} = -3.08$. The computed z is $z = \dfrac{3107 - 3200}{188/\sqrt{45}} = -3.32 \leq -3.08$, so H_o should be rejected at level .001.

72. Let p = the true proportion of mechanics who could identify the problem. Then the appropriate hypotheses are H_o: p = .75 vs H_a: p < .75, so a lower-tailed test should be used. With p_o= .75 and $\hat{p} = \dfrac{42}{72} = .583$, z = -3.28 and $P = \Phi(-3.28) = .0005$. Because this p-value is so small, the data argues strongly against H_o, so we reject it in favor of H_a.

73. We wish to test H_o: $\lambda = 4$ vs H_a: $\lambda > 4$ using the test statistic $z = \dfrac{\overline{x} - 4}{\sqrt{4/n}}$. For the given sample, n = 36 and $\overline{x} = \dfrac{160}{36} = 4.444$, so $z = \dfrac{4.444 - 4}{\sqrt{4/36}} = 1.33$. At level .02, we reject H_o if $z \geq z_{.02} \doteq 2.05$ (since $1 - \Phi(2.05) = .0202$). Because 1.33 is not ≥ 2.05, H_o should not be rejected at this level.

74. H_o: $\mu = 15$ vs H_a: $\mu > 15$. Because the sample size is less than 40, and we can assume the distribution is approximately normal, the appropriate statistic is $t = \dfrac{\overline{x} - 15}{s/\sqrt{n}} = \dfrac{17.5 - 15}{2.2/\sqrt{32}} = \dfrac{2.5}{.390} = 6.4$. Thus the p-value is "off the chart" in the 20 df column of Table A.8, and so is approximately $0 < .05$, so H_o is rejected in favor of the conclusion that the true average time exceeds 15 minutes.

75. H_o: $\sigma^2 = .25$ vs H_a: $\sigma^2 > .25$. The chi-squared critical value for 9 d.f. that captures upper-tail area .01 is 21.665. The test statistic value is $\dfrac{9(.58)^2}{.25} = 12.11$. Because 12.11 is not ≥ 21.665, H_o cannot be rejected. The uniformity specification is not contradicted.

76. The 20 df row of Table A.7 shows that $\chi^2_{.99,20} = 8.26 < 8.58$ (H_o not rejected at level .01) and $8.58 < 9.591 = \chi^2_{.975,20}$ (H_o rejected at level .025). Thus .01 < p-value < .025 and H_o cannot be rejected at level .01 (the p-value is the smallest alpha at which rejection can take place, and this exceeds .01).

77.

a $E(\overline{X} + 2.33S) = E(\overline{X}) + 2.33E(S) = \mu + 2.33\sigma$, so $\hat{\theta} = \overline{X} + 2.33S$ is approximately unbiased.

b $V(\overline{X} + 2.33S) = V(\overline{X}) + 2.33^2 V(S) = \frac{\sigma^2}{n} + 5.4289\frac{\sigma^2}{2n}$. The estimated standard error (standard deviation) is $1.927\frac{s}{\sqrt{n}}$.

c More than 99% of all soil samples have pH less than 6.75 iff the 95th percentile is less than 6.75. Thus we wish to test H_o: $\mu + 2.33\sigma = 6.75$ vs H_a: $\mu + 2.33\sigma < 6.75$. H_o will be rejected at level .01 if $z \le 2.33$. Since $z = \frac{-.047}{.0385} < 0$, H_o clearly cannot be rejected. The 95th percentile does not appear to exceed 6.75.

78.

a When H_o is true, $2\lambda_o \Sigma X_i = 2\sum \frac{X_i}{\mu_o}$ has a chi-squared distribution with df = 2n. If the alternative is H_a: $\mu > \mu_o$, large test statistic values (large Σx_i, since $\overline{x}$ is large) suggest that H_o be rejected in favor of H_a, so rejecting when $2\sum \frac{X_i}{\mu_o} \ge \chi^2_{\alpha,2n}$ gives a test with significance level α. If the alternative is H_a: $\mu < \mu_o$, rejecting when $2\sum \frac{X_i}{\mu_o} \le \chi^2_{1-\alpha,2n}$ gives a level α test. The rejection region for H_a: $\mu \ne \mu_o$ is either $2\sum \frac{X_i}{\mu_o} \ge \chi^2_{\alpha/2,2n}$ or $\le \chi^2_{1-\alpha/2,2n}$.

b H_o: $\mu = 75$ vs H_a: $\mu < 75$. The test statistic value is $\frac{2(737)}{75} = 19.65$. At level .01, H_o is rejected if $2\sum \frac{X_i}{\mu_o} \le \chi^2_{.99,20} = 8.260$. Clearly 19.65 is not in the rejection region, so H_o should not be rejected. The sample data does not suggest that true average lifetime is less than the previously claimed value.

79.

a P(type I error) = P(either $Z \geq z_\gamma$ or $Z \leq z_{\alpha-\gamma}$) (when Z is a standard normal r.v.) $= \Phi(-z_{\alpha-\gamma}) + 1 - \Phi(z_\gamma) = \alpha - \gamma + \gamma = \alpha$.

b $\beta(\mu) = P(\overline{X} \geq \mu_o + \frac{\sigma z_\gamma}{\sqrt{n}} or \overline{X} \leq \mu_o - \frac{\sigma z_{\alpha-\gamma}}{\sqrt{n}}$ when the true value is μ) =

$$\Phi\left(z_\gamma + \frac{\mu_o - \mu}{\sigma/\sqrt{n}}\right) - \Phi\left(-z_{\alpha-\gamma} + \frac{\mu_o - \mu}{\sigma/\sqrt{n}}\right)$$

c Let $\lambda = \sqrt{n}\frac{\Delta}{\sigma}$; then we wish to know when $\pi(\mu_o + \Delta) = 1 - \Phi(z_\gamma - \lambda)$ $+ \Phi(-z_{\alpha-\gamma} - \lambda) > 1 - \Phi(z_\gamma + \lambda) + \Phi(-z_{\alpha-\gamma} + \lambda) = \pi(\mu_o - \Delta)$. Using the fact that $\Phi(-c) = 1 - \Phi(c)$, this inequality becomes $\Phi(z_\gamma + \lambda) - \Phi(z_\gamma - \lambda) > \Phi(z_{\alpha-\gamma} + \lambda) - \Phi(z_{\alpha-\gamma} - \lambda)$. The l.h.s. is the area under the Z curve above the interval $(z_\gamma + \lambda, z_\gamma - \lambda)$, while the r.h.s. is the area above $(z_{\alpha-\gamma} - \lambda, z_{\alpha-\gamma} + \lambda)$. Both intervals have width 2λ, but when $z_\gamma < z_{\alpha-\gamma}$, the first interval is closer to 0 (and thus corresponds to the large area) than is the second. This happens when $\gamma > \alpha - \gamma$, i.e., when $\gamma > \alpha/2$.

80.

a $\alpha = P(X \leq 5$ when p = .9) = B(5; 10, .9) = .002, so the region (0, 1, …, 5) does specify a level .01 test.

b The first value to be placed in the upper-tailed part of a two tailed region would be 10, but P(X = 10 when p = .9) = .349, so whenever 10 is in the rejection region, $\alpha \geq .349$.

Chapter 9

Section 9.1

1.

a $E(\overline{X}-\overline{Y}) = E(\overline{X}) - E(\overline{Y}) = 4.1 - 4.5 = -.4$, irrespective of sample sizes.

b $V(\overline{X}-\overline{Y}) = V(\overline{X}) + V(\overline{Y}) = \frac{\sigma_1^2}{m} + \frac{\sigma_2^2}{n} = \frac{(1.8)^2}{100} + \frac{(2.0)^2}{100} = .0724$, and the s.d. of $\overline{X}-\overline{Y} = \sqrt{.0724} = .2691$.

c A normal curve with mean and s.d. as given in **a** and **b** (because m = n = 100, the CLT implies that both $\overline{X}$ and $\overline{Y}$ have approximately normal distributions, so $\overline{X}-\overline{Y}$ does also). The shape is not necessarily that of a normal curve when m = n = 10, because the CLT cannot be invoked. So if the two lifetime population distributions are not normal, the distribution of $\overline{X}-\overline{Y}$ will typically be quite complicated.

2. The test statistic value is $z = \frac{\bar{x}-\bar{y}}{\sqrt{\frac{s_1^2}{m}+\frac{s_2^2}{n}}}$, and H_o will be rejected if either $z \geq 1.96$ or $z \leq -1.96$. We compute $z = \frac{42{,}500 - 40{,}400}{\sqrt{\frac{2200^2}{45}+\frac{1900^2}{45}}} = \frac{2100}{433.33} = 4.85$. Since 4.85 > 1.96, reject H_o and conclude that the two brands differ with respect to true average tread lives.

3. The test statistic value is $z = \frac{(\bar{x}-\bar{y})-500}{\sqrt{\frac{s_1^2}{m}+\frac{s_2^2}{n}}}$, and H_o will be rejected at level .01 if $z \geq 2.33$. We compute $z = \frac{(43{,}500 - 36{,}800) - 500}{\sqrt{\frac{2200^2}{45}+\frac{1500^2}{45}}} = \frac{1700}{396.93} = 4.28$, which is > 2.33, so we reject H_o and conclude that the true average life for radials exceeds that for economy brand by more than 500.

4.

a The C.I. is $(\bar{x}-\bar{y}) \pm (1.96)\sqrt{\frac{s_1^2}{m}+\frac{s_2^2}{n}} = 2100 \pm 1.96(433.33) = 2100 \pm 849.33$ $= (1250.67, 2949.33)$. In the context of this problem situation, the interval is moderately wide (a consequence of the standard deviations being large), so the information about μ_1 and μ_2 is not as precise as might be desirable.

b The upper bound is $1700 + 1.645(396.93) = 1700 + 652.95 = 2352.95$.

5.

a $\sqrt{\frac{\sigma_1^2}{m}+\frac{\sigma_2^2}{n}} = \sqrt{\frac{(.04)^2}{10}+\frac{(.16)^2}{10}} = .1414$, so $z = \frac{(.64-2.05)-(-1)}{.1414} = -2.90$. At level .01, H_o is rejected if $z \le -2.33$; since –2.90 < -2.33, reject H_o.

b $P = \Phi(-2.90) = .0019$

c $\beta = 1 - \Phi\left(-2.33 - \frac{-1.2+1}{.1414}\right) = 1 - \Phi(-.92) = .8212$

d $m = n = \frac{.2(2.33+1.28)^2}{(-.2)^2} = 65.15$, so use 66.

6.

a H_o should be rejected if $z \ge 2.33$. Since $z = \frac{(18.12-16.87)}{\sqrt{\frac{2.56}{40}+\frac{1.96}{32}}} = 3.53 \ge 2.33$, H_o should be rejected at level .01.

b $\beta(1) = \Phi\left(2.33 - \frac{1-0}{.3539}\right) = \Phi(-.50) = .3085$

c $\frac{2.56}{40}+\frac{1.96}{n} = \frac{1}{(1.645+1.28)^2} = .1169 \Rightarrow \frac{1.96}{n} = .0529 \Rightarrow n = 37.06$, so use n = 38.

d Since n = 32 is not a large sample, it would no longer be appropriate to use the large sample test. A small sample t procedure should be used (section 9.2), and the appropriate conclusion would follow.

7.

1 Parameter of interest: $\mu_1 - \mu_2$ = the true difference of means for males and females on the Boredom Proneness Rating. Let μ_1 = men's average and μ_2 = women's average.

2 H$_o$: $\mu_1 - \mu_2 = 0$

3 H$_a$: $\mu_1 - \mu_2 > 0$

4 $$z = \frac{(\bar{x} - \bar{y}) - \Delta_o}{\sqrt{\frac{s_1^2}{m} + \frac{s_2^2}{n}}} = \frac{(\bar{x} - \bar{y}) - 0}{\sqrt{\frac{s_1^2}{m} + \frac{s_2^2}{n}}}$$

5 RR: $z \geq 1.645$

6 $$z = \frac{(10.40 - 9.26) - \Delta_o}{\sqrt{\frac{4.83^2}{97} + \frac{4.68^2}{148}}} = 1.83$$

7 Reject H$_o$. The data indicates the Boredom Proneness Rating is higher for males than for females.

8.

a

1 Parameter of interest: $\mu_1 - \mu_2$ = the true difference of mean tensile strength of the 1064 grade and the 1078 grade wire rod. Let μ_1 = 1064 grade average and μ_2 = 1078 grade average.

2 H$_o$: $\mu_1 - \mu_2 = -10$

3 H$_a$: $\mu_1 - \mu_2 < -10$

4 $$z = \frac{(\bar{x} - \bar{y}) - \Delta_o}{\sqrt{\frac{s_1^2}{m} + \frac{s_2^2}{n}}} = \frac{(\bar{x} - \bar{y}) - (-10)}{\sqrt{\frac{s_1^2}{m} + \frac{s_2^2}{n}}}$$

5 RR: $p - value < \alpha$

6 $$z = \frac{(107.6 - 123.6) - (-10)}{\sqrt{\frac{1.3^2}{129} + \frac{2.0^2}{129}}} = \frac{-6}{.210} = -28.57$$

7 For a lower-tailed test, the p-value = $\Phi(-28.57) \approx 0$, which is less than any α, so reject H$_o$. There is very compelling evidence that the mean tensile strength of the 1078 grade exceeds that of the 1064 grade by more than 10.

b The requested information can be provided by a 95% confidence interval for

$$\mu_1 - \mu_2 :\ (\bar{x} - \bar{y}) \pm 1.96\sqrt{\frac{s_1^2}{m} + \frac{s_2^2}{n}} = (-6) \pm 1.96(.210) = (-6.412, -5.588).$$

Chapter 9

9.

a The hypotheses are H$_o$: $\mu_1 - \mu_2 = 5$ and H$_a$: $\mu_1 - \mu_2 > 5$. At level .001, H$_o$ should be rejected if $z \ge 3.08$. Since $z = \frac{(65.6 - 59.8) - 5}{.2272} = 2.89 < 3.08$, H$_o$ cannot be rejected in favor of H$_a$ at this level, so the use of the high purity steel cannot be justified.

b $\mu_1 - \mu_2 - \Delta_o = 1$, so $\beta = \Phi\left(3.08 - \frac{1}{.2272}\right) = \Phi(-.53) = .2891$

10. $(\bar{X} - \bar{Y}) \pm z_{\alpha/2}\sqrt{\frac{s_1^2}{m} + \frac{s_2^2}{n}}$. Standard error $= \frac{s}{\sqrt{n}}$. Substitution yields $(\bar{x} - \bar{y}) \pm z_{\alpha/2}\sqrt{(SE_1)^2 + (SE_2)^2}$. Using $\alpha = .05$, $z_{\alpha/2} = 1.96$, so $(5.5 - 3.8) \pm 1.96\sqrt{(0.3)^2 + (0.2)^2} = (0.99, 2.41)$. Because we selected $\alpha = .05$, we can state that when using this method with repeated sampling, the interval calculated will bracket the true difference 95% of the time. The interval is fairly narrow, indicating precision of the estimate.

11. The C.I. is $(\bar{x} - \bar{y}) \pm 2.58\sqrt{\frac{s_1^2}{m} + \frac{s_2^2}{n}} = (-8.77) \pm 2.58\sqrt{.9104} = -8.77 \pm 2.46$ $= (-11.23, -6.31)$.

12. $\sigma_1 = \sigma_2 = .05$, d = .04, $\alpha = .01, \beta = .05$, and the test is one-tailed, so

$$n = \frac{(.0025 + .0025)(2.33 + 1.645)^2}{.0016} = 49.38$$, so use n = 50.

13. The appropriate hypotheses are H$_o$: $\theta = 0$ vs. H$_a$: $\theta < 0$, where $\theta = 2\mu_1 - \mu_2$. ($\theta < 0$ is equivalent to $2\mu_1 < \mu_2$, so normal is more than twice schizo) The estimator of θ is $\hat{\theta} = 2\bar{X} - \bar{Y}$, with $Var(\hat{\theta}) = 4Var(\bar{X}) + Var(\bar{Y}) = \frac{4\sigma_1^2}{m} + \frac{\sigma_2^2}{n}$, σ_θ is the square root of $Var(\hat{\theta})$, and $\hat{\sigma}_\theta$ is obtained by replacing each σ_i^2 with S_i^2. The test statistic is then $\frac{\hat{\theta}}{\sigma_{\hat{\theta}}}$ (since $\theta_o = 0$), and H$_o$ is rejected if $z \le -2.33$. With $\hat{\theta} = 2(2.69) - 6.35 = -.97$ and $\hat{\sigma}_\theta = \sqrt{\frac{4(2.3)^2}{43} + \frac{(4.03)^2}{45}} = .9236$, $z = \frac{-.97}{.9236} = -1.05$; Because −1.05 > -2.33, H$_o$ is not rejected.

14. As either *m* or *n* increases, σ decreases, so $\dfrac{\mu_1 - \mu_2 - \Delta_o}{\sigma}$ increases (the numerator is positive), so $\left(z_\alpha - \dfrac{\mu_1 - \mu_2 - \Delta_o}{\sigma} \right)$ decreases, so $\beta = \Phi\left(z_\alpha - \dfrac{\mu_1 - \mu_2 - \Delta_o}{\sigma} \right)$ decreases.

15. As β decreases, z_β increases, and since z_β is the numerator of n, n increases also.

16. $z = \dfrac{\overline{x} - \overline{y}}{\sqrt{\dfrac{s_1^2}{n} + \dfrac{s_2^2}{n}}} = \dfrac{.2}{\sqrt{\dfrac{2}{n}}}$. For n = 100, z = 1.41 and p-value = $2[1 - \Phi(1.41)] = .1586$.

For n = 400, z = 2.83 and p-value = .0046. From a practical point of view, the closeness of $\overline{x}$ and $\overline{y}$ suggests that there is essentially no difference between true average fracture toughness for type I and type I steels. The very small difference in sample averages has been magnified by the large sample sizes – statistical rather than practical significance. The p-value by itself would not have conveyed this message.

Section 9.2

17.

a $$\nu = \frac{\left(\frac{5^2}{10} + \frac{6^2}{10}\right)^2}{\frac{\left(\frac{5^2}{10}\right)^2}{9} + \frac{\left(\frac{6^2}{10}\right)^2}{9}} = \frac{37.21}{.694 + 1.44} = 17.43 \approx 17$$

b $$\nu = \frac{\left(\frac{5^2}{10} + \frac{6^2}{15}\right)^2}{\frac{\left(\frac{5^2}{10}\right)^2}{9} + \frac{\left(\frac{6^2}{15}\right)^2}{14}} = \frac{24.01}{.694 + .411} = 21.7 \approx 21$$

c $$\nu = \frac{\left(\frac{2^2}{10} + \frac{6^2}{15}\right)^2}{\frac{\left(\frac{2^2}{10}\right)^2}{9} + \frac{\left(\frac{6^2}{15}\right)^2}{14}} = \frac{7.84}{.018 + .411} = 18.27 \approx 18$$

d $$\nu = \frac{\left(\frac{5^2}{12} + \frac{6^2}{24}\right)^2}{\frac{\left(\frac{5^2}{12}\right)^2}{11} + \frac{\left(\frac{6^2}{24}\right)^2}{23}} = \frac{12.84}{.395 + .098} = 26.05 \approx 26$$

18. With H$_o$: $\mu_1 - \mu_2 = 0$ vs. H$_a$: $\mu_1 - \mu_2 \neq 0$, we will reject H$_o$ if $p - value < \alpha$.

$$v = \frac{\left(\frac{.164^2}{6} + \frac{.240^2}{5}\right)^2}{\frac{\left(\frac{.164^2}{6}\right)^2}{5} + \frac{\left(\frac{.240^2}{5}\right)^2}{4}} = 6.8 \approx 6$$, and the test statistic

$$t = \frac{22.73 - 21.95}{\sqrt{\frac{.164^2}{6} + \frac{.240^2}{5}}} = \frac{.78}{.1265} = 6.17$$ leads to a p-value of 2[P(t > 6.17)] < 2(.0005)

=.001, which is less than most reasonable $\alpha's$, so we reject H$_o$ and conclude that there is a difference in the densities of the two brick types.

19. For the given hypotheses, the test statistic $t = \frac{115.7 - 129.3 + 10}{\sqrt{\frac{5.03^2}{6} + \frac{5.38^2}{6}}} = \frac{-3.6}{3.007} = -1.20$,

and the d.f. is $v = \frac{(4.2168 + 4.8241)^2}{\frac{(4.2168)^2}{5} + \frac{(4.8241)^2}{5}} = 9.96$, so use d.f. = 9. We will reject H$_o$ if

$t \leq -t_{.01,9} = -2.764$; since –1.20 > -2.764, we don't reject H$_o$.

20. We want a 95% confidence interval for $\mu_1 - \mu_2$. $t_{.025,9} = 2.262$, so the interval is $-3.6 \pm 2.262(3.007) = (-10.40, 3.20)$. Because the interval is so wide, it does not appear that precise information is available.

21. Let μ_1 = the true average gap detection threshold for normal subjects, and μ_2 = the corresponding value for CTS subjects. The relevant hypotheses are H$_o$: $\mu_1 - \mu_2 = 0$ vs. H$_a$: $\mu_1 - \mu_2 < 0$, and the test statistic $t = \frac{1.71 - 2.53}{\sqrt{.0351125 + .07569}} = \frac{-.82}{.3329} = -2.46$.

Using d.f. $v = \frac{(.0351125 + .07569)^2}{\frac{(.0351125)^2}{7} + \frac{(.07569)^2}{9}} = 15.1$, or 15, the rejection region is

$t \leq -t_{.01,15} = -2.602$. Since –2.46 is not ≤ -2.602, we fail to reject H$_o$. We have insufficient evidence to claim that the true average gap detection threshold for CTS subjects exceeds that for normal subjects.

22. Let $\mu_1 =$ the true average strength for wire-brushing preparation and let $\mu_2 =$ the average strength for hand-chisel preparation. Since we are concerned about any possible difference between the two means, a two-sided test is appropriate. We test $H_0: \mu_1 - \mu_2 = 0$ vs. $H_a: \mu_1 - \mu_2 \neq 0$. We need the degrees of freedom to find the rejection region: $\nu = \dfrac{\left(\frac{1.58^2}{12} + \frac{4.01^2}{12}\right)^2}{\frac{\left(\frac{1.58^2}{12}\right)^2}{11} + \frac{\left(\frac{4.01^2}{5}\right)^2}{11}} = \dfrac{2.3964}{.0039 + .1632} = 14.33$, which we round down to 14, so we reject H_o if $|t| \geq t_{.025,14} = 2.145$. The test statistic is $t = \dfrac{19.20 - 23.13}{\sqrt{\left(\frac{1.58^2}{12} + \frac{4.01^2}{12}\right)}} = \dfrac{-3.93}{1.2442} = -3.159$, which is ≤ -2.145, so we reject H_o and conclude that there does appear to be a difference between the two population average strengths.

23.

a Normal plots

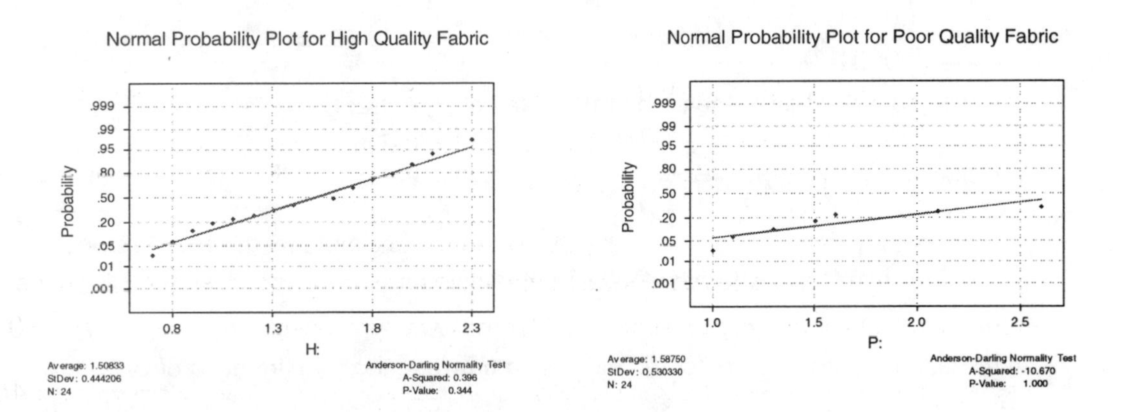

Using Minitab to generate normal probability plots, we see that both plots illustrate sufficient linearity. Therefore, it is plausible that both samples have been selected from normal population distributions.

b

Comparative Box Plot for High Quality and Poor Quality Fabric

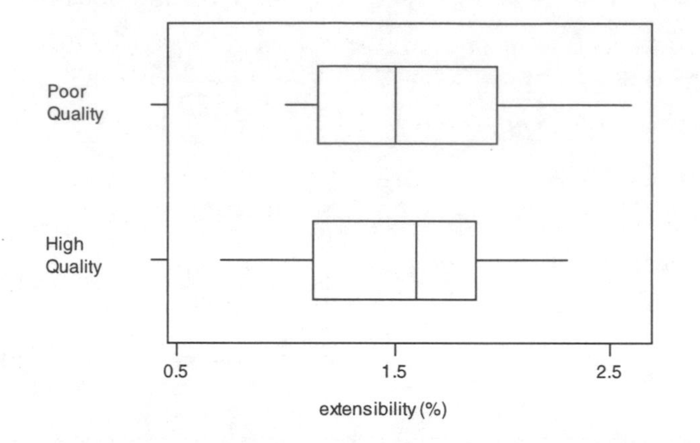

The comparative boxplot does not suggest a difference between average extensibility for the two types of fabrics.

c We test $H_0 : \mu_1 - \mu_2 = 0$ vs. $H_a : \mu_1 - \mu_2 \neq 0$. With degrees of freedom $\nu = \dfrac{(.0433265)^2}{.00017906} = 10.5$, which we round down to 10, and using significance level .05 (not specified in the problem), we reject H_o if $|t| \geq t_{.025,10} = 2.228$. The test statistic is $t = \dfrac{-.08}{\sqrt{(.0433265)}} = -.38$, which is not ≥ 2.228 in absolute value, so we cannot reject H_o. There is insufficient evidence to claim that the true average extensibility differs for the two types of fabrics.

24. A 95% confidence interval for the difference between the true firmness of zero-day apples and the true firmness of 20-day apples is $(8.74 - 4.96) \pm t_{.025,\nu}\sqrt{\dfrac{.66^2}{20} + \dfrac{.39^2}{20}}$.

We calculate the degrees of freedom $\nu = \dfrac{\left(\dfrac{.66^2}{20} + \dfrac{.39^2}{20}\right)^2}{\dfrac{\left(\frac{.66^2}{20}\right)^2}{19} + \dfrac{\left(\frac{.39^2}{20}\right)^2}{19}} = 30.83$, so we use 30 df,

and $t_{.025,30} = 2.042$, so the interval is $3.78 \pm 2.042(.17142) = (3.43, 4.13)$. Thus, with 95% confidence, we can say that the true average firmness for zero-day apples exceeds that of 20-dy apples by between 3.43 and 4.13 N.

25. We calculate the degrees of freedom $\nu = \dfrac{\left(\frac{5.5^2}{28} + \frac{7.8^2}{31}\right)^2}{\frac{\left(\frac{5.5^2}{28}\right)^2}{27} + \frac{\left(\frac{7.8^2}{31}\right)^2}{30}} = 53.95$, or about 54 (normally we would round down to 53, but this number is very close to 54 – of course for this large number of df, using either 53 or 54 won't make much difference in the critical t value) so the desired confidence interval is $(91.5 - 88.3) \pm 1.68\sqrt{\frac{5.5^2}{28} + \frac{7.8^2}{31}}$ $= 3.2 \pm 2.931 = (.269, 6.131)$. Because 0 does not lie inside this interval, we can be reasonably certain that the true difference $\mu_1 - \mu_2$ is not 0 and, therefore, that the two population means are not equal. For a 95% interval, the t value increases to about 2.01 or so, which results in the interval 3.2 ± 3.506. Since this interval does contain 0, we can no longer conclude that the means are different if we use a 95% confidence interval.

26. Let μ_1 = the true average potential drop for alloy connections and let μ_2 = the true average potential drop for EC connections. Since we are interested in whether the potential drop is higher for alloy connections, an upper tailed test is appropriate. We test $H_0 : \mu_1 - \mu_2 = 0$ vs. $H_a : \mu_1 - \mu_2 > 0$. Using the SAS output provided, the test statistic, when assuming unequal variances, is t = 3.6362, the corresponding df is 37.5, and the p-value for our upper tailed test would be ½ (two-tailed p-value) = $\frac{1}{2}(.0008) = .0004$. Our p-value of .0004 is less than the significance level of .01, so we reject H_0. We have sufficient evidence to claim that the true average potential drop for alloy connections is higher than that for EC connections.

27. The approximate degrees of freedom for this estimate are

$$\nu = \frac{\left(\frac{11.3^2}{6} + \frac{8.3^2}{8}\right)^2}{\frac{\left(\frac{11.3^2}{6}\right)^2}{5} + \frac{\left(\frac{8.3^2}{8}\right)^2}{7}} = \frac{893.59}{101.175} = 8.83$$

, which we round down to 8, so $t_{.025,8} = 2.306$ and the desired interval is $(40.3 - 21.4) \pm 2.306\sqrt{\frac{11.3^2}{6} + \frac{8.3^2}{8}} = 18.9 \pm 2.306(5.4674)$ $= 18.9 \pm 12.607 = (6.3, 31.5)$. Because 0 is not contained in this interval, there is strong evidence that $\mu_1 - \mu_2$ is not 0; i.e., we can conclude that the population means are not equal. Calculating a confidence interval for $\mu_2 - \mu_1$ would change only the order of subtraction of the sample means, but the standard error calculation would give the same result as before. Therefore, the 95% interval estimate of $\mu_2 - \mu_1$ would be (-31.5, -6.3), just the negatives of the endpoints of the original interval. Since 0 is not in this interval, we reach exactly the same conclusion as before; the population means are not equal.

28. $H_0: \mu_1 - \mu_2 = 0$ vs. $H_a: \mu_1 - \mu_2 > 0$. The test statistic is $t = \frac{18.9}{5.467} = 3.46 \approx 3.5$. With df = 8, our p-value $\approx P(t > 3.5) = .004$. Since .004 < .01, we reject H_o. The force after impact is greater for advanced players.

29. Let μ_1 = the true average compression strength for strawberry drink and let μ_2 = the true average compression strength for cola. A lower tailed test is appropriate. We test $H_0: \mu_1 - \mu_2 = 0$ vs. $H_a: \mu_1 - \mu_2 < 0$. The test statistic is $t = \frac{-14}{\sqrt{29.4 + 15}} = -2.10$.

We use degrees of freedom $\nu = \frac{(44.4)^2}{\frac{(29.4)^2}{14} + \frac{(15)^2}{14}} = \frac{1971.36}{77.8114} = 25.3$, so use df = 25.

The p-value $\approx P(t < -2.10) = .023$. This p-value indicates strong support for the alternative hypothesis. The data does suggest that the extra carbonation of cola results in a higher average compression strength.

30.

a We desire a 99% confidence interval. First we calculate the degrees of freedom:
$\nu = \frac{\left(\frac{2.2^2}{26} + \frac{4.3^2}{26}\right)^2}{\frac{\left(\frac{2.2^2}{26}\right)^2}{26} + \frac{\left(\frac{4.3^2}{26}\right)^2}{26}} = 37.24$, which we would round down to 37, except that there is no df = 37 row in Table A.5. Using 36 degrees of freedom (a more conservative choice), $t_{.005,36} = 2.719$, and the 99% C.I. is

$(33.4 - 42.8) \pm 2.719\sqrt{\frac{2.2^2}{26} + \frac{4.3^2}{26}} = -9.4 \pm 2.576 = (-11.98, -6.83)$. We are very confident that the true average load for carbon beams exceeds that for fiberglass beams by between 6.83 and 11.98 kN.

b The upper limit of the interval in part **a** does not give a 99% upper confidence bound. The 99% upper bound would be $-9.4 + 2.434(.9473) = -7.09$, meaning that the true average load for carbon beams exceeds that for fiberglass beams by at least 7.09 kN.

31.

a

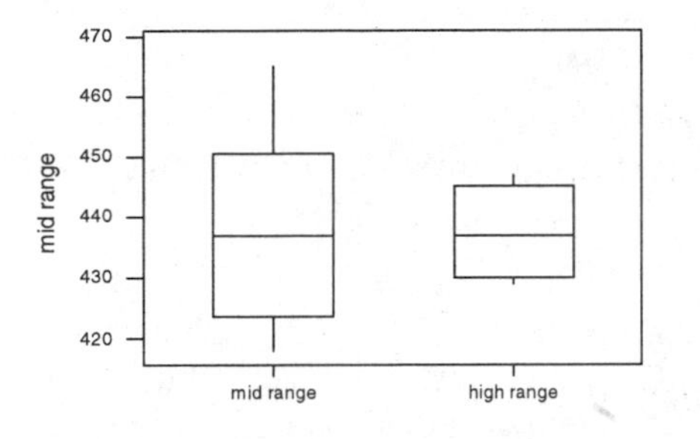

The most notable feature of these boxplots is the larger amount of variation present in the mid-range data as compared to the high-range data. Otherwise, both boxplots look reasonably symmetric and there are not outliers present.

b Using df = 23, a 95% confidence interval for $\mu_{mid-range} - \mu_{high-range}$ is $(438.3 - 437.45) \pm 2.069\sqrt{\frac{15.1^2}{17} + \frac{6.83^2}{11}} = .85 \pm 8.69 = (-7.84, 9.54)$. Since plausible values for $\mu_{mid-range} - \mu_{high-range}$ are both positive and negative (i.e., the interval spans zero) we would conclude that there is not sufficient evidence to suggest that the average value for mid-range and the average value for high-range differ.

32. Let $\mu_1 =$ the true average proportional stress limit for red oak and let $\mu_2 =$ the true average proportional stress limit for Douglas fir. We test $H_0 : \mu_1 - \mu_2 = 1$ vs. $H_a : \mu_1 - \mu_2 > 1$. The test statistic is $t = \frac{(8.48 - 6.65) - 1}{\sqrt{\frac{.79^2}{14} + \frac{1.28^2}{10}}} = \frac{1.83}{\sqrt{.2084}} = 1.818$. With degrees of freedom $v = \frac{(.2084)^2}{\frac{\left(\frac{.79^2}{14}\right)^2}{13} + \frac{\left(\frac{1.28^2}{10}\right)^2}{9}} = 2.008 \approx 2$, the p-value $\approx P(t > 1.8) = .107$. This p-value does not indicate strong support for the alternative hypothesis since we would reject H_o only at significance levels greater than .107. There is insufficient evidence to claim that true average proportional stress limit for red oak exceeds that of Douglas fir by more than 1 MPa.

33. Let $\mu_1 =$ the true average weight gain for steroid treatment and let $\mu_2 =$ the true average weight gain for the population not treated with steroids. The exercise asks if we can conclude that μ_2 exceeds μ_1 by more than 5 g., which we can restate in the equivalent form: $\mu_1 - \mu_2 < -5$. Therefore, we conduct a lower-tailed test of $H_0 : \mu_1 - \mu_2 = -5$ vs. $H_a : \mu_1 - \mu_2 < -5$. The test statistic is

$t = \dfrac{(\bar{x} - \bar{y}) - (\Delta)}{\sqrt{\dfrac{s_1^2}{m} + \dfrac{s_2^2}{n}}} = \dfrac{32.8 - 40.5 - (-5)}{\sqrt{\dfrac{2.6^2}{8} + \dfrac{2.5^2}{10}}} = \dfrac{-2.7}{1.2124} = -2.23 \approx 2.2$. The approximate d.f. is

$\nu = \dfrac{\left(\frac{2.6^2}{8} + \frac{2.5^2}{10}\right)^2}{\dfrac{\left(\frac{2.6^2}{8}\right)^2}{7} + \dfrac{\left(\frac{2.5^2}{10}\right)^2}{9}} = \dfrac{2.1609}{.1454} = 14.876$, which we round down to 14. The p-value for a lower tailed test is P(t < -2.2) = P(t > 2.2) = .022. Since this p-value is larger than the specified significance level .01, we cannot reject H_o. Therefore, this data does not support the belief that average weight gain for the control group exceeds that of the steroid group by more than 5 g.

34.

a Following the usual format for most confidence intervals: *statistic ± (critical value)(standard error),* a pooled variance confidence interval for the difference between two means is $(\bar{x} - \bar{y}) \pm t_{\alpha/2, m+n-2} \cdot s_p \sqrt{\frac{1}{m} + \frac{1}{n}}$.

b The sample means and standard deviations of the two samples are $\bar{x} = 13.90$, $s_1 = 1.225$, $\bar{y} = 12.20$, $s_2 = 1.010$. The pooled variance estimate is $s_p^2 =$ $\left(\dfrac{m-1}{m+n-2}\right)s_1^2 + \left(\dfrac{n-1}{m+n-2}\right)s_2^2 = \left(\dfrac{4-1}{4+4-2}\right)(1.225)^2 + \left(\dfrac{4-1}{4+4-2}\right)(1.010)^2$ $= 1.260$, so $s_p = 1.1227$. With df = m+n-1 = 6 for this interval, $t_{.025,6} = 2.447$ and the desired interval is $(13.90 - 12.20) \pm (2.447)(1.1227)\sqrt{\frac{1}{4} + \frac{1}{4}}$ $= 1.7 \pm 1.943 = (-.24, 3.64)$. This interval contains 0, so it does not support the conclusion that the two population means are different.

35. There are two changes that must be made to the procedure we currently use. First, the equation used to compute the value of the t test statistic is: $t = \dfrac{(\bar{x} - \bar{y}) - (\Delta)}{s_p\sqrt{\dfrac{1}{m} + \dfrac{1}{n}}}$ where s_p is defined as in Exercise 34 above. Second, the degrees of freedom = m + n – 2. Assuming equal variances in the situation from Exercise 33, we calculate s_p as follows:

$s_p = \sqrt{\left(\dfrac{7}{16}\right)(2.6)^2 + \left(\dfrac{9}{16}\right)(2.5)^2} = 2.544$. The value of the test statistic is, then,

$t = \dfrac{(32.8 - 40.5) - (-5)}{2.544\sqrt{\dfrac{1}{8} + \dfrac{1}{10}}} = -2.24 \approx -2.2$. The degrees of freedom = 16, and the p-value is P (t < -2.2) = .021. Since .021 > .01, we fail to reject H_o. This is the same conclusion reached in Exercise 33.

Section 9.3

36. $\bar{d} = 7.25$, $s_D = 11.8628$

1 Parameter of Interest: μ_D = true average difference of breaking load for fabric in unabraded or abraded condition.

2 $H_0 : \mu_D = 0$

3 $H_a : \mu_D > 0$

4 $t = \dfrac{\bar{d} - \mu_D}{s_D / \sqrt{n}} = \dfrac{\bar{d} - 0}{s_D / \sqrt{n}}$

5 RR: $t \geq t_{.01,7} = 2.998$

6 $t = \dfrac{7.25 - 0}{11.8628 / \sqrt{8}} = 1.73$

7 Fail to reject H_o. The data does not indicate a difference in breaking load for the two fabric load conditions.

37.

a This exercise calls for paired analysis. First, compute the difference between indoor and outdoor concentrations of hexavalent chromium for each of the 33 houses. These 33 differences are summarized as follows: n = 33, $\bar{d} = -.4239$, $s_d = .3868$, where d = (indoor value – outdoor value). Then $t_{.025,32} = 2.037$, and a 95% confidence interval for the population mean difference between indoor and outdoor concentration is

$-.4239 \pm (2.037)\left(\dfrac{.3868}{\sqrt{33}}\right) = -.4239 \pm .13715 = (-.5611, -.2868)$. We can be highly confident, at the 95% confidence level, that the true average concentration of hexavalent chromium outdoors exceeds the true average concentration indoors by between .2868 and .5611 nanograms/m^3.

b A 95% prediction interval for the difference in concentration for the 34th house is $\bar{d} \pm t_{.025,32}\left(s_d\sqrt{1+\frac{1}{n}}\right) = -.4239 \pm (2.037)\left(.3868\sqrt{1+\frac{1}{33}}\right) = (-1.224, .3758)$. This prediction interval means that the indoor concentration may exceed the outdoor concentration by as much as .3758 nanograms/m^3 and that the outdoor concentration may exceed the indoor concentration by a much as 1.224 nanograms/m^3, for the 34th house. Clearly, this is a wide prediction interval, largely because of the amount of variation in the differences.

38.

a The median of the "Normal" data is 46.80 and the upper and lower quartiles are 45.55 and 49.55, which yields an IQR of 49.55 – 45.55 = 4.00. The median of the "High" data is 90.1 and the upper and lower quartiles are 88.55 and 90.95, which yields an IQR of 90.95 – 88.55 = 2.40.

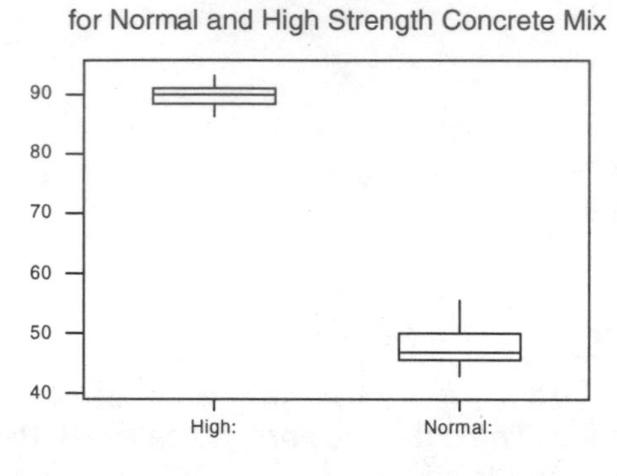

The most significant feature of these boxplots is the fact that their locations (medians) are far apart.

b This data is paired because the two measurements are taken for each of 15 test conditions. Therefore, we have to work with the differences of the two samples. A quantile of the 15 differences shows that the data follows (approximately) a straight line, indicating that it is reasonable to assume that the differences follow a normal distribution. Taking differences in the order "Normal" – "High" , we find $\bar{d} = -42.23$, and $s_d = 4.34$. With $t_{.025,14} = 2.145$, a 95% confidence interval for the difference between the population means is

$-42.23 \pm (2.145)\left(\frac{4.34}{\sqrt{15}}\right) = -42.23 \pm 2.404 = (-44.63, -39.83)$. Because 0 is not contained in this interval, we can conclude that the difference between the population means is not 0; i.e., we conclude that the two population means are not equal.

39.

a A normal probability plot shows that the data could easily follow a normal distribution.

b We test $H_0 : \mu_d = 0$ vs. $H_a : \mu_d \neq 0$, with test statistic

$t = \frac{\bar{d} - 0}{s_D / \sqrt{n}} = \frac{167.2 - 0}{228 / \sqrt{14}} = 2.74 \approx 2.7$. The two-tailed p-value is 2[P(t > 2.7)] = 2[.009] = .018. Since .018 < .05, we can reject H_o. There is strong evidence to support the claim that the true average difference between intake values measured by the two methods is not 0. There is a difference between them.

40.

a H_o will be rejected in favor of H_a if either $t \geq t_{.005,15} = 2.947$ or $t \leq -2.947$. The summary quantities are $\bar{d} = -.544$, and $s_d = .714$, so $t = \frac{-.544}{.1785} = -3.05$. Because $-3.05 \leq -2.947$, H_o is rejected in favor of H_a.

b $s_p^2 = 7.31$, $s_p = 2.70$, and $t = \frac{-.544}{.96} = -.57$, which is clearly insignificant; the incorrect analysis yields an inappropriate conclusion.

41. We test $H_0 : \mu_d = 0$ vs. $H_a : \mu_d > 0$. With $\bar{d} = 7.600$, and $s_d = 4.178$, $t = \frac{7.600 - 5}{4.178/\sqrt{9}} = \frac{2.6}{1.39} = 1.87 \approx 1.9$. With degrees of freedom n – 1 = 8, the corresponding p-value is P(t > 1.9) = .047. We would reject H_o at any alpha level greater than .047. So, at the typical significance level of .05, we would (barely) reject H_o, and conclude that the data indicates that the higher level of illumination yields a decrease of more than 5 seconds in true average task completion time.

42.

1. Parameter of interest: μ_d denotes the true average difference of spatial ability in brothers exposed to DES and brothers not exposed to DES. Let $\mu_d = \mu_{exposed} - \mu_{unexposed.}$
2. $H_0 : \mu_D = 0$
3. $H_a : \mu_D < 0$
4. $t = \frac{\bar{d} - \mu_D}{s_D / \sqrt{n}} = \frac{\bar{d} - 0}{s_D / \sqrt{n}}$
5. RR: P-value < .05, df = 8
6. $t = \frac{(12.6 - 13.7) - 0}{0.5} = -2.2$, with corresponding p-value .029 (from Table A.8)
7. Reject H_o. The data supports the idea that exposure to DES reduces spatial ability.

43.

a Although there is a "jump" in the middle of the Normal Probability plot, the data follow a reasonably straight path, so there is no strong reason for doubting the normality of the population of differences.

b A 95% lower confidence bound for the population mean difference is:

$$\bar{d} - t_{.05,14}\left(\frac{s_d}{\sqrt{n}}\right) = -38.60 - (1.761)\left(\frac{23.18}{\sqrt{15}}\right) = -38.60 - 10.54 = -49.14.$$

Therefore, with a confidence level of 95%, the population mean difference is above (–49.14).

c A 95% upper confidence bound for the corresponding population mean difference is $38.60 + 10.54 = 49.14$

44. Since the experimental design involved growing each of two types of wheat in different locations, the locations serve as the blocking variable and the appropriate analysis is a paired t test. The analysis requires we compute the difference between Sundance winter and Manitou spring wheat yields for each of the nine plot locations.

a In order to proceed, we must check the assumption that the 9 sample differences have been randomly sampled from a population that is normally distributed. A normal probability plot of the 9 differences produced by Minitab suggests that the normality assumption is reasonable, since the pattern in the plot is reasonably linear.

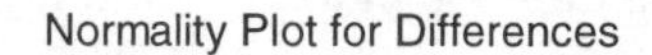

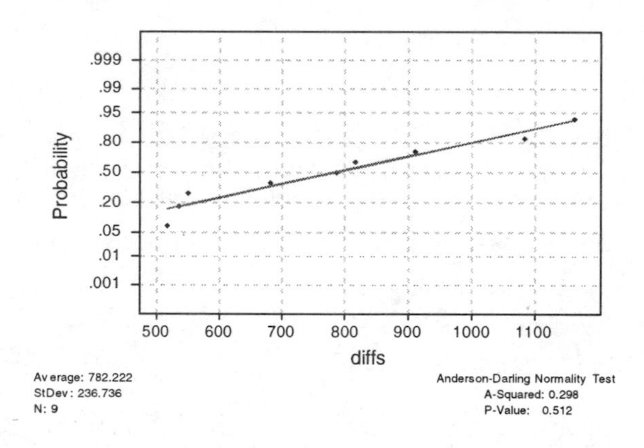

b Let μ_d denote the true average difference in yield for the winter wheat versus the spring wheat. Since we are interested in determining if the average yield for winter wheat is more than 500 kg/ha higher than for spring wheat, the relevant hypotheses are $H_0 : \mu_d = 500$ vs. $H_a : \mu_d > 500$. The descriptive statistics are n = 9, $\bar{d} = 782.2$, and $s_d = 236.7$, The corresponding test statistic is $t = \dfrac{782.2 - 500}{236.7/\sqrt{9}} = 3.58 \approx 3.6$. With degrees of freedom n – 1 = 8, the associated p-value is P(t > 3.6) = .004. Since this p-value is less than most choices for alpha, we reject H_o. We have sufficient evidence to claim that the true average yield for the winter wheat is more than 500 kg/ha higher than for the spring wheat.

45. The differences (white – black) are –7.62, -8.00, -9.09, -6.06, -1.39, -16.07, -8.40, -8.89, and –2.88, from which $\overline{d} = -7.600$, and $s_d = 4.178$. The confidence level is not specified in the problem description; for 95% confidence, $t_{.025,8} = 2.306$, and the C.I. is

$$-7.600 \pm (2.306)\left(\frac{4.178}{\sqrt{9}}\right) = -7.600 \pm 3.211 = (-10.811, -4.389).$$

46. With $(x_1, y_1) = (6,5)$, $(x_2, y_2) = (15,14)$, $(x_3, y_3) = (1,0)$, and $(x_4, y_4) = (21,20)$, $\overline{d} = 1$ and $s_d = 0$ (the d_i's are 1, 1, 1, and 1), while $s_1 = s_2 = 8.96$, so $s_p = 8.96$ and t = .16.

Section 9.4

47. H_o will be rejected if $z \le -z_{.01} = -2.33$. With $\hat{p}_1 = .150$, and $\hat{p}_2 = .300$,

$\hat{p} = \frac{30+80}{200+600} = \frac{210}{800} = .263$, and $\hat{q} = .737$. The calculated test statistic is

$$z = \frac{.150 - .300}{\sqrt{(.263)(.737)\left(\frac{1}{200} + \frac{1}{600}\right)}} = \frac{-.150}{.0359} = -4.18.$$ Because $-4.18 \le -2.33$, H_o is rejected; the proportion of those who repeat after inducement appears lower than those who repeat after no inducement.

48.

a H_o will be rejected if $z \le -1.645$. With $\hat{p}_1 = \frac{63}{300} = .2100$, and $\hat{p}_2 = \frac{75}{180} = .4167$,

$$\hat{p} = \frac{63+75}{300+180} = .2875, \quad z = \frac{.2100 - .4167}{\sqrt{(.2875)(.7125)\left(\frac{1}{300} + \frac{1}{180}\right)}} = \frac{-.2067}{.0427} = -4.84.$$

Since $-4.84 \le -1.645$, H_o is rejected.

b $\overline{p} = .275$ and $\sigma = .0432$, so power = $\Phi\left(\frac{[(-1.645)(.0421) + .2]}{.0432}\right) = \Phi(3.03)$ $= .9988$.

49.

1 Parameter of interest: $p_1 - p_2$ = true difference in proportions of those responding to two different survey covers. Let p_1 = Plain, p_2 = Picture.

2 $H_0 : p_1 - p_2 = 0$

3 $H_a : p_1 - p_2 < 0$

4 $z = \dfrac{\hat{p}_1 - \hat{p}_2}{\sqrt{\hat{p}\hat{q}\left(\frac{1}{m} + \frac{1}{n}\right)}}$

5 Reject H_o if p-value < .10

6 $z = \dfrac{\frac{104}{207} - \frac{109}{213}}{\sqrt{\left(\frac{213}{420}\right)\left(\frac{207}{420}\right)\left(\frac{1}{207} + \frac{1}{213}\right)}} = -.1910$; p-value = .4247

7 Fail to Reject H_o. The data does not indicate that plain cover surveys have a lower response rate.

50. Let $\alpha = .05$. A 95% confidence interval is $(\hat{p}_1 - \hat{p}_2) \pm z_{\alpha/2}\sqrt{\left(\frac{\hat{p}_1\hat{q}_1}{m} + \frac{\hat{p}_2\hat{q}_2}{n}\right)}$

$$= \left(\tfrac{224}{395} - \tfrac{126}{266}\right) \pm 1.96\sqrt{\left(\frac{\left(\frac{224}{395}\right)\left(\frac{171}{395}\right)}{395} + \frac{\left(\frac{126}{266}\right)\left(\frac{140}{266}\right)}{266}\right)} = .0934 \pm .0774 = (.0160, .1708).$$

51.

a $H_0 : p_1 = p_2$ will be rejected in favor of $H_a : p_1 \neq p_2$ if either $z \geq 1.645$ or $z \leq -1.645$. With $\hat{p}_1 = .193$, and $\hat{p}_2 = .182$, $\hat{p} = .188$, $z = \dfrac{.011}{.00742} = 1.48$.

Since 1.48 is not ≥ 1.645, H_o is not rejected and we conclude that no difference exists.

b Using formula (9.7) with $p_1 = .2$, $p_2 = .18$, $\alpha = .1$, $\beta = .1$, and $z_{\alpha/2} = 1.645$,

$$n = \frac{\left(1.645\sqrt{.5(.38)(1.62)} + 1.28\sqrt{.16 + .1476}\right)^2}{.0004} = 6582$$

52. Let p_1 = true proportion of irradiated bulbs that are marketable; p_2 = true proportion of untreated bulbs that are marketable; The hypotheses are $H_0 : p_1 - p_2 = 0$ vs. $H_0 : p_1 - p_2 > 0$. The test statistic is $z = \dfrac{\hat{p}_1 - \hat{p}_2}{\sqrt{\hat{p}\hat{q}\left(\frac{1}{m} + \frac{1}{n}\right)}}$. With $\hat{p}_1 = \dfrac{153}{180} = .850$, and $\hat{p}_2 = \dfrac{119}{180} = .661$, $\hat{p} = \dfrac{272}{360} = .756$, $z = \dfrac{.850 - .661}{\sqrt{(.756)(.244)\left(\frac{1}{180} + \frac{1}{180}\right)}} = \dfrac{.189}{.045} = 4.2$. The p-value = $1 - \Phi(4.2) \approx 0$, so reject H_o at any reasonable level. Radiation appears to be beneficial.

53.

a A 95% large sample confidence interval formula for $\ln(\theta)$ is

$\ln(\hat{\theta}) \pm z_{\alpha/2}\sqrt{\frac{m-x}{mx}+\frac{n-y}{ny}}$. Taking the antilogs of the upper and lower bounds gives the confidence interval for θ itself.

b $\hat{\theta} = \frac{\frac{189}{11,034}}{\frac{104}{11,037}} = 1.818$, $\ln(\hat{\theta}) = .598$, and the standard deviation is

$\sqrt{\frac{10,845}{(11,034)(189)}+\frac{10,933}{(11,037)(104)}} = .1213$, so the CI for $\ln(\theta)$ is

$.598 \pm 1.96(.1213) = (.360,.836)$. Then taking the antilogs of the two bounds gives the CI for θ to be $(1.43,2.31)$.

54.

a The "after" success probability is $p_1 + p_3$ while the "before" probability is $p_1 + p_2$, so $p_1 + p_3 > p_1 + p_2$ becomes $p_3 > p_2$; thus we wish to test $H_0 : p_3 = p_2$ versus $H_a : p_3 > p_2$.

b The estimator of $(p_1 + p_3) - (p_1 + p_2)$ is $\frac{(X_1 + X_3) - (X_1 + X_2)}{n} = \frac{X_3 - X_2}{n}$.

c When H_o is true, $p_2 = p_3$, so $Var\left(\frac{X_3 - X_2}{n}\right) = \frac{p_2 + p_3}{n}$, which is estimated by

$\frac{\hat{p}_2 + \hat{p}_3}{n}$. The Z statistic is then $\frac{\frac{X_3 - X_2}{n}}{\sqrt{\frac{\hat{p}_2 + \hat{p}_3}{n}}} = \frac{X_3 - X_2}{\sqrt{X_2 + X_3}}$.

d The computed value of Z is $\frac{200-150}{\sqrt{200+150}} = 2.68$, so $P = 1 - \Phi(2.68) = .0037$. At level .01, H_o can be rejected but at level .001 H_o would not be rejected.

55. $\hat{p}_1 = \frac{15+7}{40} = .550$, $\hat{p}_2 = \frac{29}{42} = .690$, and the 95% C.I. is

$(.550 - .690) \pm 1.96(.106) = -.14 \pm .21 = (-.35,.07)$.

56. Using $p_1 = q_1 = p_2 = q_2 = .5$, $L = 2(1.96)\sqrt{\left(\frac{.25}{n}+\frac{.25}{n}\right)} = \frac{2.7719}{\sqrt{n}}$, so L = .1 requires n = 769.

Section 9.5

57.

a From Table A.9, column 5, row 8, $F_{.01,5,8} = 3.69$.

b From column 8, row 5, $F_{.01,8,5} = 4.82$.

c $F_{.95,5,8} = \frac{1}{F_{.05,8,5}} = .207$.

d $F_{.95,8,5} = \frac{1}{F_{.05,5,8}} = .271$

e $F_{.01,10,12} = 4.30$

f $F_{.99,10,12} = \frac{1}{F_{.01,12,10}} = \frac{1}{4.71} = .212$.

g $F_{.05,6,4} = 6.16$, so $P(F \le 6.16) = .95$.

h Since $F_{.99,10,5} = \frac{1}{5.64} = .177$, $P(.177 \le F \le 4.74) = P(F \le 4.74) - P(F \le .177)$ $= .95 - .01 = .94$.

58.

a Since the given f value of 4.75 falls between $F_{.05,5,10} = 3.33$ and $F_{.01,5,10} = 5.64$, we can say that the upper-tailed p-value is between .01 and .05.

b Since the given f of 2.00 is less than $F_{.10,5,10} = 2.52$, the p-value > .10.

c The two tailed p-value = $2P(F \ge 5.64) = 2(.01) = .02$.

d For a lower tailed test, we must first use formula 9.9 to find the critical values:
$F_{.90,5,10} = \frac{1}{F_{.10,10,5}} = .3030$, $F_{.95,5,10} = \frac{1}{F_{.05,10,5}} = .2110$, $F_{.99,5,10} = \frac{1}{F_{.01,10,5}} = .0995$.
Since $.0995 < f = .200 < .2110$, $.01 <$ p-value $< .05$ (but obviously closer to .05).

e There is no column for numerator d.f. of 35 in Table A.9, however looking at both df = 30 and df = 40 columns, we see that for denominator df = 20, our f value is between $F_{.01}$ and $F_{.001}$. So we can say $.001 <$ p-value $< .01$.

59. $H_0 : \sigma_1 = \sigma_2$ will be rejected in favor of $H_a : \sigma_1 \neq \sigma_2$ if either $f \geq F_{.05,19,19} \approx 2.18$ or if $f \leq \frac{1}{2.18} = .459$. Since $f = \frac{(40.5)^2}{(32.1)^2} = 1.59$, which is neither ≥ 2.18 nor $\leq .459$, H_o is not rejected. The data does not suggest a difference in the two variances.

60. With σ_1 = true standard deviation for not-fused specimens and σ_2 = true standard deviation for fused specimens, we test $H_0 : \sigma_1 = \sigma_2$ vs. $H_a : \sigma_1 > \sigma_2$. The calculated test statistic is $f = \frac{(277.3)^2}{(205.9)^2} = 1.814$. With numerator d.f. = m – 1 = 10 – 1 = 9, and denominator d.f. = n – 1 = 8 – 1 = 7, $f = 1.814 < 2.72 = F_{.10,9,7}$. We can say that the p-value > .10, which is obviously > .01, so we cannot reject H_o. There is not sufficient evidence that the standard deviation of the strength distribution for fused specimens is smaller than that of not-fused specimens.

61. Let σ_1^2 = variance in weight gain for low-dose treatment, and σ_2^2 = variance in weight gain for control condition. We wish to test $H_0 : \sigma_1^2 = \sigma_2^2$ vs. $H_a : \sigma_1^2 > \sigma_2^2$. The test statistic is $f = \frac{s_1^2}{s_2^2}$, and we reject H_o at level .05 if $f > F_{.05,19,22} \approx 2.08$.

$f = \frac{(54)^2}{(32)^2} = 2.85 \geq 20.8$, so reject H_o at level .05. The data does suggest that there is more variability in the low-dose weight gains.

62. $H_0 : \sigma_1 = \sigma_2$ will be rejected in favor of $H_a : \sigma_1 \neq \sigma_2$ if either $f \leq F_{.975,47,44} \approx .56$ or if $f \geq F_{.025,47,44} \approx 1.8$. Because $f = 1.22$, H_o is not rejected. The data does not suggest a difference in the two variances.

63. $P\left(F_{1-\alpha/2,m-1,n-1} \le \dfrac{S_1^2/\sigma_1^2}{S_2^2/\sigma_2^2} \le F_{\alpha/2,m-1,n-1}\right) = 1-\alpha$. The set of inequalities inside the parentheses is clearly equivalent to $\dfrac{S_2^2 F_{1-\alpha/2,m-1,n-1}}{S_1^2} \le \dfrac{\sigma_2^2}{\sigma_1^2} \le \dfrac{S_2^2 F_{\alpha/2,m-1,n-1}}{S_1^2}$.

Substituting the sample values s_1^2 and s_2^2 yields the confidence interval for $\dfrac{\sigma_2^2}{\sigma_1^2}$, and taking the square root of each endpoint yields the confidence interval for $\dfrac{\sigma_2}{\sigma_1}$. m = n = 4, so we need $F_{.05,3,3} = 9.28$ and $F_{.95,3,3} = \dfrac{1}{9.28} = .108$. Then with s_1 = .160 and s_2 = .074, the C. I. for $\dfrac{\sigma_2^2}{\sigma_1^2}$ is (.023, 1.99), and for $\dfrac{\sigma_2}{\sigma_1}$ is (.15, 1.41).

64. A 95% upper bound for $\dfrac{\sigma_2}{\sigma_1}$ is $\sqrt{\dfrac{s_2^2 F_{.05,9,9}}{s_1^2}} = \sqrt{\dfrac{(3.59)^2(3.18)}{(.79)^2}} = 8.10$. We are confident that the ratio of the standard deviation of triacetate porosity distribution to that of the cotton porosity distribution is at most 8.10.

Supplementary Exercises

65. We test $H_0: \mu_1 - \mu_2 = 0$ vs. $H_a: \mu_1 - \mu_2 \ne 0$. The test statistic is

$$t = \frac{(\bar{x}-\bar{y})-(\Delta)}{\sqrt{\frac{s_1^2}{m}+\frac{s_2^2}{n}}} = \frac{807-757}{\sqrt{\frac{27^2}{10}+\frac{41^2}{10}}} = \frac{50}{\sqrt{241}} = \frac{50}{15.524} = 3.22$$

. The approximate d.f. is

$$\nu = \frac{(241)^2}{\frac{(72.9)^2}{9}+\frac{(168.1)^2}{9}} = 15.6$$

, which we round down to 15. The p-value for a two-tailed test is approximately 2P(t > 3.22) = 2(.003) = .006. This small of a p-value gives strong support for the alternative hypothesis. The data indicates a significant difference.

66.

a

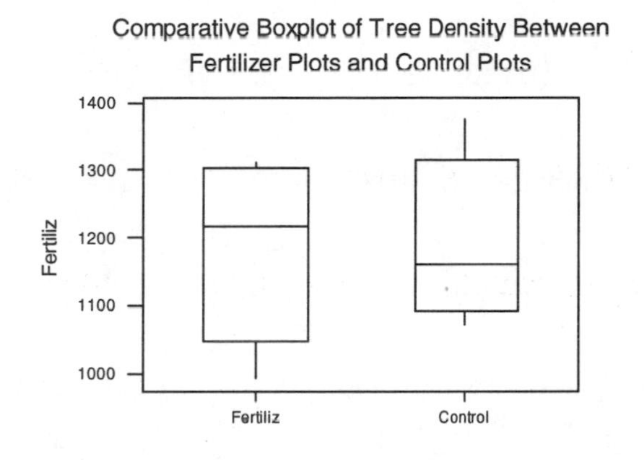

Although the median of the fertilizer plot is higher than that of the control plots, the fertilizer plot data appears negatively skewed, while the opposite is true for the control plot data.

b A test of $H_0: \mu_1 - \mu_2 = 0$ vs. $H_a: \mu_1 - \mu_2 \neq 0$ yields a t value of -.20, and a two-tailed p-value of .85. (d.f. = 13). We would fail to reject H_o; the data does not indicate a significant difference in the means.

c With 95% confidence we can say that the true average difference between the tree density of the fertilizer plots and that of the control plots is somewhere between –144 and 120. Since this interval contains 0, 0 is a plausible value for the difference, which further supports the conclusion based on the p-value.

67. Let p_1 = true proportion of returned questionnaires that included no incentive; p_2 = true proportion of returned questionnaires that included an incentive. The hypotheses are $H_0: p_1 - p_2 = 0$ vs. $H_0: p_1 - p_2 < 0$. The test statistic is $z = \dfrac{\hat{p}_1 - \hat{p}_2}{\sqrt{\hat{p}\hat{q}\left(\frac{1}{m} + \frac{1}{n}\right)}}$.

$\hat{p}_1 = \dfrac{75}{110} = .682$, and $\hat{p}_2 = \dfrac{66}{98} = .673$. At this point we notice that since $\hat{p}_1 > \hat{p}_2$, the numerator of the z statistic will be > 0, and since we have a lower tailed test, the p-value will be > .5. We fail to reject H_o. This data does not suggest that including an incentive increases the likelihood of a response.

68. Summary quantities are m = 24, $\bar{x} = 103.66$, s_1 = 3.74, n = 11, $\bar{y} = 101.11$, s_2 = 3.60. We use the pooled t interval based on 24 + 11 – 2 = 33 d.f.; 95% confidence requires $t_{.025,33} = 2.03$. With $s_p^2 = 13.68$ and $s_p = 3.70$, the confidence interval is

$2.55 \pm (2.03)(3.70)\sqrt{\frac{1}{24} + \frac{1}{11}} = 2.55 \pm 2.73 = (-.18, 5.28)$. We are confident that the difference between true average dry densities for the two sampling methods is between -.18 and 5.28. Because the interval contains 0, we cannot say that there is a significant difference between them.

69. The center of any confidence interval for $\mu_1 - \mu_2$ is always $\bar{x}_1 - \bar{x}_2$, so $\bar{x}_1 - \bar{x}_2 = \frac{-473.3 + 1691.9}{2} = 609.3$. Furthermore, half of the width of this interval is $\frac{1691.9 - (-473.3)}{2} = 1082.6$. Equating this value to the expression on the right of the 95% confidence interval formula, $1082.6 = (1.96)\sqrt{\frac{s_1^2}{n_1} + \frac{s_2^2}{n_2}}$, we find $\sqrt{\frac{s_1^2}{n_1} + \frac{s_2^2}{n_2}} = \frac{1082.6}{1.96} = 552.35$. For a 90% interval, the associated z value is 1.645, so the 90% confidence interval is then $609.3 \pm (1.645)(552.35) = 609.3 \pm 908.6$ $= (-299.3, 1517.9)$.

70.

a A 95% lower confidence bound for the true average strength of joints with a side coating is $\bar{x} - t_{.025,9}\left(\frac{s}{\sqrt{n}}\right) = 63.23 - (1.833)\left(\frac{5.96}{\sqrt{10}}\right) = 63.23 - 3.45 = 59.78$. That is, with a confidence level of 95%, the average strength of joints with a side coating is at least 59.78 (Note: this bound is valid only if the distribution of joint strength is normal.)

b A 95% lower prediction bound for the strength of a single joint with a side coating is $\bar{x} - t_{.025,9}\left(s\sqrt{1 + \frac{1}{n}}\right) = 63.23 - (1.833)\left(5.96\sqrt{1 + \frac{1}{10}}\right) = 63.23 - 11.46 = 51.77$. That is, with a confidence level of 95%, the strength of a single joint with a side coating would be at least 51.77.

c For a confidence level of 95%, a two-sided tolerance interval for capturing at least 95% of the strength values of joints with side coating is $\bar{x} \pm$ (tolerance critical value)s. The tolerance critical value is obtained from Table A.6 with 95% confidence, k = 95%, and n = 10. Thus, the interval is $63.23 \pm (3.379)(5.96) = 63.23 \pm 20.14 = (43.09, 83.37)$. That is, we can be highly confident that at least 95% of all joints with side coatings have strength values between 43.09 and 83.37.

d A 95% confidence interval for the difference between the true average strengths for the two types of joints is $(80.95-63.23)\pm t_{.025,v}\sqrt{\frac{(9.59)^2}{10}+\frac{(5.96)^2}{10}}$. The approximate degrees of freedom is $v=\frac{\left(\frac{91.9681}{10}+\frac{35.5216}{10}\right)^2}{\frac{\left(\frac{91.9681}{10}\right)^2}{9}+\frac{\left(\frac{35.5216}{10}\right)^2}{9}}=15.05$, so we use 15 d.f., and $t_{.025,15}=2.131$. The interval is , then,
$17.72\pm(2.131)(3.57)=17.72\pm7.61=(10.11,25.33)$. With 95% confidence, we can say that the true average strength for joints without side coating exceeds that of joints with side coating by between 10.11 and 25.33 lb-in./in.

71. m = n = 40, $\bar{x}=3975.0$, s_1 = 245.1, $\bar{y}=2795.0$, s_2 = 293.7. The large sample 99% confidence interval for $\mu_1-\mu_2$ is $(3975.0-2795.0)\pm2.58\sqrt{\frac{245.1^2}{40}+\frac{293.7^2}{40}}$ $(1180.0)\pm1560.5\approx(1024,1336)$. The value 0 is not contained in this interval so we can state that, with very high confidence, the value of $\mu_1-\mu_2$ is not 0, which is equivalent to concluding that the population means are not equal.

72. This exercise calls for a paired analysis. First compute the difference between the amount of cone penetration for commutator and pinion bearings for each of the 17 motors. These 17 differences are summarized as follows: n = 17, $\bar{d}=-4.18$, $s_d=35.85$, where d = (commutator value – pinion value). Then $t_{.025,16}=2.120$, and the 95% confidence interval for the population mean difference between penetration for the commutator armature bearing and penetration for the pinion bearing is:
$-4.18\pm(2.120)\left(\frac{35.85}{\sqrt{17}}\right)=-4.18\pm18.43=(-22.61,14.25)$. We would have to say that the population mean difference has not been precisely estimated. The bound on the error of estimation is quite large. In addition, the confidence interval spans zero. Because of this, we have insufficient evidence to claim that the population mean penetration differs for the two types of bearings.

73. Since we can assume that the distributions from which the samples were taken are normal, we use the two-sample t test. Let μ_1 denote the true mean headability rating for aluminum killed steel specimens and μ_2 denote the true mean headability rating for silicon killed steel. Then the hypotheses are $H_0 : \mu_1 - \mu_2 = 0$ vs. $H_a : \mu_1 - \mu_2 \neq 0$.

The test statistic is $t = \frac{-.66}{\sqrt{.03888 + .047203}} = \frac{-.66}{\sqrt{.086083}} = -2.25$. The approximate degrees of freedom $\nu = \frac{(.086083)^2}{\frac{(.03888)^2}{29} + \frac{(.047203)^2}{29}} = 57.5$, so we use 57. The two-tailed p-value $\approx 2(.014) = .028$, which is less than the specified significance level, so we would reject H_o. The data supports the article's authors' claim.

74. Let μ_1 denote the true average tear length for Brand A and let μ_2 denote the true average tear length for Brand B. The relevant hypotheses are $H_0 : \mu_1 - \mu_2 = 0$ vs. $H_a : \mu_1 - \mu_2 > 0$. Assuming both populations have normal distributions, the two-sample t test is appropriate. m = 16, $\bar{x} = 74.0$, s_1 = 14.8, n = 14, $\bar{y} = 61.0$, s_2 = 12.5,

so the approximate d.f. is $\nu = \frac{\left(\frac{14.8^2}{16} + \frac{12.5^2}{14}\right)^2}{\frac{\left(\frac{14.8^2}{16}\right)^2}{15} + \frac{\left(\frac{12.5^2}{14}\right)^2}{13}} = 27.97$, which we round down to 27.

The test statistic is $t = \frac{74.0 - 61.0}{\sqrt{\frac{14.8^2}{16} + \frac{12.5^2}{14}}} \approx 2.6$. From Table A.7, the p-value = P(t > 2.6) = .007. At a significance level of .05, H_o is rejected and we conclude that the average tear length for Brand A is larger than that of Brand B.

75.

a The relevant hypotheses are $H_0 : \mu_1 - \mu_2 = 0$ vs. $H_a : \mu_1 - \mu_2 \neq 0$. Assuming both populations have normal distributions, the two-sample t test is appropriate. m = 11, $\bar{x} = 98.1$, s_1 = 14.2, n = 15, $\bar{y} = 129.2$, s_2 = 39.1. The test statistic is

$t = \frac{-31.1}{\sqrt{18.3309 + 101.9207}} = \frac{-31.1}{\sqrt{120.252}} = -2.84$. The approximate degrees of freedom $\nu = \frac{(120.252)^2}{\frac{(18.3309)^2}{10} + \frac{(101.9207)^2}{14}} = 18.64$, so we use 18. From Table A.7, the two-tailed p-value $\approx 2(.006) = .012$. No, obviously, the results are different.

b For the hypotheses $H_0: \mu_1 - \mu_2 = -25$ vs. $H_a: \mu_1 - \mu_2 < -25$, the test statistic changes to $t = \frac{-31.1-(-25)}{\sqrt{120.252}} = -.556$. With degrees of freedom 18, the p-value $\approx P(t < -.6) = .278$. Since the p-value is greater than any sensible choice of α, we fail to reject H_o. There is insufficient evidence that the true average strength for males exceeds that for females by more than 25N.

76.

a The relevant hypotheses are $H_0: \mu_1^* - \mu_2^* = 0$ (which is equivalent to saying $\mu_1 - \mu_2 = 0$) versus $H_a: \mu_1^* - \mu_2^* \neq 0$ (which is the same as saying $\mu_1 - \mu_2 \neq 0$). The pooled t test is based on d.f. = m + n – 2 = 8 + 9 – 2 = 15. The pooled variance is $s_p^2 = \left(\frac{m-1}{m+n-2}\right)s_1^2 + \left(\frac{n-1}{m+n-2}\right)s_2^2$

$\left(\frac{8-1}{8+9-2}\right)(4.9)^2 + \left(\frac{9-1}{8+9-2}\right)(4.6)^2 = 22.49$, so $s_p = 4.742$. The test statistic is $t = \frac{\bar{x}^* - \bar{y}^*}{s_p\sqrt{\frac{1}{m}+\frac{1}{n}}} = \frac{18.0-11.0}{4.742\sqrt{\frac{1}{8}+\frac{1}{9}}} = 3.04 \approx 3.0$. From Table A.7, the p-value associated with t = 3.0 is 2P(t > 3.0) = 2(.004) = .008. At significance level .05, H_o is rejected and we conclude that there is a difference between μ_1^* and μ_2^*, which is equivalent to saying that there is a difference between μ_1 and μ_2.

b No. The mean of a lognormal distribution is $\mu = e^{\mu^* + (\sigma^*)^2/2}$, where μ^* and σ^* are the parameters of the lognormal distribution (i.e., the mean and standard deviation of ln(x)). So when $\sigma_1^* = \sigma_2^*$, then $\mu_1^* = \mu_2^*$ would imply that $\mu_1 = \mu_2$. However, when $\sigma_1^* \neq \sigma_2^*$, then even if $\mu_1^* = \mu_2^*$, the two means μ_1 and μ_2 (given by the formula above) would not be equal.

77. This is paired data, so the paired t test is employed. The relevant hypotheses are $H_0: \mu_d = 0$ vs. $H_a: \mu_d < 0$, where μ_d denotes the difference between the population average control strength minus the population average heated strength. The observed differences (control – heated) are: -.06, .01, -.02, 0, and -.05. The sample mean and standard deviation of the differences are $\bar{d} = -.024$ and $s_d = .0305$. The test statistic is $t = \frac{-.024}{.0305/\sqrt{5}} = -1.76 \approx -1.8$. From Table A.7, with d.f. = 5 – 1 = 4, the lower tailed p-value associated with t = -1.8 is P(t < -1.8) = P(t > 1.8) = .073. At significance level .05, H_o should not be rejected. Therefore, this data does not show that the heated average strength exceeds the average strength for the control population.

78. Let μ_1 denote the true average ratio for young men and μ_2 denote the true average ratio for elderly men. The relevant hypotheses are $H_0 : \mu_1 - \mu_2 = 0$ vs. $H_a : \mu_1 - \mu_2 > 0$. The value of the test statistic is $t = \frac{(7.47 - 6.71)}{\sqrt{\frac{(.22)^2}{13} + \frac{(.28)^2}{12}}} = 7.5$. The d.f. = 20 and the p-value is P(t > 7.5) ≈ 0. Since the p-value is $< \alpha = .05$, we reject H$_0$. We have sufficient evidence to claim that the true average ratio for young men exceeds that for elderly men.

79.

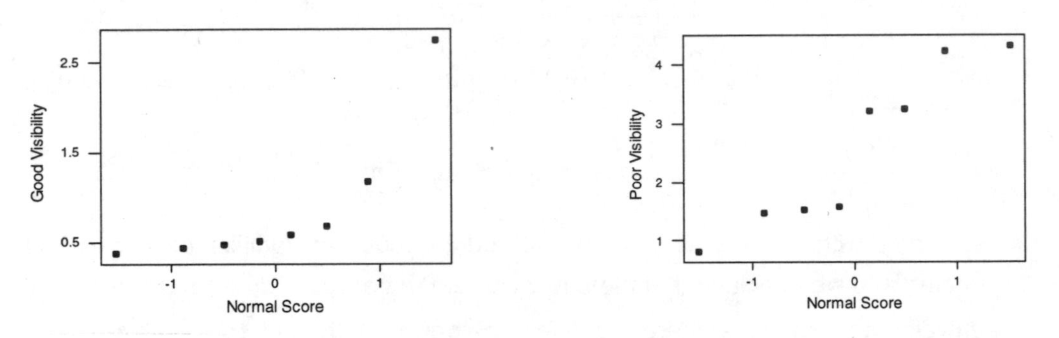

A normal probability plot indicates the data for good visibility does not follow a normal distribution, thus a t-test is not appropriate for this small a sample size.

80. The relevant hypotheses would be $\mu_M = \mu_F$ versus $\mu_M \neq \mu_F$ for both the distress and delight indices. The reported p-value for the test of mean differences on the distress index was less than 0.001. This indicates a statistically significant difference in the mean scores, with the mean score for women being higher. The reported p-value for the test of mean differences on the delight index was > 0.05. This indicates a lack of statistical significance in the difference of delight index scores for men and women.

81.

a With n denoting the second sample size, the first is m = 3n. We then wish $20 = 2(2.58)\sqrt{\frac{900}{3n} + \frac{400}{n}}$, which yields n = 47, m = 141.

b We wish to find the n which minimizes $2(z_{\alpha/2})\sqrt{\frac{900}{400-n}+\frac{400}{n}}$, or equivalently, the n which minimizes $\frac{900}{400-n}+\frac{400}{n}$. Taking the derivative with respect to n and equating to 0 yields $900(400-n)^{-2}-400n^{-2}=0$, whence $9n^2=4(400-n)^2$, or $5n^2+3200n-640{,}000=0$. This yields n = 160, m = 400 – n = 240.

82. Let p_1 = true survival rate at $11^\circ C$; p_2 = true survival rate at $30^\circ C$; The hypotheses are $H_0: p_1 - p_2 = 0$ vs. $H_a: p_1 - p_2 \neq 0$. The test statistic is $z=\frac{\hat{p}_1-\hat{p}_2}{\sqrt{\hat{p}\hat{q}\left(\frac{1}{m}+\frac{1}{n}\right)}}$. With

$\hat{p}_1=\frac{73}{91}=.802$, and $\hat{p}_2=\frac{102}{110}=.927$, $\hat{p}=\frac{175}{201}=.871$, $\hat{q}=.129$.

$z=\frac{.802-.927}{\sqrt{(.871)(.129)\left(\frac{1}{91}+\frac{1}{110}\right)}}=\frac{-.125}{.0320}=-3.91$. The p-value = $\Phi(-3.91)<\Phi(-3.49)=.0003$, so reject H_o at any reasonable level. The two survival rates appear to differ.

83.

a We test $H_0:\mu_1-\mu_2=0$ vs. $H_a:\mu_1-\mu_2\neq 0$. Assuming both populations have normal distributions, the two-sample t test is appropriate. The approximate degrees of freedom $\nu=\frac{(.042721)^2}{\frac{(.0325125)^2}{7}+\frac{(.0102083)^2}{11}}=11.4$, so we use df = 11.

$t_{.0005,11}=4.437$, so we reject H_o if $t\geq 4.437$ or $t\leq -4.437$ The test statistic is $t=\frac{.68}{\sqrt{.042721}}\approx 3.3$, which is not ≥ 4.437, so we cannot reject H_o. At significance level .001, the data does not indicate a difference in true average insulin-binding capacity due to the dosage level.

b P-value = 2P(t > 3.3) = 2 (.004) = .008 which is > .001.

84. $\hat{\sigma}^2=\frac{\left[(n_1-1)S_1^2+(n_2-1)S_2^2+(n_3-1)S_3^2+(n_4-1)S_4^2\right]}{n_1+n_2+n_3+n_4-4}$

$E(\hat{\sigma}^2)=\frac{\left[(n_1-1)\sigma_1^2+(n_2-1)\sigma_2^2+(n_3-1)\sigma_3^2+(n_4-1)\sigma_4^2\right]}{n_1+n_2+n_3+n_4-4}=\sigma^2$. The estimate for the given data is $=\frac{[15(.4096)+17(.6561)+7(.2601)+11(.1225)]}{50}=.409$

85. $\Delta_0 = 0$, $\sigma_1 = \sigma_2 = 10$, d = 1, $\sigma = \sqrt{\frac{200}{n}} = \frac{14.142}{\sqrt{n}}$, so $\beta = \Phi\left(1.645 - \frac{\sqrt{n}}{14.142}\right)$, giving β = .9015, .8264, .0294, and .0000 for n = 25, 100, 2500, and 10,000 respectively. If the μ_i's referred to true average IQ's resulting from two different conditions, $\mu_1 - \mu_2 = 1$ would have little practical significance, yet very large sample sizes would yield statistical significance in this situation.

86. $H_0 : \mu_1 - \mu_2 = 0$ is tested against $H_a : \mu_1 - \mu_2 \neq 0$ using the two-sample t test, rejecting H_o at level .05 if either $t \geq t_{.025,15} = 2.131$ or if $t \leq -2.131$. With $\bar{x} = 11.20$, $s_1 = 2.68$, $\bar{y} = 9.79$, $s_2 = 3.21$, and m = n = 8, s_p = 2.96, and t = .95, so H_o is not rejected. In the situation described, the effect of carpeting would be mixed up with any effects due to the different types of hospitals, so no separate assessment could be made. The experiment should have been designed so that a separate assessment could be obtained (e.g., a randomized block design).

87. $H_0 : p_1 = p_2$ will be rejected at level α in favor of $H_a : p_1 \neq p_2$ if either $z \geq z_{.025} = 1.96$ or $z \leq -1.96$. With $\hat{p}_1 = .3438$, $\hat{p}_2 = .3125$, and $\hat{p} = .3267$, $z = \frac{.0313}{.0502} = .62$, so H_o cannot be rejected and there appears to be no difference.

88. The computed value of Z is $z = \frac{34 - 46}{\sqrt{34 + 46}} = -1.34$. A lower tailed test would be appropriate, so the p-value $= \Phi(-1.34) = .0901 > .05$, so we would not judge the drug to be effective.

89.

a Let μ_1 and μ_2 denote the true average weights for operations 1 and 2, respectively. The relevant hypotheses are $H_0 : \mu_1 - \mu_2 = 0$ vs. $H_a : \mu_1 - \mu_2 \neq 0$. The value of the test statistic is

$$t = \frac{(1402.24 - 1419.63)}{\sqrt{\frac{(10.97)^2}{30} + \frac{(9.96)^2}{30}}} = \frac{-17.39}{\sqrt{4.011363 + 3.30672}} = \frac{-17.39}{\sqrt{7.318083}} = -6.43.$$ The

d.f. $\nu = \frac{(7.318083)^2}{\frac{(4.011363)^2}{29} + \frac{(3.30672)^2}{29}} = 57.5$, so use df = 57. $t_{.025,57} \approx 2.000$, so we can reject H_o at level .05. The data indicates that there is a significant difference between the true mean weights of the packages for the two operations.

b $H_0 : \mu_1 = 1400$ will be tested against $H_a : \mu_1 > 1400$ using a one-sample t test with test statistic $t = \frac{\bar{x} - 1400}{s_1/\sqrt{m}}$. With degrees of freedom = 29, we reject H_o if $t > t_{.05,29} = 1.699$. The test statistic value is $t = \frac{1402.24 - 1400}{10.97/\sqrt{30}} = \frac{2.24}{2.00} = 1.1$.

Because 1.1 < 1.699, H_o is not rejected. True average weight does not appear to exceed 1400.

90. $Var(\bar{X} - \bar{Y}) = \frac{\lambda_1}{m} + \frac{\lambda_2}{n}$ and $\hat{\lambda}_1 = \bar{X}$, $\hat{\lambda}_2 = \bar{Y}$, $\hat{\lambda} = \frac{m\bar{X} + n\bar{Y}}{m+n}$, giving $Z = \frac{\bar{X} - \bar{Y}}{\sqrt{\frac{\hat{\lambda}}{m} + \frac{\hat{\lambda}}{n}}}$.

With $\bar{x} = 1.616$ and $\bar{y} = 2.557$, z = -5.3 and p-value = $2(\Phi(-5.3)) < .0006$, so we would certainly reject $H_0 : \lambda_1 = \lambda_2$ in favor of $H_a : \lambda_1 \neq \lambda_2$.

91. $\hat{\lambda}_1 = \bar{x} = 1.62$, $\hat{\lambda}_2 = \bar{y} = 2.56$, $\sqrt{\frac{\hat{\lambda}_1}{m} + \frac{\hat{\lambda}_2}{n}} = 1.77$, and the confidence interval is $-.94 \pm (1.96)(1.77) = -.94 \pm .35 = (-1.29, -.59)$.

Chapter 10

Section 10.1

1. H_o will be rejected if $f \geq F_{.05,4,15} = 3.06$ (since I – 1 = 4, and I (J – 1) = (5)(3) = 15).

   The computed value of F is $f = \dfrac{2573.3}{1394.2} = 1.85$. Since 1.85 is not ≥ 3.06, H_o is not rejected. The data does not indicate a difference in the mean tensile strengths of the different types of copper wires.

2.

Type of Box	$\overline{x}$	s
1	713.00	46.55
2	756.93	40.34
3	698.07	37.20
4	682.02	39.87

Grand mean = 712.51

$$MSTr = \frac{6}{4-1}\left[(713.00-712.51)^2 + (756.93-712.51)^2 + (698.07-712.51)^2 + (682.02-712.51)^2\right] = 6{,}223.0604$$

$$MSE = \frac{1}{4}\left[(46.55)^2 + (40.34)^2 + (37.20)^2 + (39.87)^2\right] = 1{,}691.9188$$

$$f = \frac{MSTr}{MSE} = \frac{6{,}223.0604}{1{,}691.9188} = 3.678$$

$F_{.05,3,20} = 3.10$

3.678 > 3.10, so reject H_o. There is a difference in compression strengths among the four box types.

Chapter 10

3. With μ_i = true average lumen output for brand i bulbs, we wish to test $H_0 : \mu_1 = \mu_2 = \mu_3$ versus H_a : at least two μ_i's are unequal.

$MSTr = \hat{\sigma}_B^2 = \frac{591.2}{2} = 295.60$, $MSE = \hat{\sigma}_W^2 = \frac{4773.3}{21} = 227.30$, so

$f = \frac{MSTr}{MSE} = \frac{295.60}{227.30} = 1.30$ For finding the p-value, we need degrees of freedom I – 1 = 2 and I (J – 1) = 21. In the 2nd row and 21st column of Table A.9, we see that $1.30 < F_{.10,2,21} = 2.57$, so the p-value > .10. Since .10 is not < .05 , we cannot reject H_o. There are no differences in the average lumen outputs among the three brands of bulbs.

4. $x_{\bullet\bullet} = IJ\bar{x}_{\bullet\bullet} = 32(5.19) = 166.08$, so $SST = 911.91 - \frac{(166.08)^2}{32} = 49.95$.

$SSTr = 8\left[(4.39 - 5.19)^2 + \ldots + (6.36 - 5.19)^2\right] = 20.38$, so

$SSE = 49.95 - 20.38 = 29.57$. Then $f = \frac{20.38/3}{29.57/28} = 6.43$. Since

$6.43 \geq F_{.05,2,28} = 2.95$, $H_0 : \mu_1 = \mu_2 = \mu_3 = \mu_4$ is rejected at level .05. There are differences between at least two average flight times for the four treatments.

5. μ_i = true mean modulus of elasticity for grade i (i = 1, 2, 3). We test $H_0 : \mu_1 = \mu_2 = \mu_3$ vs. H_a : at least two μ_i's are unequal. Reject H_o if $f \geq F_{.01,2,27} = 5.49$. The grand mean = 1.5367,

$$MSTr = \frac{10}{2}\left[(1.63 - 1.5367)^2 + (1.56 - 1.5367)^2 + (1.42 - 1.5367)^2\right] = .1143$$

$MSE = \frac{1}{3}\left[(.27)^2 + (.24)^2 + (.26)^2\right] = .0660$, $f = \frac{MSTr}{MSE} = \frac{.1143}{.0660} = 1.73$. Fail to reject H_o. The three grades do not appear to differ.

6.

Source	Df	SS	MS	F
Treatments	3	509.112	169.707	10.85
Error	36	563.134	15.643	
Total	39	1,072.256		

$F_{.01,3,36} \approx F_{.01,3,30} = 4.51$. The computed test statistic value of 10.85 exceeds 4.51, so reject H_o in favor of H_a: at least two of the four means differ.

7.

Source	Df	SS	MS	F
Treatments	3	75,081.72	25,027.24	1.70
Error	16	235,419.04	14,713.69	
Total	19	310,500.76		

The hypotheses are $H_0 : \mu_1 = \mu_2 = \mu_3 = \mu_4$ vs. H_a : at least two μ_i's are unequal. $1.70 < F_{.10,3,16} = 2.46$, so p-value > .10, and we fail to reject H_o.

8. The summary quantities are $x_{1\bullet} = 2332.5$, $x_{2\bullet} = 2576.4$, $x_{3\bullet} = 2625.9$, $x_{4\bullet} = 2851.5$, $x_{5\bullet} = 3060.2$, $x_{\bullet\bullet} = 13,446.5$, so CF = 5,165,953.21, SST = 75,467.58, SSTr = 43,992.55, SSE = 31,475.03, $MSTr = \frac{43,992.55}{4} = 10,998.14$,

$MSE = \frac{31,475.03}{30} = 1049.17$ and $f = \frac{10,998.14}{1049.17} = 10.48$. (These values should be displayed in an ANOVA table as requested.) Since $10.48 \geq F_{.01,4,30} = 4.02$, $H_0 : \mu_1 = \mu_2 = \mu_3 = \mu_4 = \mu_5$ is rejected. There are differences in the true average axle stiffness for the different plate lengths.

9. The summary quantities are $x_{1\bullet} = 34.3$, $x_{2\bullet} = 39.6$, $x_{3\bullet} = 33.0$, $x_{4\bullet} = 41.9$,

$x_{\bullet\bullet} = 148.8$, $\Sigma\Sigma x_{ij}^2 = 946.68$, so $CF = \frac{(148.8)^2}{24} = 922.56$,

$SST = 946.68 - 922.56 = 24.12$, $SSTr = \frac{(34.3)^2 + ... + (41.9)^2}{6} - 922.56 = 8.98$,

$SSE = 24.12 - 8.98 = 15.14$.

Source	Df	SS	MS	F
Treatments	3	8.98	2.99	3.95
Error	20	15.14	.757	
Total	23	24.12		

Since $3.10 = F_{.05,3,20} < 3.95 < 4.94 = F_{.01,3,20}$, $.01 < p-value < .05$ and H_o is rejected at level .05.

Chapter 10

10.

a $E(\overline{X}_{\bullet\bullet}) = \dfrac{\Sigma E(\overline{X}_{i\bullet})}{I} = \dfrac{\Sigma \mu_i}{I} = \mu$.

b $E(\overline{X}_{i\bullet}^2) = Var(\overline{X}_{i\bullet}) + [E(\overline{X}_{i\bullet})]^2 = \dfrac{\sigma^2}{J} + \mu_i^2$.

c $E(\overline{X}_{\bullet\bullet}^2) = Var(\overline{X}_{\bullet\bullet}) + [E(\overline{X}_{\bullet\bullet})]^2 = \dfrac{\sigma^2}{IJ} + \mu^2$.

d $E(SSTr) = E[J\Sigma\overline{X}_{i\bullet}^2 - IJ\overline{X}_{\bullet\bullet}^2] = J\sum\left(\dfrac{\sigma^2}{J + \mu_i^2}\right) - IJ\left(\dfrac{\sigma^2}{IJ + \mu^2}\right)$

$= I\sigma^2 + J\Sigma\mu_i^2 - \sigma^2 - IJ\mu^2 = (I-1)\sigma^2 + J\Sigma(\mu_i - \mu)^2$, so

$E(MSTr) = \dfrac{E(SSTr)}{I-1} = E[J\Sigma\overline{X}_{i\bullet}^2 - IJ\overline{X}_{\bullet\bullet}^2] = \sigma^2 + J\sum\dfrac{(\mu_i - \mu)^2}{I-1}$.

e When H_o is true, $\mu_1 = \ldots = \mu_i = \mu$, so $\Sigma(\mu_i - \mu)^2 = 0$ and $E(MSTr) = \sigma^2$.
When H_o is false, $\Sigma(\mu_i - \mu)^2 > 0$, so $E(MSTr) > \sigma^2$ (on average, MSTr overestimates σ^2).

Section 10.2

11. $Q_{.05,5,15} = 4.37$, $w = 4.37\sqrt{\dfrac{272.8}{4}} = 36.09$.

3	1	4	2	5
437.5	462.0	469.3	512.8	532.1

The brands seem to divide into two groups: 1, 3, and 4; and 2 and 5; with no significant differences within each group but all between group differences are significant.

12.

3	1	4	2	5
437.5	462.0	469.3	512.8	532.1

Brands 2 and 5 do not differ significantly from one another, but both differ significantly from brands 1, 3, and 4. While brands 3 and 4 do differ significantly, there is not enough evident to indicate a significant difference between 1 and 3 or 1 and 4.

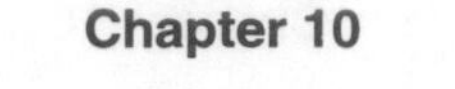

13.

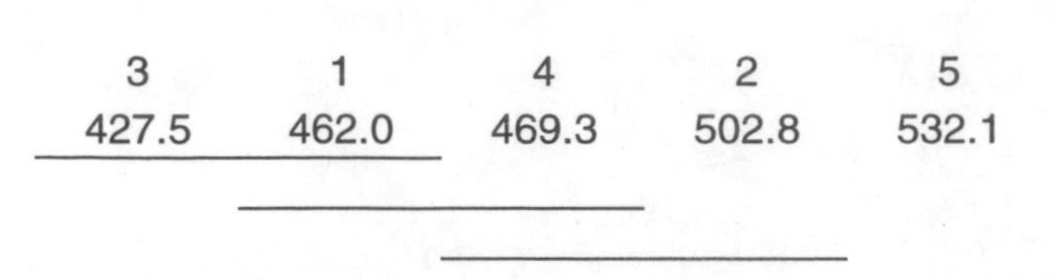

3	1	4	2	5
427.5	462.0	469.3	502.8	532.1

Brand 1 does not differ significantly from 3 or 4, 2 does not differ significantly from 4 or 5, 3 does not differ significantly from1, 4 does not differ significantly from 1 or 2, 5 does not differ significantly from 2, but all other differences (e.g., 1 with 2 and 5, 2 with 3, etc.) do appear to be significant.

14. I = 4, J = 8, so $Q_{.05,4,28} \approx 3.87$, $w = 3.87\sqrt{\frac{1.06}{8}} = 1.41$.

1	2	3	4
4.39	4.52	5.49	6.36

Treatment 4 appears to differ significantly from both 1 and 2, but there are no other significant differences.

15. $Q_{.01,4,36} = 4.75$, $w = 4.75\sqrt{\frac{15.64}{10}} = 5.94$.

2	1	3	4
24.69	26.08	29.95	33.84

Treatment 4 appears to differ significantly from both 1 and 2, but there are no other significant differences.

16.

a Since the largest standard deviation (s_4 = 44.51) is only slightly more than twice the smallest (s_3 = 20.83) it is plausible that the population variances are equal (see text p. 406).

b The relevant hypotheses are $H_0 : \mu_1 = \mu_2 = \mu_3 = \mu_4 = \mu_5$ vs. H_a :at least two μ_i's differ. With the given f of 10.48 and associated p-value of 0.000, we can reject H_o and conclude that there is a difference in axial stiffness for the different plate lengths.

c

4	6	8	10	12
333.21	368.06	375.13	407.36	437.17

There is no significant difference in the axial stiffness for lengths 4, 6, and 8, and for lengths 6, 8, and 10, yet 4 and 10 differ significantly. Length 12 differs from 4, 6, and 8, but does not differ from 10.

17. $\theta = \Sigma c_i \mu_i$ where $c_1 = c_2 = .5$ and $c_3 = -1$, so $\hat{\theta} = .5\bar{x}_{1\bullet} + .5\bar{x}_{2\bullet} - \bar{x}_{3\bullet} = -.396$ and $\Sigma c_i^2 = 1.50$. With $t_{.025,6} = 2.447$ and MSE = .03106, the CI is (from 10.5 on page 418)

$$-.396 \pm (2.447)\sqrt{\frac{(.03106)(1.50)}{3}} = -.396 \pm .305 = (-.701, -.091).$$

18.

a Let μ_i = true average growth when hormone #*i* is applied. $H_0 : \mu_1 = ... = \mu_5$ will be rejected in favor of H_a :at least two μ_i's differ if $f \geq F_{.05,4,15} = 3.06$. With

$\frac{x_{\bullet\bullet}^2}{IJ} = \frac{(278)^2}{20} = 3864.20$ and $\Sigma\Sigma x_{ij}^2 = 4280$, SST = 415.80.

$\frac{\Sigma x_{i\bullet}^2}{J} = \frac{(51)^2 + (71)^2 + (70)^2 + (46)^2 + (40)^2}{4} = 4064.50$, so SSTr = 4064.50 – 3864.20 = 200.3, and SSE = 415.80 – 200.30 = 215.50. Thus

$MSTr = \frac{200.3}{4} = 50.075$, $MSE = \frac{215.5}{15} = 14.3667$, and $f = \frac{50.075}{14.3667} = 3.49$.

Because $3.49 \geq 3.06$, reject H_o. There appears to be a difference in the average growth with the application of the different growth hormones.

b $Q_{.05,5,15} = 4.37$, $w = 4.37\sqrt{\frac{14.3667}{4}} = 8.28$. The sample means are, in increasing order, 10.00, 11.50, 12.75, 17.50, and 17.75. The most extreme difference is 17.75 – 10.00 = 7.75 which doesn't exceed 8.28, so no differences are judged significant. Tukey's method and the F test are at odds.

19. MSTr = 140, error d.f. = 12, so $f = \frac{140}{SSE/12} = \frac{1680}{SSE}$ and $F_{.05,2,12} = 3.89$.

$w = Q_{.05,3,12}\sqrt{\frac{MSE}{J}} = 3.77\sqrt{\frac{SSE}{60}} = .4867\sqrt{SSE}$. Thus we wish $\frac{1680}{SSE} > 3.89$ (significance of f) and $.4867\sqrt{SSE} > 10$ (= 20 – 10, the difference between the extreme $\bar{x}_{i\bullet}$'s - so no significant differences are identified). These become $431.88 > SSE$ and $SSE > 422.16$, so SSE = 425 will work.

20. Now MSTr = 125, so $f = \frac{1500}{SSE}$, $w = .4867\sqrt{SSE}$ as before, and the inequalities become $385.60 > SSE$ and $SSE > 422.16$. Clearly no value of SSE can satisfy both inequalities.

21.

a Grand mean = 222.167, MSTr = 38,015.1333, MSE = 1,681.8333, and f = 22.6. The hypotheses are $H_0 : \mu_1 = ... = \mu_6$ vs. H_a :at least two μ_i's differ. Reject H_o if $f \ge F_{.01,5,78}$ (but since there is no table value for $\nu_2 = 78$, use $f \ge F_{.01,5,60} = 3.34$) With $22.6 \ge 3.34$, we reject H_o. the data indicates there is a dependence on injection regimen.

b Assume $t_{.005,78} \approx 2.645$

i) Confidence interval for $\mu_1 - \frac{1}{5}(\mu_2 + \mu_3 + \mu_4 + \mu_5 + \mu_6)$:

$$\Sigma c_i \bar{x}_i \pm t_{\alpha/2, I(J-1)}\sqrt{\frac{MSE(\Sigma c_i^2)}{J}}$$

$$= -67.4 \pm (2.645)\sqrt{\frac{1,681.8333(1.2)}{14}} = (-99.16, -35.64).$$

ii) Confidence interval for $\frac{1}{4}(\mu_2 + \mu_3 + \mu_4 + \mu_5) - \mu_6$:

$$= 61.75 \pm (2.645)\sqrt{\frac{1,681.8333(1.25)}{14}} = (29.34, 94.16)$$

Chapter 10

Section 10.3

22. Summary quantities are $x_{1\bullet} = 291.4$, $x_{2\bullet} = 221.6$, $x_{3\bullet} = 203.4$, $x_{4\bullet} = 227.5$, $x_{\bullet\bullet} = 943.9$, $CF = 49{,}497.07$, $\Sigma\Sigma x_{ij}^2 = 50{,}078.07$, from which $SST = 581$,

$$SSTr = \frac{(291.4)^2}{5} + \frac{(221.6)^2}{4} + \frac{(203.4)^2}{4} + \frac{(227.5)^2}{5} - 49{,}497.07$$

$= 49{,}953.57 - 49{,}497.07 = 456.50$, and $SSE = 124.50$. Thus

$MSTr = \dfrac{456.50}{3} = 152.17$, $MSE = \dfrac{124.50}{18-4} = 8.89$, and f = 17.12. Because $17.12 \geq F_{.05,3,14} = 3.34$, $H_0 : \mu_1 = \ldots = \mu_4$ is rejected at level .05. There is a difference in yield of tomatoes for the four different levels of salinity.

23. $J_1 = 5$, $J_2 = 4$, $J_3 = 4$, $J_4 = 5$, $\bar{x}_{1\bullet} = 58.28$, $\bar{x}_{2\bullet} = 55.40$, $\bar{x}_{3\bullet} = 50.85$, $\bar{x}_{4\bullet} = 45.50$, MSE = 8.89. With $W_{ij} = Q_{.05,4,14} \cdot \sqrt{\dfrac{MSE}{2}\left(\dfrac{1}{J_i} + \dfrac{1}{J_j}\right)} = 4.11\sqrt{\dfrac{8.89}{2}\left(\dfrac{1}{J_i} + \dfrac{1}{J_j}\right)}$,

$\bar{x}_{1\bullet} - \bar{x}_{2\bullet} \pm W_{12} = (2.88) \pm (5.81)$; $\quad \bar{x}_{1\bullet} - \bar{x}_{3\bullet} \pm W_{13} = (7.43) \pm (5.81)$*;

$\bar{x}_{1\bullet} - \bar{x}_{4\bullet} \pm W_{14} = (12.78) \pm (5.48)$*; $\quad \bar{x}_{2\bullet} - \bar{x}_{3\bullet} \pm W_{23} = (4.55) \pm (6.13)$;

$\bar{x}_{2\bullet} - \bar{x}_{4\bullet} \pm W_{24} = (9.90) \pm (5.81)$*; $\quad \bar{x}_{3\bullet} - \bar{x}_{4\bullet} \pm W_{34} = (5.35) \pm (5.81)$;

*Indicates an interval that doesn't include zero, corresponding to μ's that are judged significantly different.

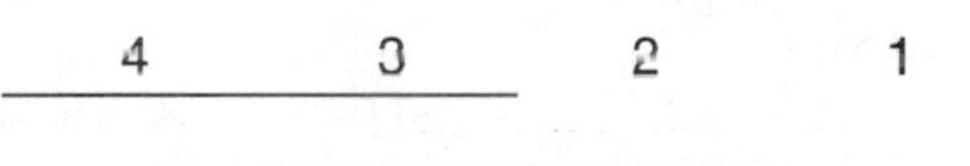

This underscoring pattern does not have a very straightforward interpretation.

24.

Source	Df	SS	MS	F
Groups	3-1=2	152.18	76.09	5.56
Error	74-3=71	970.96	13.68	
Total	74-1=73	1123.14		

Since $5.56 \geq F_{.01,2,71} \approx 4.94$, reject $H_0 : \mu_1 = \mu_2 = \mu_3$ at level .01.

25.

Diet	J_i	$\bar{x}_{i\bullet}$	$s_{i\bullet}$	$x_{i\bullet}$
Red Maple '74	13	1.134	.0252	14.742
Red Oak/Red Maple	10	1.148	.0253	11.480
Red Maple '75	20	1.159	.0179	23.180
Red Oak	16	1.191	.0200	19.056
Red Oak/White Pine	16	1.217	.0160	19.472
	75			87.930

Where $X_{i\bullet} = J_i \cdot \bar{X}_{i\bullet}$. Also, $X_{i\bullet}^2 = S_i^2(J_i - 1) + \frac{(X_{i\bullet})^2}{J_i}$ (from computational formula for variance).

a $\bar{x}_{\bullet\bullet} = \frac{87.93}{75} = 1.1724$, $SSTr = \frac{(14.742)^2}{13} + \frac{(11.480)^2}{10} + \frac{(23.180)^2}{20} + \frac{(19.056)^2}{16} + \frac{(19.472)^2}{16} - \frac{(87.93)^2}{75} = .066076$,

$$MSTr = \frac{SSTr}{I-1} = \frac{.066076}{4} = .016519.$$

b $SSE = (12)(.0252)^2 + (9)(.0253)^2 + (19)(.0179)^2 + (15)(.0200)^2 + (15)(.0160)^2 = .029309$, and the $MSE = \frac{SSE}{n-I} = \frac{.029309}{70} = .0004187$.

c $f = \frac{MSTr}{MSE} = \frac{.016519}{.0004187} = 39.45$. Using d.f = 4, 60 (because there is no table for 70) we find that $39.45 > 5.31 = F_{.001,4,60}$, so the p-value < .001, which is less than the significance level of .05, so we reject H_o. The data does suggest differences among the means.

d $W_{ij} = Q_{\alpha,I,n-I} \cdot \sqrt{\frac{MSE}{2}\left(\frac{1}{J_i}+\frac{1}{J_j}\right)}$; $Q_{.05,5,70}$ not tabled so use $Q_{.05,5,60} = 3.98$.

w_{ij}	$\bar{x}_j - \bar{x}_i$	
$w_{12} = .024$	.014	Not significantly different
$w_{13} = .020$	.025	Significantly different
$w_{23} = .020$	.011	Not significantly different
$w_{24} = .023$	.048	Significantly different
$w_{34} = .019$	.032	Significantly different
$w_{45} = .020$	.026	Significantly different

1 2 3 4 5

(underline under 1 2; underline under 2 3)

26.

a

i:	1	2	3	4	5	6	
J_i:	4	5	4	4	5	4	
$x_{i\bullet}$:	56.4	64.0	55.3	52.4	85.7	72.4	$x_{\bullet\bullet} = 386.2$
$\bar{x}_{i\bullet}$:	14.10	12.80	13.83	13.10	17.14	18.10	$\Sigma\Sigma x_j^2 = 5850.20$

Thus SST = 113.64, SSTr = 108.19, SSE = 5.45, MSTr = 21.64, MSE = .273, f = 79.3. Since $79.3 \geq F_{.01,5,20} = 4.10$, $H_0 : \mu_1 = \ldots = \mu_6$ is rejected.

b The modified Tukey intervals are as follows: (The first number is $\bar{x}_{i\bullet} - \bar{x}_{j\bullet}$ and the second is $W_{ij} = Q_{.01} \cdot \sqrt{\frac{MSE}{2}\left(\frac{1}{J_i}+\frac{1}{J_j}\right)}$.)

Pair	Interval	Pair	Interval	Pair	Interval
1,2	1.30 ± 1.37	2,3	-1.03 ± 1.37	3,5	-3.31 ± 1.37 *
1,3	$.27 \pm 1.44$	2,4	$-.30 \pm 1.37$	3,6	-4.27 ± 1.44 *
1,4	1.00 ± 1.44	2,5	-4.34 ± 1.29 *	4,5	-4.04 ± 1.37 *
1,5	-3.04 ± 1.37 *	2,6	-5.30 ± 1.37 *	4,6	-5.00 ± 1.44 *
1,6	-4.00 ± 1.44 *	3,4	$.37 \pm 1.44$	5,6	$-.96 \pm 1.37$

Asterisks identify pairs of means that are judged significantly different from one another.

c The 99% t confidence interval is $\Sigma c_i \bar{x}_{i\bullet} \pm t_{.005,I(J-1)}\sqrt{\dfrac{MSE(\Sigma c_i^2)}{J_i}}$.

$\Sigma c_i \bar{x}_{i\bullet} = \frac{1}{4}\bar{x}_{1\bullet} + \frac{1}{4}\bar{x}_{2\bullet} + \frac{1}{4}\bar{x}_{3\bullet} + 14\bar{x}_{4\bullet} - 12\bar{x}_{5\bullet} - \frac{1}{2}\bar{x}_{6\bullet} = -4.16$, $\dfrac{(\Sigma c_i^2)}{J_i} = .1719$,

MSE = .273, $t_{.005,20} = 2.845$. The resulting interval is

$-4.16 \pm (2.845)\sqrt{(.273)(.1719)} = -4.16 \pm .62 = (-4.78, -3.54)$. The interval in the answer section is a Scheffe' interval, and is substantially wider than the t interval.

27.

a Let μ_i = true average folacin content for specimens of brand I. The hypotheses to be tested are $H_0 : \mu_1 = \mu_2 = \mu_3 = \mu_4$ vs. H_a : at least two μ_i's differ.

$\Sigma\Sigma x_{ij}^2 = 1246.88$ and $\dfrac{x_{\bullet\bullet}^2}{n} = \dfrac{(168.4)^2}{24} = 1181.61$, so SST = 65.27.

$\dfrac{\Sigma x_{i\bullet}^2}{J_i} = \dfrac{(57.9)^2}{7} + \dfrac{(37.5)^2}{5} + \dfrac{(38.1)^2}{6} + \dfrac{(34.9)^2}{6} = 1205.10$, so

$SSTr = 1205.10 - 1181.61 = 23.49$.

Source	Df	SS	MS	F
Treatments	3	23.49	7.83	3.75
Error	20	41.78	2.09	
Total	23	65.27		

With numerator df = 3 and denominator = 20,

$F_{.05,3,20} = 3.10 < 3.75 < F_{.01,3,20} = 4.94$, so $.01 < p-value < .05$, and since the p-value < .05, we reject H_o. At least one of the pairs of brands of green tea have different average folacin content.

b With $\bar{x}_{i\bullet}$ = 8.27, 7.50, 6.35, and 5.82 for I = 1, 2, 3, 4, we calculate the residuals $x_{ij} - \bar{x}_{i\bullet}$ for all observations. A normal probability plot appears below, and indicates that the distribution of residuals could be normal, so the normality assumption is plausible.

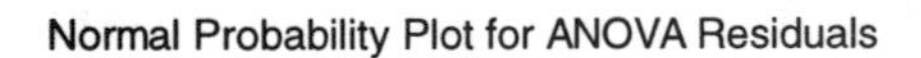

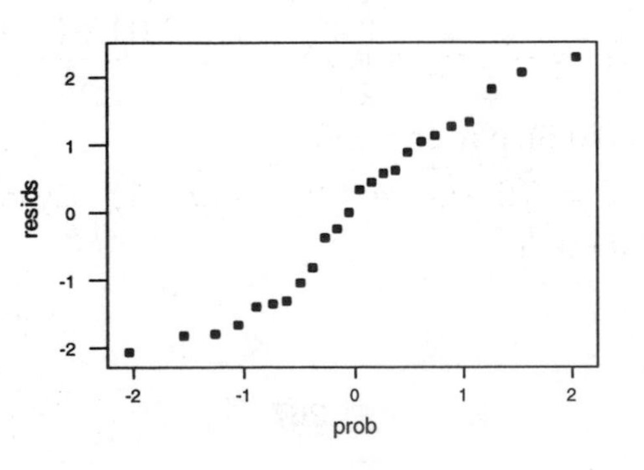

c $Q_{.05,4,20} = 3.96$ and $W_{ij} = 3.96 \cdot \sqrt{\frac{2.09}{2}\left(\frac{1}{J_i} + \frac{1}{J_j}\right)}$, so the Modified Tukey intervals are:

Pair	Interval	Pair	Interval
1,2	$.77 \pm 2.37$	2,3	1.15 ± 2.45
1,3	1.92 ± 2.25	2,4	1.68 ± 2.45
1,4	2.45 ± 2.25 *	3,4	$.53 \pm 2.34$

4 3 2 1

Only Brands 1 and 4 are different from each other.

28. $$SSTr = \sum_i \left\{\sum_j (\overline{X}_{i\cdot} - \overline{X}_{\cdot\cdot})^2\right\} = \sum_i J_i(\overline{X}_{i\cdot} - \overline{X}_{\cdot\cdot})^2 = \sum_i J_i \overline{X}_{i\cdot}^2 - 2\overline{X}_{\cdot\cdot}\sum_i J_i \overline{X}_{i\cdot} + \overline{X}_{\cdot\cdot}^2 \sum_i J_i$$
$$= \sum_i J_i \overline{X}_{i\cdot}^2 - 2\overline{X}_{\cdot\cdot} X_{\cdot\cdot} + n\overline{X}_{\cdot\cdot}^2 = \sum_i J_i \overline{X}_{i\cdot}^2 - 2n\overline{X}_{\cdot\cdot}^2 + n\overline{X}_{\cdot\cdot}^2 = \sum_i J_i \overline{X}_{i\cdot}^2 - n\overline{X}_{\cdot\cdot}^2.$$

29. $$E(SSTr) = E\left(\sum_i J_i \overline{X}_{i\cdot}^2 - n\overline{X}_{\cdot\cdot}^2\right) = \Sigma J_i E(\overline{X}_{i\cdot}^2) - nE(\overline{X}_{\cdot\cdot}^2)$$
$$= \Sigma J_i \left[Var(\overline{X}_{i\cdot}) + (E(\overline{X}_{i\cdot}))^2\right] - n\left[Var(\overline{X}_{\cdot\cdot}) + (E(\overline{X}_{\cdot\cdot}))^2\right]$$
$$= \Sigma J_i \left[\frac{\sigma^2}{J_i} + \mu_i^2\right] - n\left[\frac{\sigma^2}{n} + \frac{(\Sigma J_i \mu_i)^2}{n}\right]$$
$$= (I-1)\sigma^2 + \Sigma J_i(\mu + \alpha_i)^2 - [\Sigma J_i(\mu + \alpha_i)]^2$$
$$= (I-1)\sigma^2 + \Sigma J_i \mu^2 + 2\mu\Sigma J_i \alpha_i + \Sigma J_i \alpha_i^2 - [\mu \Sigma J_i]^2$$
$= (I-1)\sigma^2 + \Sigma J_i \alpha_i^2$, from which E(MSTr) is obtained through division by $(I-1)$.

30.

a $\alpha_1 = \alpha_2 = 0$, $\alpha_3 = -1$, $\alpha_4 = 1$, so $\Phi^2 = \frac{2(0^2 + 0^2 + (-1)^2 + 1^2)}{1} = 4$, $\Phi = 2$, and from figure (10.5), power $\approx .90$.

b $\Phi^2 = .5J$, so $\Phi = .707\sqrt{J}$ and $\nu_2 = 4(J-1)$. By inspection of figure (10.5), J = 9 looks to be sufficient.

c $\mu_1 = \mu_2 = \mu_3 = \mu_4, \mu_5 = \mu_1 + 1$, so $\mu = \mu_1 + \frac{1}{5}$, $\alpha_1 = \alpha_2 = \alpha_3 = \alpha_4 = -\frac{1}{5}$,
$\alpha_4 = \frac{4}{5}$, $\Phi^2 = \dfrac{2\left(20/25\right)}{1} = 1.60$ $\Phi = 1.26$, $\nu_1 = 4$, $\nu_2 = 45$. By inspection of figure (10.6), power $\approx .55$.

31. With $\sigma = 1$ (any other σ would yield the same Φ), $\alpha_1 = -1$, $\alpha_2 = \alpha_3 = 0$, $\alpha_4 = 1$,
$\Phi^2 = \dfrac{.25\left(5(-1)^2 + 5(0)^2 + 5(0)^2 + 5(1)^2\right)}{1} = 2.5$, $\Phi = 1.58$, $\nu_1 = 3$, $\nu_2 = 14$, and power $\approx .62$.

32. With Poisson data, the ANOVA should be done using $y_{ij} = \sqrt{x_{ij}}$. This gives
$y_{1\bullet} = 15.43$, $y_{2\bullet} = 17.15$, $y_{3\bullet} = 19.12$, $y_{4\bullet} = 20.01$, $y_{\bullet\bullet} = 71.71$, $\Sigma\Sigma y_{ij}^2 = 263.79$, CF = 257.12, SST = 6.67, SSTr = 2.52, SSE = 4.15, MSTr = .84, MSE = .26, f = 3.23. Since $F_{.01,3,16} = 5.29$, H_o cannot be rejected. The expected number of flaws per reel does not seem to depend upon the brand of tape.

33. $g(x) = x\left(1 - \dfrac{x}{n}\right) = nu(1-u)$ where $u = \dfrac{x}{n}$, so $h(x) = \int [u(1-u)]^{-1/2} du$. From a table of integrals, this gives $h(x) = \arcsin\left(\sqrt{u}\right) = \arcsin\left(\sqrt{\dfrac{x}{n}}\right)$ as the appropriate transformation.

34. $E(MSTr) = \sigma^2 + \dfrac{1}{I-1}\left(n - \dfrac{IJ^2}{n}\right)\sigma_A^2 = \sigma^2 + \dfrac{n-J}{I-1}\sigma_A^2 = \sigma^2 + J\sigma_A^2$

Supplementary Exercises

35.

a $H_0 : \mu_1 = \mu_2 = \mu_3 = \mu_4$ vs. H_a : at least two μ_i's differ ; 3.68 is not $\geq F_{.01,3,20} = 4.94$, thus fail to reject H_o. The means do not appear to differ.

b We reject H_o when the p-value < alpha. Since .029 is not < .01, we still fail to reject H_o.

36.

a $H_0 : \mu_1 = ... = \mu_5$ will be rejected in favor of H_a :at least two μ_i's differ if $f \geq F_{.05,4,40} = 2.61$. With $\bar{x}_{\bullet\bullet} = 30.82$, straightforward calculation yields $MSTr = \frac{221.112}{4} = 55.278$, $MSE = \frac{80.4591}{5} = 16.1098$, and $f = \frac{55.278}{16.1098} = 3.43$. Because $3.43 \geq 2.61$, H_o is rejected. There is a difference among the five teaching methods with respect to true mean exam score.

b The format of this test is identical to that of part **a**. The calculated test statistic is $f = \frac{33.12}{20.109} = 1.65$. Since $1.65 < 2.61$, H_o is not rejected. The data suggests that with respect to true average retention scores, the five methods are not different from one another.

37. Let μ_i = true average amount of motor vibration for each of five bearing brands. Then the hypotheses are $H_0 : \mu_1 = ... = \mu_5$ vs. H_a :at least two μ_i's differ. The ANOVA table follows:

Source	Df	SS	MS	F
Treatments	4	30.855	7.714	8.44
Error	25	22.838	0.914	
Total	29	53.694		

$8.44 > F_{.001,4,25} = 6.49$, so p-value < .001, which is also < .05, so we reject H_o. At least two of the means differ from one another. The Tukey multiple comparison is appropriate. $Q_{.05,5,25} = 4.15$ (from Minitab output. Using Table A.10, approximate with $Q_{.05,5,24} = 4.17$). $W_{ij} = 4.15\sqrt{.914/6} = 1.620$.

Pair	$\bar{x}_{i\bullet} - \bar{x}_{j\bullet}$	Pair	$\bar{x}_{i\bullet} - \bar{x}_{j\bullet}$
1,2	-2.267*	2,4	1.217
1,3	0.016	2,5	2.867*
1,4	-1.050	3,4	-1.066
1,5	0.600	3,5	0.584
2,3	2.283*	4,5	1.650*

*Indicates significant pairs.

5 3 1 4 2

38. $x_{1\bullet} = 15.48$, $x_{2\bullet} = 15.78$, $x_{3\bullet} = 12.78$, $x_{4\bullet} = 14.46$, $x_{5\bullet} = 14.94$ $x_{\bullet\bullet} = 73.44$, so $CF = 179.78$, SST = 3.62, SSTr = 180.71 – 179.78 = .93, SSE = 3.62 - .93 = 2.69.

Source	Df	SS	MS	F
Treatments	4	.93	.233	2.16
Error	25	2.69	.108	
Total	29	3.62		

$F_{.05,4,25} = 2.76$. Since 2.16 is not ≥ 2.76, do not reject H_o at level .05.

39. $\hat{\theta} = 2.58 - \frac{2.63 + 2.13 + 2.41 + 2.49}{4} = .165$, $t_{.025,25} = 2.060$, MSE = .108, and $\Sigma c_i^2 = (1)^2 + (-.25)^2 + (-.25)^2 + (-.25)^2 + (-.25)^2 = 1.25$, so a 95% confidence interval for θ is $.165 \pm 2.060\sqrt{\frac{(.108)(1.25)}{6}} = .165 \pm .309 = (-.144, .474)$. This interval does include zero, so 0 is a plausible value for θ.

40. $\mu_1 = \mu_2 = \mu_3, \mu_4 = \mu_5 = \mu_1 - \sigma$, so $\mu = \mu_1 - \frac{2}{5}\sigma$, $\alpha_1 = \alpha_2 = \alpha_3 = \frac{2}{5}\sigma$, $\alpha_4 = \alpha_5 = -\frac{3}{5}\sigma$. Then $\Phi^2 = \frac{J}{I}\sum \frac{\alpha_i^2}{\sigma^2}$

$$= \frac{6}{5}\left[\frac{3\left(\frac{2}{5}\sigma\right)^2}{\sigma^2} + \frac{2\left(-\frac{3}{5}\sigma\right)^2}{\sigma^2}\right] = 1.632$$

and $\Phi = 1.28$, $v_1 = 4$, $v_2 = 25$. By inspection of figure (10.6), power $\approx .48$, so $\beta \approx .52$.

41. This is a random effects situation. $H_0 : \sigma_A^2 = 0$ states that variation in laboratories doesn't contribute to variation in percentage. H_o will be rejected in favor of H_a if $f \geq F_{.05,3,8} = 4.07$. SST = 86,078.9897 – 86,077.2224 = 1.7673, SSTr = 1.0559, and SSE = .7114. Thus $f = \frac{1.0559/3}{.7114/8} = 3.96$, which is not ≥ 4.07, so H_o cannot be rejected at level .05. Variation in laboratories does not appear to be present.

42.

a μ_i = true average CFF for the three iris colors. Then the hypotheses are $H_0 : \mu_1 = \mu_2 = \mu_3$ vs. H_a :at least two $\mu_i's$ differ. SST = 13,659.67 – 13,598.36 = 61.31, $SSTR = \left(\frac{(204.7)^2}{8} + \frac{(134.6)^2}{5} + \frac{(169.0)^2}{6}\right) - 13{,}598.36 = 23.00$ The ANOVA table follows:

Source	Df	SS	MS	F
Treatments	2	23.00	11.50	4.803
Error	16	38.31	2.39	
Total	18	61.31		

Because $F_{.05,2,16} = 3.63 < 4.803 < F_{.01,2,16} = 6.23$, .01 < p-value < .05, so we reject H_o. There are differences in CFF based on iris color.

b $Q_{.05,3,16} = 3.65$ and $W_{ij} = 3.65 \cdot \sqrt{\frac{2.39}{2}\left(\frac{1}{J_i} + \frac{1}{J_j}\right)}$, so the Modified Tukey intervals are:

Pair	$(\bar{x}_{i\bullet} - \bar{x}_{j\bullet}) \pm W_{ij}$
1,2	-1.33 ± 2.27
1,3	-2.58 ± 2.15 *
2,3	-1.25 ± 2.42

Brown	Green	Blue
25.59	26.92	28.17

The CFF is only significantly different for Brown and Blue iris color.

43. $\sqrt{(I-1)(MSE)(F_{.05,I-1,n-I})} = \sqrt{(2)(2.39)(3.63)} = 4.166$. For $\mu_1 - \mu_2$, $c_1 = 1$, $c_2 = -1$, and $c_3 = 0$, so $\sqrt{\sum \frac{c_i^2}{J_i}} = \sqrt{\frac{1}{8} + \frac{1}{5}} = .570$. Similarly, for $\mu_1 - \mu_3$, $\sqrt{\sum \frac{c_i^2}{J_i}} = \sqrt{\frac{1}{8} + \frac{1}{6}} = .540$; for $\mu_2 - \mu_3$, $\sqrt{\sum \frac{c_i^2}{J_i}} = \sqrt{\frac{1}{5} + \frac{1}{6}} = .606$, and for $.5\mu_2 + .5\mu_2 - \mu_3$, $\sqrt{\sum \frac{c_i^2}{J_i}} = \sqrt{\frac{.5^2}{8} + \frac{.5^2}{5} + \frac{(-1)^2}{6}} = .498$.

(continued)

Contrast	Estimate	Interval
$\mu_1 - \mu_2$	25.59 – 26.92 = -1.33	$(-1.33) \pm (.570)(4.166) = (-3.70, 1.04)$
$\mu_1 - \mu_3$	25.59 – 28.17 = -2.58	$(-2.58) \pm (.540)(4.166) = (-4.83, -.33)$
$\mu_2 - \mu_3$	26.92 – 28.17 = -1.25	$(-1.25) \pm (.606)(4.166) = (-3.77, 1.27)$
$.5\mu_2 + .5\mu_2 - \mu_3$	-1.92	$(-1.92) \pm (.498)(4.166) = (-3.99, 0.15)$

44.

Source	Df	SS	MS	F	$F_{.05}$
Treatments	3	24,937.63	8312.54	1117.8	4.07
Error	8	59.49	7.44		
Total	11	24,997.12			

Because $1117.8 \geq 4.07$, $H_0: \mu_1 = \mu_2 = \mu_3 = \mu_4$ is rejected. $Q_{.05,4,8} = 4.53$, so $w = 4.53\sqrt{\frac{7.44}{3}} = 7.13$. The four sample means are $\bar{x}_{4\bullet} = 29.92$, $\bar{x}_{1\bullet} = 33.96$, $\bar{x}_{3\bullet} = 115.84$, and $\bar{x}_{2\bullet} = 129.30$. Only $\bar{x}_{1\bullet} - \bar{x}_{4\bullet} < 7.13$, so all means are judged significantly different from one another except for μ_4 and μ_1 (corresponding to PCM and OCM).

45. $Y_{ij} - \bar{Y}_{\bullet\bullet} = c(X_{ij} - \bar{X}_{\bullet\bullet})$ and $\bar{Y}_{i\bullet} - \bar{Y}_{\bullet\bullet} = c(\bar{X}_{i\bullet} - \bar{X}_{\bullet\bullet})$, so each sum of squares involving Y will be the corresponding sum of squares involving X multiplied by c^2. Since F is a ratio of two sums of squares, c^2 appears in both the numerator and denominator so cancels, and F computed from Y_{ij}'s = F computed from X_{ij}'s.

46. The ordered residuals are –6.67, -5.67, -4, -2.67, -1, -1, 0, 0, 0, .33, .33, .33, 1, 1, 2.33, 4, 5.33, 6.33. The corresponding z percentiles are –1.91, -1.38, -1.09, -.86, -.67, -.51, -.36, -.21, -.07, .07, .21, .36, .51, .67, .86, 1.09, 1.38, and 1.91. The resulting plot of the respective pairs (the Normal Probability Plot) is reasonably straight, and thus there is no reason to doubt the normality assumption.

Chapter 11

Section 11.1

1.

a $MSA = \frac{30.6}{4} = 7.65$, $MSE = \frac{59.2}{12} = 4.93$, $f_A = \frac{7.65}{4.93} = 1.55$. Since 1.55 is not $\geq F_{.05,4,12} = 3.26$, don't reject H_{oA}. There is no difference in true average tire lifetime due to different makes of cars.

b $MSB = \frac{44.1}{3} = 14.70$, $f_B = \frac{14.70}{4.93} = 2.98$. Since 2.98 is not $\geq F_{.05,3,12} = 3.49$, don't reject H_{oB}. There is no difference in true average tire lifetime due to different brands of tires.

2.

a $x_{1\bullet} = 163$, $x_{2\bullet} = 152$, $x_{3\bullet} = 142$, $x_{4\bullet} = 146$, $x_{\bullet 1} = 215$, $x_{\bullet 2} = 188$, $x_{\bullet 3} = 200$, $x_{\bullet\bullet} = 603$, $\Sigma\Sigma x_{ij}^2 = 30{,}599$, $CF = \frac{(603)^2}{12} = 30{,}300.75$, so SST = 298.25,

$SSA = \frac{1}{3}\left[(163)^2 + (152)^2 + (142)^2 + (146)^2\right] - 30{,}300.75 = 83.58$,

$SSB = 30{,}392.25 - 30{,}300.75 = 91.50$, $SSE = 298.25 - 83.58 - 91.50 = 123.17$.

Source	Df	SS	MS	F
A	3	83.58	27.86	1.36
B	2	91.50	45.75	2.23
Error	6	123.17	20.53	
Total	11	298.25		

$F_{.05,3,6} = 4.76$, $F_{.05,2,6} = 5.14$. Since neither f is greater than the appropriate critical value, neither H_{oA} nor H_{oB} is rejected.

b $\hat{\mu} = \bar{x}_{\bullet\bullet} = 50.25$, $\hat{\alpha}_1 = \bar{x}_{1\bullet} - \bar{x}_{\bullet\bullet} = 4.08$, $\hat{\alpha}_2 = .42$, $\hat{\alpha}_3 = -2.92$, $\hat{\alpha}_4 = -1.58$, $\hat{\beta}_1 = \bar{x}_{\bullet 1} - \bar{x}_{\bullet\bullet} = 3.50$, $\hat{\beta}_2 = -3.25$, $\hat{\beta}_3 = -.25$.

3. $x_{1\bullet} = 927$, $x_{2\bullet} = 1301$, $x_{3\bullet} = 1764$, $x_{4\bullet} = 2453$, $x_{\bullet 1} = 1347$, $x_{\bullet 2} = 1529$, $x_{\bullet 3} = 1677$, $x_{\bullet 4} = 1892$, $x_{\bullet\bullet} = 6445$, $\Sigma\Sigma x_{ij}^2 = 2{,}969{,}375$,

$CF = \dfrac{(6445)^2}{16} = 2{,}596{,}126.56$, $SSA = 324{,}082.2$, $SSB = 39{,}934.2$,

$SST = 373{,}248.4$, $SSE = 9232.0$

a

Source	Df	SS	MS	F
A	3	324,082.2	108,027.4	105.3
B	3	39,934.2	13,311.4	13.0
Error	9	9232.0	1025.8	
Total	15	373,248.4		

Since $F_{.01,3,9} = 6.99$, both H_{oA} and H_{oB} are rejected.

b $Q_{.01,4,9} = 5.96$, $w = 5.96\sqrt{\dfrac{1025.8}{4}} = 95.4$

i:	1	2	3	4
$\bar{x}_{i\bullet}$:	231.75	325.25	441.00	613.25

All levels of Factor A (gas rate) differ significantly except for 1 and 2

c $w = 95.4$, as in **b**

i:	1	2	3	4
$\bar{x}_{\bullet j}$:	336.75	382.25	419.25	473

Only levels 1 and 4 appear to differ significantly.

4.

a After subtracting 400, $x_{1\bullet} = 151$, $x_{2\bullet} = 137$, $x_{3\bullet} = 125$, $x_{4\bullet} = 124$, $x_{\bullet 1} = 183$, $x_{\bullet 2} = 169$, $x_{\bullet 3} = 185$, $x_{\bullet\bullet} = 537$, $SSA = 159.98$, $SSB = 38.00$, $SST = 238.25$, $SSE = 40.67$.

Source	Df	SS	MS	f	$F_{.05}$
A	3	159.58	53.19	7.85	4.76
B	2	38.00	19.00	2.80	5.14
Error	6	40.67	6.78		
Total	11	238.25			

b Since $7.85 \geq 4.76$, reject H_{oA}: $\alpha_1 = \alpha_2 = \alpha_3 = \alpha_4 = 0$: The amount of coverage depends on the paint brand.

c Since 2.80 is not ≥ 5.14, do not reject H_{oA}: $\beta_1 = \beta_2 = \beta_3 = 0$. The amount of coverage does not depend on the roller brand.

d Because H_{oB} was not rejected. Tukey's method is used only to identify differences in levels of factor A (brands of paint). $Q_{.05,4,6} = 4.90$, w = 7.37.

i:	4	3	2	1
$\bar{x}_{i\bullet}$:	231.75	325.25	441.00	613.25

Brands 3 and 4 differ significantly from brand 1

5.

Source	Df	SS	MS	f
Angle	3	58.16	19.3867	2.5565
Connector	4	246.97	61.7425	8.1419
Error	12	91.00	7.5833	
Total	19	396.13		

$H_0 : \alpha_1 = \alpha_2 = \alpha_3 = \alpha_4 = 0;$ H_a : at least one α_i is not zero.
$f_A = 2.5565 < F_{.01,3,12} = 5.95$, so fail to reject H_o. The data fails to indicate any effect due to the angle of pull.

6.

a $MSA = \frac{11.7}{2} = 5.85$, $MSE = \frac{25.6}{8} = 3.20$, $f = \frac{5.85}{3.20} = 1.83$, which is not significant at level .05.

b Otherwise extraneous variation associated with houses would tend to interfere with our ability to assess assessor effects. If there really was a difference between assessors, house variation might have hidden such a difference. Alternatively, an observed difference between assessors might have been due just to variation among houses and the manner in which assessors were allocated to homes.

7.

a CF = 140,454, SST = 3476, $SSTr = \frac{(905)^2 + (913)^2 + (936)^2}{18} - 140,454 = 28.78$,
$SSBl = \frac{430,295}{3} - 140,454 = 2977.67$, SSE = 469.55, MSTr = 14.39, MSE = 13.81, $f_{Tr} = 1.04$, which is clearly insignificant when compared to $F_{.05,2,51}$.

b $f_{Bl} = 12.68$, which is significant, and suggests substantial variation among subjects. If we had not controlled for such variation, it might have affected the analysis and conclusions.

8.

a $x_{1\bullet} = 4.34$, $x_{2\bullet} = 4.43$, $x_{3\bullet} = 8.53$, $x_{\bullet\bullet} = 17.30$, $SST = 3.8217$, $SSTr = 1.1458$, $SSBl = \frac{32.8906}{3} - 9.9763 = .9872$, $SSE = 1.6887$, $MSTr = .5729$, $MSE = .0938$, f = 6.1. Since $6.1 \ge F_{.05,2,18} = 3.55$, H_{oA} is rejected; there appears to be a difference between anesthetics.

b $Q_{.05,3,18} = 3.61$, w = .35. $\bar{x}_{1\bullet} = .434$, $\bar{x}_{2\bullet} = .443$, $\bar{x}_{3\bullet} = .853$, so both anesthetic 1 and anesthetic 2 appear to be different from anesthetic 3 but not from one another.

9.

Source	Df	SS	MS	f
Treatment	3	81.1944	27.0648	22.36
Block	8	66.5000	8.3125	6.87
Error	24	29.0556	1.2106	
Total	35	176.7500		

$F_{.05,3,24} = 3.01$. Reject H_o. There is an effect due to treatments.

$$Q_{.05,4,24} = 3.90;\ w = (3.90)\sqrt{\frac{1.2106}{9}} = 1.43$$

1	4	3	2
8.56	9.22	10.78	12.44

10.

Source	Df	SS	MS	f
Method	2	23.23	11.61	8.69
Batch	9	86.79	9.64	7.22
Error	18	24.04	1.34	
Total	29	134.07		

$F_{.01,2,18} = 6.01 < 8.69 < F_{.001,2,18} = 10.39$, so .001 < p-value < .01, which is significant. At least two of the curing methods produce differing average compressive strengths. (With p-value < .001, there are differences between batches as well.)

$$Q_{.05,3,18} = 3.61;\ w = (3.61)\sqrt{\frac{1.34}{10}} = 1.32$$

Method A	Method B	Method C
29.49	31.31	31.40

Methods B and C produce strengths that are not significantly different, but Method A produces strengths that are different (less) than those of both B and C.

Chapter 11

11. The residual, percentile pairs are (-0.1225, -1.73), (-0.0992, -1.15), (-0.0825, -0.81), (-0.0758, -0.55), (-0.0750, -0.32), (0.0117, -0.10), (0.0283, 0.10), (0.0350, 0.32), (0.0642, 0.55), (0.0708, 0.81), (0.0875, 1.15), (0.1575, 1.73).

Normal Probability Plot

residuals

z-percentile

The pattern is sufficiently linear, so normality is plausible.

12. $MSB = \frac{113.5}{4} = 28.38$, $MSE = \frac{25.6}{8} = 3.20$, $f_B = 8.87$, $F_{.01,4,8} = 7.01$, and since $8.87 \geq 7.01$, we reject H_o and conclude that $\sigma_B^2 > 0$.

13.

a With $Y_{ij} = X_{ij} + d$, $\overline{Y}_{i\bullet} = \overline{X}_{i\bullet} + d$, $\overline{Y}_{\bullet j} = \overline{X}_{\bullet j} + d$, $\overline{Y}_{\bullet\bullet} = \overline{X}_{\bullet\bullet} + d$, so all quantities inside the parentheses in (11.5) remain unchanged when the Y quantities are substituted for the corresponding X's (e.g., $\overline{Y}_{i\bullet} - \overline{Y}_{\bullet\bullet} = \overline{X}_{i\bullet} - \overline{X}_{\bullet\bullet}$, etc.).

b With $Y_{ij} = cX_{ij}$, each sum of squares for Y is the corresponding SS for X multiplied by c^2. However, when F ratios are formed the c^2 factors cancel, so all F ratios computed from Y are identical to those computed from X. If $Y_{ij} = cX_{ij} + d$, the conclusions reached from using the Y's will be identical to those reached using the X's.

14. $$E(\overline{X}_{i\bullet} - \overline{X}_{\bullet\bullet}) = E(\overline{X}_{i\bullet}) - E(\overline{X}_{\bullet\bullet}) = \frac{1}{J}E\left(\sum_j X_{ij}\right) - \frac{1}{IJ}E\left(\sum_i\sum_j X_{ij}\right)$$

$$= \frac{1}{J}\sum_j(\mu + \alpha_i + \beta_j) - \frac{1}{IJ}\sum_i\sum_j(\mu + \alpha_i + \beta_j)$$

$$= \mu + \alpha_i + \frac{1}{J}\sum_j \beta_j - \mu - \frac{1}{I}\sum_i \alpha_i - \frac{1}{J}\sum_j \beta_j = \alpha_i$$, as desired.

15.

a $\Sigma\alpha_i^2 = 24$, so $\Phi^2 = \left(\frac{3}{4}\right)\left(\frac{24}{16}\right) = 1.125$, $\Phi = 1.06$, $\nu_1 = 3$, $\nu_2 = 6$, and from figure 10.5, power $\approx .2$. For the second alternative, $\Phi = 1.59$, and power $\approx .43$.

b $\Phi^2 = \left(\frac{1}{J}\right)\sum\frac{\beta_j^2}{\sigma^2} = \left(\frac{4}{5}\right)\left(\frac{20}{16}\right) = 1.00$, so $\Phi = 1.00$, $\nu_1 = 4$, $\nu_2 = 12$, and power $\approx .3$.

Section 11.2

16.

a

Source	Df	SS	MS	f
A	2	30,763.0	15,381.50	3.79
B	3	34,185.6	11,395.20	2.81
AB	6	43,581.2	7263.53	1.79
Error	24	97,436.8	4059.87	
Total	35	205,966.6		

b $f_{AB} = 1.79$ which is not $\geq F_{.05,6,24} = 2.51$, so H_{oAB} cannot be rejected, and we conclude that no interaction is present.

c $f_A = 3.79$ which is $\geq F_{.05,2,24} = 3.40$, so H_{oA} is rejected at level .05.

d $f_B = 2.81$ which is not $\geq F_{.05,3,24} = 3.01$, so H_{oB} is not rejected.

e $Q_{.05,3,24} = 3.53$, $w = 3.53\sqrt{\frac{4059.87}{12}} = 64.93$.

3	1	2
3960.02	4010.88	4029.10

(Underlines: 3960.02–4010.88; 4010.88–4029.10)

Only times 2 and 3 yield significantly different strengths.

17.

a

Source	Df	SS	MS	f	$F_{.05}$
Sand	2	705	352.5	3.76	4.26
Fiber	2	1,278	639.0	6.82*	4.26
Sand&Fiber	4	279	69.75	0.74	3.63
Error	9	843	93.67		
Total	17	3,105			

There appears to be an effect due to carbon fiber addition.

b

Source	Df	SS	MS	f	$F_{.05}$
Sand	2	106.78	53.39	6.54*	4.26
Fiber	2	87.11	43.56	5.33*	4.26
Sand&Fiber	4	8.89	2.22	.27	3.63
Error	9	73.50	8.17		
Total	17	276.28			

There appears to be an effect due to both sand and carbon fiber addition to casting hardness.

c

Sand%	Fiber%	$\bar{x}$
0	0	62
15	0	68
30	0	69.5
0	.25	69
15	.25	71.5
30	.25	73
0	.50	68
15	.50	71.5
30	.50	74

The plot below indicates some effect due to sand and fiber addition with no significant interaction. This agrees with the statistical analysis in part **b**.

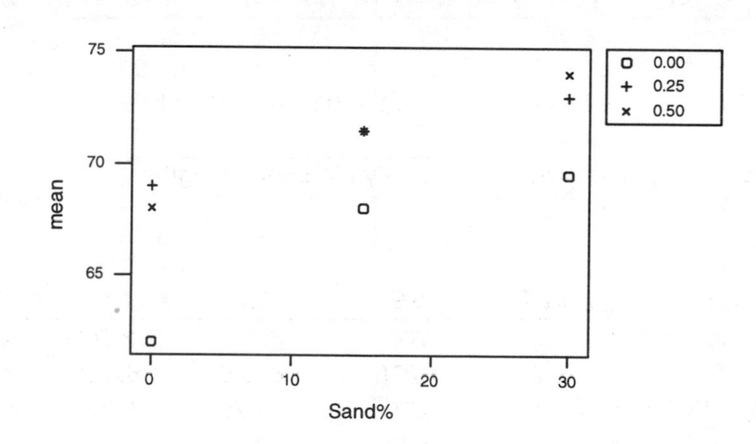

18.

Source	Df	SS	MS	f	$F_{.05}$	$F_{.01}$
Formulation	1	2,253.44	2,253.44	376.2**	4.75	9.33
Speed	2	230.81	115.41	19.27**	3.89	6.93
Formulation & Speed	2	18.58	9.29	1.55	3.89	6.93
Error	12	71.87	5.99			
Total	17	2,574.7				

a There appears to be no interaction between the two factors.

b Both formulation and speed appear to have a highly statistically significant effect on yield.

c Let formulation = Factor A and speed = Factor B.

For Factor A: $\mu_{1\bullet} = 187.03$ $\quad \mu_{2\bullet} = 164.66$

For Factor B: $\mu_{\bullet 1} = 177.83$ $\quad \mu_{\bullet 2} = 170.82$ $\quad \mu_{\bullet 3} = 178.88$

For Interaction: $\mu_{11} = 189.47$ $\quad \mu_{12} = 180.6$ $\quad \mu_{13} = 191.03$

$\mu_{21} = 166.2$ $\quad \mu_{22} = 161.03$ $\quad \mu_{33} = 166.73$

overall mean: $\mu = 175.84$

$\alpha_i = \mu_{i\bullet} - \mu$: $\quad \alpha_1 = 11.19$ $\quad \alpha_2 = -11.18$

$\beta_j = \mu_{\bullet j} - \mu$: $\quad \beta_1 = 1.99$ $\quad \beta_2 = -5.02$ $\quad \beta_3 = 3.04$

$y_{ij} = \mu_{ij} - (\mu + \alpha_i + \beta_j)$:

$y_{11} = .45$ $\quad y_{12} = -1.41$ $\quad y_{13} = .96$

$y_{21} = -.45$ $\quad y_{22} = 1.39$ $\quad y_{23} = -.97$

d

Observed	Fitted	Residual	Observed	Fitted	Residual
189.7	189.47	0.23	161.7	161.03	0.67
188.6	189.47	-0.87	159.8	161.03	-1.23
190.1	189.47	0.63	161.6	161.03	0.57
165.1	166.2	-1.1	189.0	191.03	-2.03
165.9	166.2	-0.3	193.0	191.03	1.97
167.6	166.2	1.4	191.1	191.03	0.07
185.1	180.6	4.5	163.3	166.73	-3.43
179.4	180.6	-1.2	166.6	166.73	-0.13
177.3	180.6	-3.3	170.3	166.73	3.57

e

i	Residual	Percentile	z-percentile
1	-3.43	2.778	-1.91
2	-3.30	8.333	-1.38
3	-2.03	13.889	-1.09
4	-1.23	19.444	-0.86
5	-1.20	25.000	-0.67
6	-1.10	30.556	-0.51
7	-0.87	36.111	-0.36
8	-0.30	41.667	-0.21
9	-0.13	47.222	-0.07
10	0.07	52.778	0.07
11	0.23	58.333	0.21
12	0.57	63.889	0.36
13	0.63	69.444	0.51
14	0.67	75.000	0.67
15	1.40	80.556	0.86
16	1.97	86.111	1.09
17	3.57	91.667	1.38
18	4.50	97.222	1.91

Normal Probability Plot of ANOVA Residuals

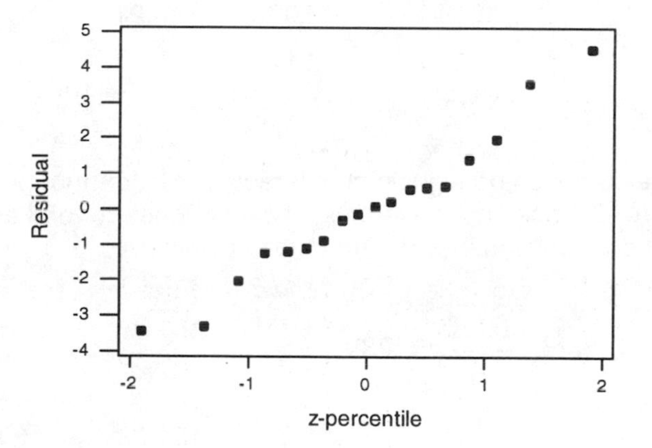

The residuals appear to be normally distributed.

19.

a

$x_{ij\bullet}$		j: 1	2	3	$x_{i\bullet\bullet}$
i	1	16.44	17.27	16.10	49.81
	2	16.24	17.00	15.91	49.15
	3	16.80	17.37	16.20	50.37
$x_{\bullet j\bullet}$		49.48	51.64	48.21	$x_{\bullet\bullet\bullet} = 149.33$
					CF = 1238.8583

Thus SST = 1240.1525 – 1238.8583 = 1.2942,

$$SSE = 1240.1525 - \frac{2479.9991}{2} = .1530,$$

$$SSA = \frac{(49.81)^2 + (49.15)^2 + (50.37)^2}{6} - 1238.8583 = .1243,\ SSB = 1.0024$$

Source	Df	SS	MS	f	$F_{.01}$
A	2	.1243	.0622	3.66	8.02
B	2	1.0024	.5012	29.48*	8.02
AB	4	.0145	.0036	.21	6.42
Error	9	.1530	.0170		
Total	17	1.2942			

H_{oAB} cannot be rejected, so no significant interaction; H_{oA} cannot be rejected, so varying levels of NaOH does not have a significant impact on total acidity; H_{oB} is rejected: type of coal does appear to affect total acidity.

b $Q_{.01,3,9} = 5.43$, $w = 5.43\sqrt{\frac{.0170}{6}} = .289$

j:	3	1	2
$\bar{x}_{\bullet j\bullet}$	8.035	8.247	8.607

(8.035 and 8.247 are underscored together.)

Coal 2 is judged significantly different from both 1 and 3, but these latter two don't differ significantly from each other.

20. $x_{11\bullet} = 855$, $x_{12\bullet} = 905$, $x_{13\bullet} = 845$, $x_{21\bullet} = 705$, $x_{22\bullet} = 735$, $x_{23\bullet} = 675$, $x_{1\bullet\bullet} = 2605$, $x_{2\bullet\bullet} = 2115$, $x_{\bullet 1\bullet} = 1560$, $x_{\bullet 2\bullet} = 1640$, $x_{\bullet 3\bullet} = 1520$, $x_{\bullet\bullet\bullet} = 4720$, $\Sigma\Sigma\Sigma x_{ijk}^2 = 1{,}253{,}150$, CF = 1,237,688.89, $\Sigma\Sigma x_{ij\bullet}^2 = 3{,}756{,}950$, which yields the accompanying ANOVA table.

Source	Df	SS	MS	f	$F_{.01}$
A	1	13,338.89	13,338.89	192.09*	9.93
B	2	1244.44	622.22	8.96*	6.93
AB	2	44.45	22.23	.32	6.93
Error	12	833.33	69.44		
Total	17	15,461.11			

Clearly, $f_{AB} = .32$ is insignificant, so H_{oAB} is not rejected. Both H_{oA} and H_{oB} are both rejected, since they are both greater than the respective critical values. Both phosphor type and glass type significantly affect the current necessary to produce the desired level of brightness.

21.

a $SST = 12{,}280{,}103 - \dfrac{(19{,}143)^2}{30} = 64{,}954.70$,

$SSE = 12{,}280{,}103 - \dfrac{(24{,}529{,}699)}{2} = 15{,}253.50$,

$SSA = \dfrac{122{,}380{,}901}{10} - \dfrac{(19{,}143)^2}{30} = 22{,}941.80$, $SSB = 22{,}765.53$,

$SSAB = 64{,}954.70 - [22{,}941.80 + 22{,}765.53 + 15{,}253.50] = 3993.87$

Source	Df	SS	MS	f
A	2	22,941.80	11,470.90	$\frac{11{,}470.90}{499.23} = 22.98$
B	4	22,765.53	5691.38	$\frac{5691.38}{499.23} = 11.40$
AB	8	3993.87	499.23	.49
Error	15	15,253.50	1016.90	
Total	29	64,954.70		

b $f_{AB} = .49$ is clearly not significant. Since $22.98 \geq F_{.05,2,8} = 4.46$, H_{oA} is rejected; since $11.40 \geq F_{.05,4,8} = 3.84$, H_{oB} is also rejected. Thus we conclude that the different cement factors affect flexural strength differently and that batch variability contributes to variation in flexural strength.

22. The relevant null hypotheses are $H_{0A}: \alpha_1 = \alpha_2 = \alpha_3 = \alpha_4 = 0$; $H_{0B}: \sigma_B^2 = 0$; $H_{0AB}: \sigma_G^2 = 0$.

$$SST = 11{,}499{,}492 - \frac{(16{,}598)^2}{24} = 20{,}591.83,$$

$$SSE = 11{,}499{,}492 - \frac{(22{,}982{,}552)}{2} = 8216.0,$$

$$SSA = \left[\frac{(4112)^2 + (4227)^2 + (4122)^2 + (4137)^2}{6}\right] - \frac{(16{,}598)^2}{24} = 1387.5,$$

$$SSB = \left[\frac{(5413)^2 + (5621)^2 + (5564)^2}{8}\right] - \frac{(16{,}598)^2}{24} = 2888.08,$$

$$SSAB = 20{,}591.83 - [8216.0 + 1387.5 + 2888.08] = 8216.25$$

Source	Df	SS	MS	f	$F_{.05}$
A	3	1387.5	462.5	$\frac{MSA}{MSAB} = .34$	4.76
B	2	2888.08	1444.04	$\frac{MSB}{MSAB} = 1.07$	5.14
AB	6	8100.25	1350.04	$\frac{MSAB}{MSE} = 1.97$	3.00
Error	12	8216.0	684.67		
Total	23	20,591.83			

Interaction between brand and writing surface has no significant effect on the lifetime of the pen, and since neither f_A nor f_B is greater than its respective critical value, we can conclude that neither the surface nor the brand of pen has a significant effect on the writing lifetime.

23. Summary quantities include $x_{1\bullet\bullet} = 9410$, $x_{2\bullet\bullet} = 8835$, $x_{3\bullet\bullet} = 9234$, $x_{\bullet 1\bullet} = 5432$, $x_{\bullet 2\bullet} = 5684$, $x_{\bullet 3\bullet} = 5619$, $x_{\bullet 4\bullet} = 5567$, $x_{\bullet 3\bullet} = 5177$, $x_{\bullet\bullet\bullet} = 27{,}479$, $CF = 16{,}779{,}898.69$, $\Sigma x_{i\bullet\bullet}^2 = 251{,}872{,}081$, $\Sigma x_{\bullet j\bullet}^2 = 151{,}180{,}459$, resulting in the accompanying ANOVA table.

Source	Df	SS	MS	f
A	2	11,573.38	5786.69	$\frac{MSA}{MSAB} = 26.70$
B	4	17,930.09	4482.52	$\frac{MSB}{MSAB} = 20.68$
AB	8	1734.17	216.77	$\frac{MSAB}{MSE} = 1.38$
Error	30	4716.67	157.22	
Total	44	35,954.31		

Since $1.38 < F_{.01,8,30} = 3.17$, H_{oG} cannot be rejected, and we continue: $26.70 \geq F_{.01,2,8} = 8.65$, and $20.68 \geq F_{.01,4,8} = 7.01$, so both H_{oA} and H_{oB} are rejected. Both capping material and the different batches affect compressive strength of concrete cylinders.

24.

a $$E(\overline{X}_{i..} - \overline{X}_{...}) = \frac{1}{JK}\sum_j\sum_k E(X_{ijk}) - \frac{1}{IJK}\sum_i\sum_j\sum_k E(X_{ijk})$$
$$= \frac{1}{JK}\sum_j\sum_k(\mu + \alpha_i + \beta_j + \gamma_{ij}) - \frac{1}{IJK}\sum_i\sum_j\sum_k(\mu + \alpha_i + \beta_j + \gamma_{ij}) = \mu + \alpha_i - \mu = \alpha_i$$

b $$E(\hat{\gamma}_{ij}) = \frac{1}{K}\sum_k E(X_{ijk}) - \frac{1}{JK}\sum_j\sum_k E(X_{ijk}) - \frac{1}{IK}\sum_i\sum_k E(X_{ijk}) + \frac{1}{IJK}\sum_i\sum_j\sum_k E(X_{ijk})$$
$$= \mu + \alpha_i + \beta_j + \gamma_{ij} - (\mu + \alpha_i) - (\mu + \beta_j) + \mu = \gamma_{ij}$$

25. With $\theta = \alpha_i - \alpha_i'$, $\hat{\theta} = \overline{X}_{i..} - \overline{X}_{i'..} = \frac{1}{JK}\sum_j\sum_k(X_{ijk} - X_{i'jk})$, and since $i \neq i'$, $X_{ijk} and X_{i'jk}$ are independent for every j, k. Thus $Var(\hat{\theta}) = Var(\overline{X}_{i..}) + Var(\overline{X}_{i'..}) = \frac{\sigma^2}{JK} + \frac{\sigma^2}{JK} = \frac{2\sigma^2}{JK}$ (because $Var(\overline{X}_{i..}) = Var(\overline{\varepsilon}_{i..})$ and $Var(\varepsilon_{ijk}) = \sigma^2$) so $\hat{\sigma}_{\hat{\theta}} = \sqrt{\frac{2MSE}{JK}}$. The appropriate number of d.f. is IJ(K – 1), so the C.I. is $(\overline{x}_{i..} - \overline{x}_{i'..}) \pm t_{\alpha/2,IJ(K-1)}\sqrt{\frac{2MSE}{JK}}$. For the data of exercise 19, $\overline{x}_{2..} = 49.15$, $\overline{x}_{3..} = 50.37$, MSE = .0170, $t_{.025,9} = 2.262$, J = 3, K = 2, so the C.I. is
$$(49.15 - 50.37) \pm 2.262\sqrt{\frac{.0370}{6}} = -1.22 \pm .17 = (-1.39, -1.05).$$

26.

a $\dfrac{E(MSAB)}{E(MSE)} = 1 + \dfrac{K\sigma_G^2}{\sigma^2} = 1$ if $\sigma_G^2 = 0$ and > 1 if $\sigma_G^2 > 0$, so $\dfrac{MSAB}{MSE}$ is the appropriate F ratio.

b $\dfrac{E(MSA)}{E(MSAB)} = \dfrac{\sigma^2 + K\sigma_G^2 + JK\sigma_A^2}{\sigma^2 + K\sigma_G^2} = 1 + \dfrac{JK\sigma_A^2}{\sigma^2 + K\sigma_G^2} = 1$ if $\sigma_A^2 = 0$ and > 1 if $\sigma_A^2 > 0$, so $\dfrac{MSA}{MSAB}$ is the appropriate F ratio.

Section 11.3

27.

a

Source	Df	SS	MS	f	$F_{.05}$
A	2	14,144.44	7072.22	61.06	3.35
B	2	5,511.27	2755.64	23.79	3.35
c	2	244,696.39	122.348.20	1056.24	3.35
AB	4	1,069.62	267.41	2.31	2.73
AC	4	62.67	15.67	.14	2.73
BC	4	331.67	82.92	.72	2.73
ABC	8	1,080.77	135.10	1.17	2.31
Error	27	3,127.50	115.83		
Total	53	270,024.33			

b the computed f-statistics for all four interaction terms are less than the tabled values for statistical significance at the level .05. This indicates that none of the interactions are statistically significant.

c The computed f-statistics for all three main effects exceed the tabled value for significance at level .05. All three main effects are statistically significant.

d $Q_{.05,3,27}$ is not tabled, use $Q_{.05,3,24} = 3.53$, $w = 3.53\sqrt{\dfrac{115.83}{(3)(3)(2)}} = 8.95$. All three levels differ significantly from each other.

28.

Source	Df	SS	MS	f	$F_{.01}$
A	3	19,149.73	6,383.24	2.70	4.72
B	2	2,589,047.62	1,294,523.81	546.79	5.61
C	1	157,437.52	157,437.52	66.50	7.82
AB	6	53,238.21	8,873.04	3.75	3.67
AC	3	9,033.73	3,011.24	1.27	4.72
BC	2	91,880.04	45,940.02	19.40	5.61
ABC	6	6,558.46	1,093.08	.46	3.67
Error	24	56,819.50	2,367.48		
Total	47	2,983,164.81			

The statistically significant interactions are AB and BC. Factor A appears to be the least significant of all the factors. It does not have a significant main effect and the significant interaction (AB) is only slightly greater than the tabled value at significance level .01

29. I = 3, J = 2, K = 4, L = 4; $SSA = JKL\sum(\bar{x}_{i...} - \bar{x}_{....})^2$; $SSB = IKL\sum(\bar{x}_{.j..} - \bar{x}_{....})^2$;
$SSC = IJL\sum(\bar{x}_{..k.} - \bar{x}_{....})^2$.

For level A: $\bar{x}_{1...} = 3.781$ $\bar{x}_{2...} = 3.625$ $\bar{x}_{3...} = 4.469$

For level B: $\bar{x}_{.1..} = 4.979$ $\bar{x}_{.2..} = 2.938$

For level C: $\bar{x}_{..1.} = 3.417$ $\bar{x}_{..2.} = 5.875$ $\bar{x}_{..3.} = .875$ $\bar{x}_{..4.} = 5.667$

$\bar{x}_{....} = 3.958$

SSA = 12.907; SSB = 99.976; SSC = 393.436

a

Source	Df	SS	MS	f	$F_{.05}$*
A	2	12.907	6.454	1.04	3.15
B	1	99.976	99.976	16.09	4.00
C	3	393.436	131.145	21.10	2.76
AB	2	1.646	.823	.13	3.15
AC	6	71.021	11.837	1.90	2.25
BC	3	1.542	.514	.08	2.76
ABC	6	9.805	1.634	.26	2.25
Error	72	447.500	6.215		
Total	95	1,037.833			

*use 60 df for denominator of tabled F.

b No interaction effects are significant at level .05

c Factor B and C main effects are significant at the level .05

d $Q_{.05,4,72}$ is not tabled, use $Q_{.05,4,60} = 3.74$, $w = 3.74\sqrt{\dfrac{6.215}{(3)(2)(4)}} = 1.90$.

Machine:	3	1	4	2
Mean:	.875	3.417	5.667	5.875

30.

a

b

Source	Df	SS	MS	f	$F_{.05}$
A	3	.22625	.075417	77.35	9.28
B	1	.000025	.000025	.03	10.13
C	1	.0036	.0036	3.69	10.13
AB	3	.004325	.0014417	1.48	9.28
AC	3	.00065	.000217	.22	9.28
BC	1	.000625	.000625	.64	10.13
ABC	3	.002925	.000975		
Error	--	--	--		
Total	15	.2384			

The only statistically significant effect at the level .05 is the factor A main effect: levels of nitrogen.

c $Q_{.05,4,3} = 6.82$; $w = 6.82\sqrt{\dfrac{.002925}{(2)(2)}} = .1844$.

1	2	3	4
1.1200	1.3025	1.3875	1.4300

31.

$x_{ij.}$	B_1	B_2	B_3
A_1	210.2	224.9	218.1
A_2	224.1	229.5	221.5
A_3	217.7	230.0	202.0
$x_{.j.}$	652.0	684.4	641.6

$x_{i.k}$	A$_1$	A$_2$	A$_3$
C$_1$	213.8	222.0	205.0
C$_2$	225.6	226.5	223.5
C$_3$	213.8	226.6	221.2
$x_{i..}$	653.2	675.1	649.7

$x_{.jk}$	C$_1$	C$_2$	C$_3$
B$_1$	213.5	220.5	218.0
B$_2$	214.3	246.1	224.0
B$_3$	213.0	209.0	219.6
$x_{..k}$	640.8	675.6	661.6

$\Sigma\Sigma x_{ij.}^2 = 435{,}382.26$ $\Sigma\Sigma x_{i.k}^2 = 435{,}156.74$ $\Sigma\Sigma x_{.jk}^2 = 435{,}666.36$

$\Sigma x_{.j.}^2 = 1{,}305{,}157.92$ $\Sigma x_{i..}^2 = 1{,}304{,}540.34$ $\Sigma x_{..k}^2 = 1{,}304{,}774.56$

Also, $\Sigma\Sigma\Sigma x_{ijk}^2 = 145{,}386.40$, $x = 1978$, CF = 144,906.81, from which we obtain the ANOVA table displayed in the problem statement. $F_{.01,4,8} = 7.01$, so the AB and BC interactions are significant (as can b seen from the p-values) and tests for main effects are not appropriate.

32.

a Since $\frac{E(MSABC)}{E(MSE)} = \frac{\sigma^2 + L\sigma_{ABC}^2}{\sigma^2} = 1$ if $\sigma_{ABC}^2 = 0$ and > 1 if $\sigma_{ABC}^2 > 0$, $\frac{MSABC}{MSE}$ is the appropriate F ratio for testing $H_0 : \sigma_{ABC}^2 = 0$. Similarly, $\frac{MSC}{MSE}$ is the F ratio for testing $H_0 : \sigma_C^2 = 0$; $\frac{MSAB}{MSABC}$ is the F ratio for testing $H_0 : all\ \gamma_{ij}^{AB} = 0$; and $\frac{MSA}{MSAC}$ is the F ratio for testing $H_0 : all\ \alpha_i = 0$.

b

Source	Df	SS	MS	f	$F_{.01}$
A	1	14,318.24	14,318.24	$\frac{MSA}{MSAC}=19.85$	98.50
B	3	9656.4	3218.80	$\frac{MSB}{MSBC}=6.24$	9.78
C	2	2270.22	1135.11	$\frac{MSC}{MSE}=3.15$	5.61
AB	3	3408.93	1136.31	$\frac{MSAB}{MSABC}=2.41$	9.78
AC	2	1442.58	721.29	$\frac{MSAC}{MSABC}=2.00$	5.61
BC	6	3096.21	516.04	$\frac{MSBC}{MSE}=1.43$	3.67
ABC	6	2832.72	472.12	$\frac{MSABC}{MSE}=1.31$	3.67
Error	24	8655.60	360.65		
Total	47				

At level .01, no H_o's can be rejected, so there appear to be no interaction or main effects present.

33.

Source	Df	SS	MS	f
A	6	67.32	11.02	
B	6	51.06	8.51	
C	6	5.43	.91	.61
Error	30	44.26	1.48	
Total	48	168.07		

Since .61 < $F_{.05,6,30}=2.42$, treatment was not effective.

34.

	1	2	3	4	5	6
$x_{i..}$	144	205	272	293	85	98
$x_{.j.}$	171	199	147	221	177	182
$x_{..k}$	180	161	186	171	169	230

Thus $x_{...}=1097$, $CF=\frac{(1097)^2}{36}=33{,}428.03$, $\Sigma\Sigma x_{ij(k)}^2=42{,}219$, $\Sigma x_{i..}^2=239{,}423$, $\Sigma x_{.j.}^2=203{,}745$, $\Sigma x_{..k}^2=203.619$

Source	Df	SS	MS	f
A	5	6475.80	1295.16	
B	5	529.47	105.89	
C	5	508.47	101.69	1.59
Error	20	1277.23	63.89	
Total	35	8790.97		

Since 1.59 is not $\geq F_{.05,5,20} = 2.71$, H_{oC} is not rejected; shelf space does not appear to affect sales.

35.

	1	2	3	4	5	
$x_{i..}$	40.68	30.04	44.02	32.14	33.21	$\Sigma x_{i..}^2 = 6630.91$
$x_{.j.}$	29.19	31.61	37.31	40.16	41.82	$\Sigma x_{.j.}^2 = 6605.02$
$x_{..k}$	36.59	36.67	36.03	34.50	36.30	$\Sigma x_{..k}^2 = 6489.92$

$x_{...} = 180.09$, CF = 1297.30, $\Sigma\Sigma x_{ij(k)}^2 = 1358.60$

36.

Source	Df	SS	MS	f	$F_{.01}$*
A (laundry treatment)	3	39.171	13.057	16.23	3.95
B (pen type)	2	.665	.3325	.41	4.79
C (Fabric type)	5	21.508	4.3016	5.35	3.17
AB	6	1.432	.2387	.30	2.96
AC	15	15.953	1.0635	1.32	2.19
BC	10	1.382	.1382	.17	2.47
ABC	30	9.016	.3005	.37	1.86
Error	144	115.820	.8043		
Total	215	204.947			

*Because a denominator degrees of freedom for 144 is not tabled, use 120.

At the level .01, there are two statistically significant main effects (laundry treatment and fabric type). There are no statistically significant interactions.

37.

Source	Df	MS	f	$F_{.01}$*
A	2	2207.329	2259.29	5.39
B	1	47.255	48.37	7.56
C	2	491.783	503.36	5.39
D	1	.044	.05	7.56
AB	2	15.303	15.66	5.39
AC	4	275.446	281.93	4.02
AD	2	.470	.48	5.39
BC	2	2.141	2.19	5.39
BD	1	.273	.28	7.56
CD	2	.247	.25	5.39
ABC	4	3.714	3.80	4.02
ABD	2	4.072	4.17	5.39
ACD	4	.767	.79	4.02
BCD	2	.280	.29	5.39
ABCD	4	.347	.355	4.02
Error	36	.977		
Total	71			

*Because denominator d.f. for 36 is not tabled, use d.f. = 30

SST = (71)(93.621) = 6,647.091. Computing all other sums of squares and adding them up = 6,645.702. Thus SSABCD = 6,647.091 – 6,645.702 = 1.389 and

$$MSABCD = \frac{1.389}{4} = .347.$$

At level .01 the statistically significant main effects are A, B, C. The interaction AB and AC are also statistically significant. No other interactions are statistically significant.

Chapter 11

Section 11.4

38.

a

Treatment Condition	$x_{ijk.}$	1	2	Effect Contrast	$SS = \frac{(contrast)^2}{16}$
$(1) = x_{111.}$	404.2	839.2	1991.0	3697.0	
$a = x_{211.}$	435.0	1151.8	1706.0	164.2	1685.1
$b = x_{121.}$	549.6	717.6	83.4	583.4	21,272.2
$ab = x_{221.}$	602.2	988.4	80.8	24.2	36.6
$c = x_{112.}$	339.2	30.8	312.6	-285.0	5076.6
$ac = x_{212.}$	378.4	52.6	270.8	-2.6	.4
$bc = x_{122.}$	473.4	39.2	21.8	-41.8	109.2
$abc = x_{222.}$	515.0	41.6	2.4	-19.4	23.5

$$\Sigma\Sigma\Sigma\Sigma x_{ijkl}^2 = 882{,}573.38; \quad SST = 882{,}573.38 - \frac{(3697)^2}{16} = 28{,}335.3$$

b The important effects are those with small associated p-values, indicating statistical significance. Those effects significant at level .05 (i.e., p-value < .05) are the three main effects and the speed by distance interaction.

39.

Condition	Total	1	2	Contrast	$SS = \frac{(contrast)^2}{24}$
111	315	927	2478	5485	
211	612	1551	3007	1307	A = 71,177.04
121	584	1163	680	1305	B = 70,959.38
221	967	1844	627	199	AB = 1650.04
112	453	297	624	529	C = 11,660.04
212	710	383	681	-53	AC = 117.04
122	737	257	86	57	BC = 135.38
222	1107	370	113	27	ABC = 30.38

a $$\hat{\beta}_1 = \bar{x}_{.2..} - \bar{x}_{....} = \frac{584+967+737+1107-315-612-453-710}{24} = 54.38$$

$$\hat{\gamma}_{11}^{AC} = \frac{315-612+584-967-453+710-737+1107}{24} = 2.21;$$

$$\hat{\gamma}_{21}^{AC} = -\hat{\gamma}_{11}^{AC} = 2.21.$$

b Factor SS's appear above. With $CF = \frac{5485^2}{24} = 1{,}253{,}551.04$ and $\Sigma\Sigma\Sigma\Sigma x_{ijkl}^2 = 1{,}411{,}889$, SST = 158,337.96, from which SSE = 2608.7. The ANOVA table appears in the answer section. $F_{.05,1,16} = 4.49$, from which we see that the AB interaction and al the main effects are significant.

40.

a In the accompanying ANOVA table, effects are listed in the order implied by Yates' algorithm. $\Sigma\Sigma\Sigma\Sigma\Sigma x_{ijklm}^2 = 4783.16$, $x_{.....} = 388.14$, so $SST = 4783.16 - \frac{368.14^2}{32} = 72.56$ and SSE = 72.56 – (sum of all other SS's) = 35.85.

Source	Df	SS	MS	f
A	1	.17	.17	< 1
B	1	1.94	1.94	< 1
C	1	3.42	3.42	1.53
D	1	8.16	8.16	3.64
AB	1	.26	.26	< 1
AC	1	.74	.74	< 1
AD	1	.02	.02	< 1
BC	1	13.08	13.08	5.84
BD	1	.91	.91	< 1
CD	1	.78	.78	< 1
ABC	1	.78	.78	< 1
ABD	1	6.77	6.77	3.02
ACD	1	.62	.62	< 1
BCD	1	1.76	1.76	< 1
ABCD	1	.00	.00	< 1
Error	16	35.85	2.24	
Total	31			

b $F_{.05,1,16} = 4.49$, so none of the interaction effects is judged significant, and only the D main effect is significant.

41. $\Sigma\Sigma\Sigma\Sigma\Sigma x_{ijklm}^2 = 3{,}308{,}143$, $x_{.....} = 11{,}956$, so $CF = \frac{(11{,}956)^2}{48} = 2{,}979{,}535.02$, and SST = 328,607.98. Each SS is $\frac{(effectcontrast)^2}{48}$ and SSE is obtained by subtraction. The ANOVA table appears in the answer section. $F_{.05,1,32} \approx 4.15$, a value exceeded by the F ratios for AB interaction and the four main effects.

42. $\Sigma\Sigma\Sigma\Sigma\Sigma x_{ijklm}^2 = 32{,}917{,}817$, $x_{.....} = 39{,}371$, $SS = \frac{(contrast)^2}{48}$, and error d.f. = 32.

Effect	MS	f	Effect	MS	f
A	16,170.02	3.42	BD	3519.19	< 1
B	332,167.69	70.17	CD	4700.52	< 1
C	43,140.02	9.11	ABC	1210.02	< 1
D	20,460.02	4.33	ABD	15,229.69	3.22
AB	1989.19	< 1	ACD	1963.52	< 1
AC	776.02	< 1	BCD	10,354.69	2.19
AD	16,170.02	3.42	ABCD	1692.19	< 1
BC	3553.52	< 1	Error	4733.69	

$F_{.01,1,32} \approx 7.5$, so only the B and C main effects are judged significant at the 1% level.

43.

Condition/ Effect	$SS = \frac{(contrast)^2}{16}$	f	Condition/ Effect	$SS = \frac{(contrast)^2}{16}$	f
(1)	--		D	414.123	1067.33
A	.436	1.12	AD	.017	< 1
B	.099	< 1	BD	.456	< 1
AB	.497	1.28	ABD	.055	--
C	.109	< 1	CD	2.190	5.64
AC	.078	< 1	ACD	1.020	--
BC	1.404	3.62	BCD	.133	--
ABC	.051	--	ABCD	.681	--

SSE = .051 + .055 + 1.020 + .133 + .681 = 1.940, d.f. = 5, so MSE = .388. $F_{.05,1,5} = 6.61$, so only the D main effect is significant.

44.

a The eight treatment conditions which have even number of letters in common with abcd and thus go in the first (principle) block are (1), ab, ac, bc, ad, bd, cd, and abd; the other eight conditions are placed in the second block.

b and **c**

$x_{....} = 1290$, $\Sigma\Sigma\Sigma\Sigma x_{ijkl}^2 = 105{,}160$, so SST = 1153.75. The two block totals are 639 and 651, so $SSBl = \frac{639^2}{8} + \frac{651^2}{8} - \frac{1290^2}{16} = 9.00$, which is identical (as it must be here) to SSABCD computed from Yates algorithm.

Condition/Effect	Block	$SS = \frac{(contrast)^2}{16}$	f
(1)	1	--	
A	2	25.00	1.93
B	2	9.00	< 1
AB	1	12.25	< 1
C	2	49.00	3.79
AC	1	2.25	< 1
BC	1	.25	< 1
ABC	2	9.00	--
D	2	930.25	71.90
AD	1	36.00	2.78
BD	1	25.00	1.93
ABD	2	20.25	--
CD	1	4.00	< 1
ACD	2	20.25	--
BCD	2	2.25	--
ABCD=Blocks	1	9.00	--
Total		1153.75	

$SSE = 9.0 + 20.25 + 20.25 + 2.25 = 51.75$; d.f. = 4, so MSE = 12.9375, $F_{.05,1,4} = 7.71$, so only the D main effect is significant.

45.

a The allocation of treatments to blocks is as given in the answer section, with block #1 containing all treatments having an even number of letters in common with both ab and cd, etc.

b $x_{.....} = 16{,}898$, so $SST = 9{,}035{,}054 - \frac{16{,}898^2}{32} = 111{,}853.88$. The eight $block \times replication$ totals are 2091 (= 618 + 421 + 603 + 449, the sum of the four observations in block #1 on replication #1), 2092, 2133, 2145, 2113, 2080, 2122, and 2122, so $SSBl = \frac{2091^2}{4} + \ldots + \frac{2122^2}{4} - \frac{16{,}898^2}{32} = 898.88$. The remaining SS's as well as all F ratios appear in the ANOVA table in the answer section. With $F_{.01,1,12} = 9.33$, only the A and B main effects are significant.

46. The result is clearly true if either defining effect is represented by either a single letter (e.g., A) or a pair of letters (e.g. AB). The only other possibilities are for both to be "triples" (e.g. ABC or ABD, all of which must have two letters in common.) or one a triple and the other ABCD. But the generalized interaction of ABC and ABD is CD, so a two-factor interaction is confounded, and the generalized interaction of ABC and ABCD is D, so a main effect is confounded.

47. See the answer section.

48.

a The treatment conditions in the observed group are (in standard order) (1), ab, ac, bc, ad, bd, cd, and abcd. The alias pairs are {A, BCD}, {B, ACD}, {C, ABD}, {D, ABC}, {AB, CD}, {AC, BD}, and {AD, BC}.

b

	A	B	C	D	AB	AC	AD
(1) = 19.09	-	-	-	-	+	+	+
Ab = 20.11	+	+	-	-	+	-	-
Ac = 21.66	+	-	+	-	-	+	-
Bc = 20.44	-	+	+	-	-	-	+
Ad = 13.72	+	-	-	+	-	-	+
Bd = 11.26	-	+	-	+	-	+	-
Cd = 11.72	-	-	+	+	+	-	-
Abcd = 12.29	+	+	+	+	+	+	+
Contrast	5.27	-2.09	1.93	-32.31	-3.87	-1.69	.79
SS	3.47	.55	.47	130.49	1.87	.36	.08
f	4.51	< 1	< 1	169.47	SSE=2.31	MSE=.770	

$F_{.05,1,3} = 10.13$, so only the D main effect is judged significant.

49.

		A	B	C	D	E	AB	AC	AD	AE	BC	BD	BE	CD	CE	DE
a	70.4	+	-	-	-	-	-	-	-	-	+	+	+	+	+	+
b	72.1	-	+	-	-	-	-	+	+	+	-	-	-	+	+	+
c	70.4	-	-	+	-	-	+	-	+	+	-	+	+	-	-	+
abc	73.8	+	+	+	-	-	+	+	-	-	+	-	-	-	-	+
d	67.4	-	-	-	+	-	+	+	-	+	+	-	+	-	+	-
abd	67.0	+	+	-	+	-	+	-	+	-	-	+	-	-	+	-
acd	66.6	+	-	+	+	-	-	+	+	-	-	-	+	+	-	-
bcd	66.8	-	+	+	+	-	-	-	-	+	+	+	-	+	-	-
e	68.0	-	-	-	-	+	+	+	+	-	+	+	-	+	-	-
abe	67.8	+	+	-	-	+	+	-	-	+	-	-	+	+	-	-
ace	67.5	+	-	+	-	+	-	+	-	+	-	+	-	-	+	-
bce	70.3	-	+	+	-	+	-	-	+	-	+	-	+	-	+	-
ade	64.0	+	-	-	+	+	-	-	+	+	+	-	-	-	-	+
bde	67.9	-	+	-	+	+	-	+	-	-	-	+	+	-	-	+
cde	65.9	-	-	+	+	+	+	-	-	-	-	-	-	+	+	+
abcde	68.0	+	+	+	+	+	+	+	+	+	+	+	+	+	+	+

Thus $SSA = \dfrac{(70.4 - 72.1 - 70.4 + ... + 68.0)^2}{16} = 2.250$, SSB = 7.840, SSC = .360, SSD = 52.563, SSE = 10.240, SSAB = 1.563, SSAC = 7.563, SSAD = .090, SSAE = 4.203, SSBC = 2.103, SSBD = .010, SSBE = .123, SSCD = .010, SSCE = .063, SSDE = 4.840, Error SS = sum of two factor SS's = 20.568, Error MS = 2.057, $F_{.01,1,10} = 10.04$, so only the D main effect is significant.

Supplementary Exercises

50.

Source	Df	SS	MS	f
Treatment	4	14.962	3.741	36.7
Block	8	9.696		
Error	32	3.262	.102	
Total	44	27.920		

$H_0 : \alpha_1 = \alpha_2 = \alpha_3 = \alpha_4 = \alpha_5 = 0$ will be rejected if $f = \dfrac{MSTr}{MSE} \ge F_{.05,4,32} = 2.67$.

Because $36.7 \ge 2.67$, H_o is rejected. We conclude that expected smoothness score does depend somehow on the drying method used.

51.

Source	Df	SS	MS	f
A	1	322.667	322.667	980.38
B	3	35.623	11.874	36.08
AB	3	8.557	2.852	8.67
Error	16	5.266	.329	
Total	23	372.113		

We first test the null hypothesis of no interactions ($H_0 : \gamma_{ij} = 0$ for all I, j). H_o will be rejected at level .05 if $f_{AB} = \frac{MSAB}{MSE} \geq F_{.05,3,16} = 3.24$. Because $8.67 \geq 3.24$, H_o is rejected. Because we have concluded that interaction is present, tests for main effects are not appropriate.

52. Let X_{ij} = the amount of clover accumulation when the ith sowing rate is used in the jth plot = $\mu + \alpha_i + \beta_j + e_{ij}$. $H_0 : \alpha_1 = \alpha_2 = \alpha_3 = \alpha_4 = 0$ will be rejected if

$$f = \frac{MSTr}{MSE} \geq F_{\alpha, I-1,(I-1)(J-1)} = F_{.05,3,9} = 3.86$$

Source	Df	SS	MS	f
Treatment	3	3,141,153.5	1,040,751.17	2.28
Block	3	19,470,550.0		
Error	9	4,141,165.5	460,129.50	
Total	15	26,752,869.0		

Because $2.28 < 3.86$, H_o is not rejected. Expected accumulation does not appear to depend on sowing rate.

53. Let A = spray volume, B = belt speed, C = brand.

Condition	Total	1	2	Contrast	$SS = \frac{(contrast)^2}{16}$
(1)	76	129	289	592	21,904.00
A	53	160	303	22	30.25
B	62	143	13	48	144.00
AB	98	160	9	134	1122.25
C	88	-23	31	14	12.25
AC	55	36	17	-4	1.00
BC	59	-33	59	-14	12.25
ABC	101	42	75	16	16.00

The ANOVA table is as follows:

Effect	Df	MS	f
A	1	30.25	6.72
B	1	144.00	32.00
AB	1	1122.25	249.39
C	1	12.25	2.72
AC	1	1.00	.22
BC	1	12.25	2.72
ABC	1	16.00	3.56
Error	8	4.50	
Total	15		

$F_{.05,1,8} = 5.32$, so all of the main effects are significant at level .05, but none of the interactions are significant.

54. We use Yates' method for calculating the sums of squares, and for ease of calculation, we divide each observation by 1000.

Condition	Total	1	2	Contrast	$SS = \frac{(contrast)^2}{8}$
(1)	23.1	66.1	213.5	317.2	-
A	43.0	147.4	103.7	20.2	51.005
B	71.4	70.2	24.5	44.6	248.645
AB	76.0	33.5	-4.3	-12.0	18.000
C	37.0	19.9	81.3	-109.8	1507.005
AC	33.2	4.6	-36.7	-28.8	103.68
BC	17.0	-3.8	-15.3	-118.0	1740.5
ABC	16.5	-.5	3.3	18.6	43.245

We assume that there is no three-way interaction, so the MSABC becomes the MSE for ANOVA:

Source	df	MS	f
A	1	51.005	1.179
B	1	248.645	5.750*
AB	1	18.000	< 1
C	1	1507.005	34.848*
AC	1	103.68	2.398
BC	1	1740.5	40.247*
Error	1	43.245	
Total	8		

With $F_{.05,1,8} = 5.32$, the B and C main effects are significant at the .05 level, as well as the BC interaction. We conclude that although binder type (A) is not significant, both amount of water (B) and the land disposal scenario (C) affect the leaching characteristics under study., and there is some interaction between the two factors.

55.

a

Effect	%Iron	1	2	3	Effect Contrast	SS
	7	18	37	174	684	
A	11	19	137	510	144	1296
B	7	62	169	50	36	81
AB	12	75	341	94	0	0
C	21	79	9	14	272	4624
AC	41	90	41	22	32	64
BC	27	165	47	2	12	9
ABC	48	176	47	-2	-4	1
D	28	4	1	100	336	7056
AD	51	5	13	172	44	121
BD	33	20	11	32	8	4
ABD	57	21	11	0	0	0
CD	70	23	1	12	72	324
ACD	95	24	1	0	-32	64
BCD	77	25	1	0	-12	9
ABCD	99	22	-3	-4	-4	1

We use $estimate = \dfrac{contrast}{2^p}$ when n = 1 (see p 472 of text) to get

$\hat{\alpha}_1 = \dfrac{144}{2^4} = \dfrac{144}{16} = 9.00$, $\hat{\beta}_1 = \dfrac{36}{16} = 2.25$, $\hat{\delta}_1 = \dfrac{272}{16} = 17.00$, $\hat{\gamma}_1 = \dfrac{336}{16} = 21.00$.

Similarly, $\left(\widehat{\alpha\beta}\right)_{11} = 0$, $\left(\widehat{\alpha\delta}\right)_{11} = 2.00$, $\left(\widehat{\alpha\gamma}\right)_{11} = 2.75$, $\left(\widehat{\beta\delta}\right)_{11} = .75$, $\left(\widehat{\beta\gamma}\right)_{11} = .50$, and $\left(\widehat{\delta\gamma}\right)_{11} = 4.50$.

b

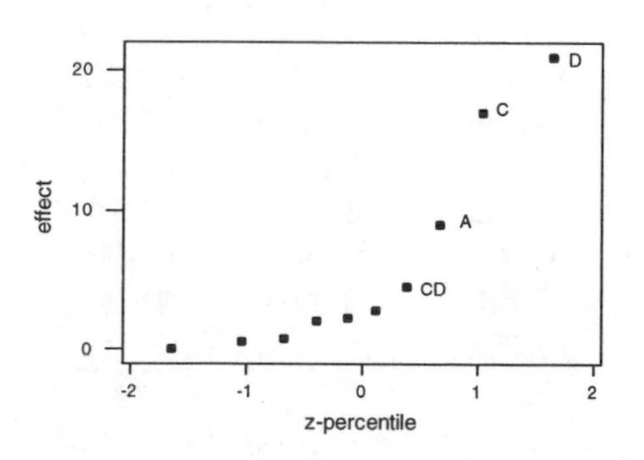

The plot suggests that main effects A, C, and D are quite important, and perhaps the interaction CD as well. (See answer section for comment)

56. The summary quantities are:

	$x_{ij\bullet}$	j = 1	2	3	$x_{i\bullet\bullet}$
i	1	6.2	4.0	5.8	16.0
	2	7.6	6.2	6.4	20.2
	$x_{\bullet j\bullet}$	13.8	10.2	12.2	$x_{\bullet\bullet\bullet} = 36.2$

$CF = \frac{(36.2)^2}{30} = 43.6813$, $\Sigma\Sigma\Sigma x_{ijk}^2 = 45.560$, so $SST = 45.560 - 43.6813 = 1.8787$,

$SSE = 45.560 - \frac{225.24}{5} = .5120$, $SSA = \frac{(16.0)^2 + (20.2)^2}{15} - CF = .5880$,

$SSB = \frac{(13.8)^2 + (10.2)^2 + (12.2)^2}{10} - CF = .6507$, and by subtraction, SSAB = .128

```
Analysis of Variance for Average Bud Rating
Source          DF        SS          MS          F
Health           1      0.5880      0.5880      27.56
pH               2      0.6507      0.3253      15.25
Interaction      2      0.1280      0.0640       3.00
Error           24      0.5120      0.0213
Total           29      1.8787
```

Since 3.00 is not $\geq F_{.05,2,24} = 3.40$, we fail to reject the no interactions hypothesis, and we continue: $27.56 \geq F_{.05,1,24} = 4.26$, and $15.25 \geq F_{.05,2,24} = 3.40$, so we conclude that both the health of the seedling and its pH level have an effect on the average rating.

57. The ANOVA table is:

Source	df	SS	MS	f	$F_{.01}$
A	2	34,436	17,218	436.92	5.49
B	2	105,793	52,897	1342.30	5.49
C	2	516,398	258,199	6552.04	5.49
AB	4	6,868	1,717	43.57	4.11
AC	4	10,922	2,731	69.29	4.11
BC	4	10,178	2,545	64.57	4.11
ABC	8	6,713	839	21.30	3.26
Error	27	1,064	39		
Total	53	692,372			

All calculated f values are greater than their respective tabled values, so all effects, including the interaction effects, are significant at level .01.

58.

Source	df	SS	MS	f	$F_{.05}$
A(pressure)	1	6.94	6.940	11.57*	4.26
B(time)	3	5.61	1.870	3.12*	3.01
C(concen.)	2	12.33	6.165	10.28*	3.40
AB	3	4.05	1.350	2.25	3.01
AC	2	7.32	3.660	6.10*	3.40
BC	6	15.80	2.633	4.39*	2.51
ABC	6	4.37	.728	1.21	2.51
Error	24	14.40	.600		
Total	47	70.82			

There appear to be no three-factor interactions. However both AC and BC two-factor interactions appear to be present.

59. Based on the p-values in the ANOVA table, statistically significant factors at the level .01 are adhesive type and cure time. The conductor material does not have a statistically significant effect on bond strength. There are no significant interactions.

60.

Source	df	SS	MS	f	$F_{.05}$
A (diet)	2	18,138	9.69.0	28.9*	≈ 3.32
B (temp.)	2	5,182	2591.0	8.3*	≈ 3.32
Interaction	4	1,737	434.3	1.4	≈ 2.69
Error	36	11,291	313.6		
Total	44	36,348			

Interaction appears to be absent. However, since both main effect f values exceed the corresponding F critical values, both diet and temperature appear to affect expected energy intake.

61. $SSA = \sum_i \sum_j (\overline{X}_{i...} - \overline{X}_{....})^2 = \frac{1}{N}\Sigma X_{i...}^2 - \frac{X_{....}^2}{N}$, with similar expressions for SSB, SSC, and SSD, each having N – 1 df.

$SST = \sum_i \sum_j (X_{ij(kl)} - \overline{X}_{....})^2 = \sum_i \sum_j X_{ij(kl)}^2 - \frac{X_{....}^2}{N}$ with $N^2 - 1$ df, leaving $N^2 - 1 - 4(N-1)$ df for error.

	1	2	3	4	5	Σx^2
$x_{i...}$:	482	446	464	468	434	1,053,916
$x_{.j..}$:	470	451	440	482	451	1,053,626
$x_{..k.}$:	372	429	484	528	481	1,066,826
$x_{...l}$:	340	417	466	537	534	1,080,170

Also, $\Sigma\Sigma x_{ij(kl)}^2 = 220{,}378$, $x_{....} = 2294$, and CF = 210,497.44

Source	df	SS	MS	f	$F_{.05}$
A	4	285.76	71.44	.594	3.84
B	4	227.76	56.94	.473	3.84
C	4	2867.76	716.94	5.958*	3.84
D	4	5536.56	1384.14	11.502*	3.84
Error	8	962.72	120.34		
Total	24				

H_{oA} and H_{oB} cannot be rejected, while H_{oC} and H_{oD} are rejected.

Chapter 12

Section 12.1

1.

a Stem and Leaf display of temp:

```
17|0
17|23          stem = tens
17|445         leaf = ones
17|67
17|
18|0000011
18|2222
18|445
18|6
18|8
```

180 appears to be a typical value for this data. The distribution is reasonably symmetric in appearance and somewhat bell-shaped. The variation in the data is fairly small since the range of values (188 – 170 = 18) is fairly small compared to the typical value of 180.

```
0|889
1|0000         stem = ones
1|3            leaf = tenths
1|4444
1|66
1|8889
2|11
2|
2|5
2|6
2|
3|00
```

For the ratio data, a typical value is around 1.6 and the distribution appears to be positively skewed. The variation in the data is large since the range of the data (3.08 - .84 = 2.24) is very large compared to the typical value of 1.6. The two largest values could be outliers.

b The efficiency ratio is not uniquely determined by temperature since there are several instances in the data of equal temperatures associated with different efficiency ratios. For example, the five observations with temperatures of 180 each have different efficiency ratios.

c A scatter plot of the data appears below. The points exhibit quite a bit of variation and do not appear to fall close to any straight line or simple curve.

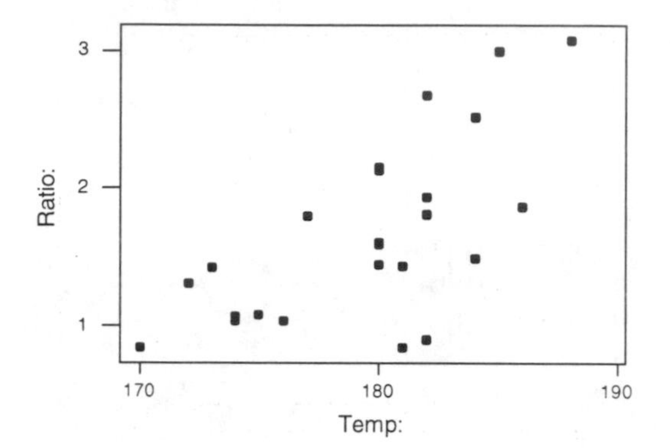

2. Scatter plots for the emissions vs age:

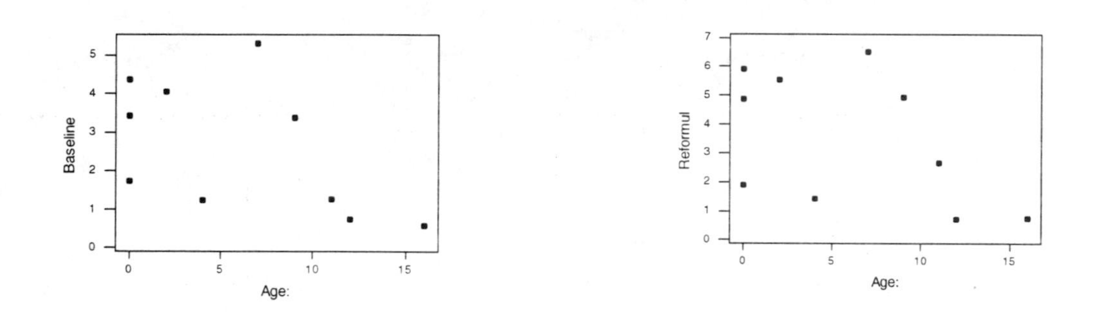

With this data the relationship between the age of the lawn mower and its NO_x emissions seems somewhat dubious. One might have expected to see that as the age of the lawn mower increased the emissions would also increase. We certainly do not see such a pattern. Age does not seem to be a particularly useful predictor of NO_x emission.

3. A scatter plot of the data appears below. The points fall very close to a straight line with an intercept of approximately 0 and a slope of about 1. This suggests that the two methods are producing substantially the same concentration measurements.

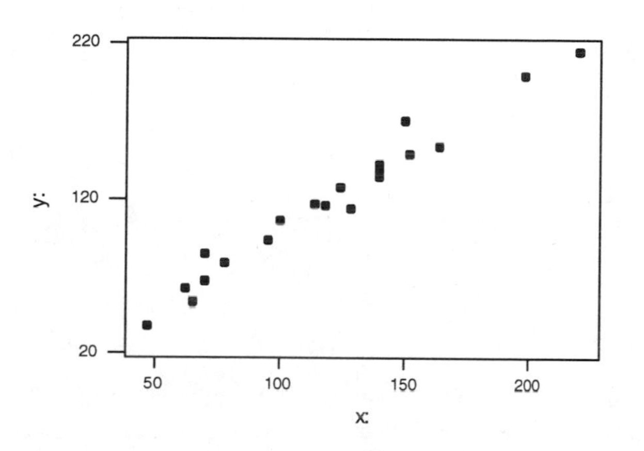

4.

a Box plots of both variables:

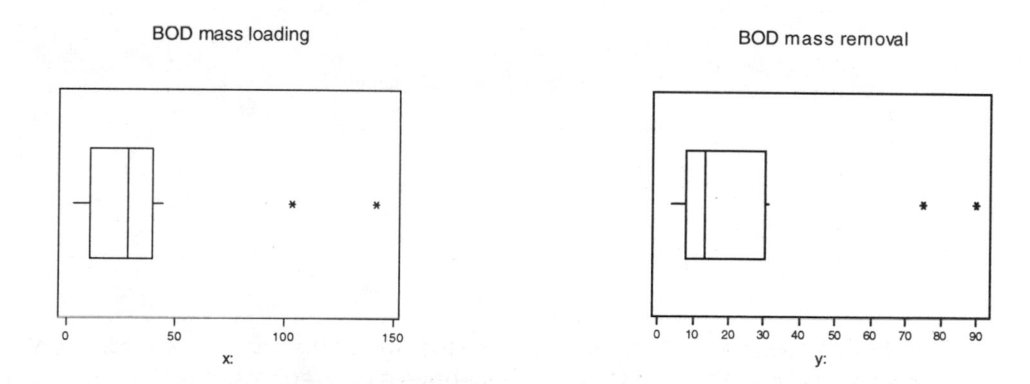

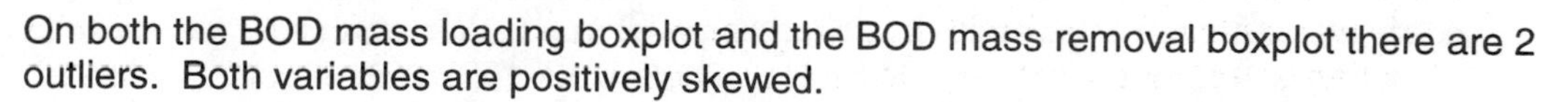
On both the BOD mass loading boxplot and the BOD mass removal boxplot there are 2 outliers. Both variables are positively skewed.

b Scatter plot of the data:

BOD mass loading (x) vs BOD mass removal (y)

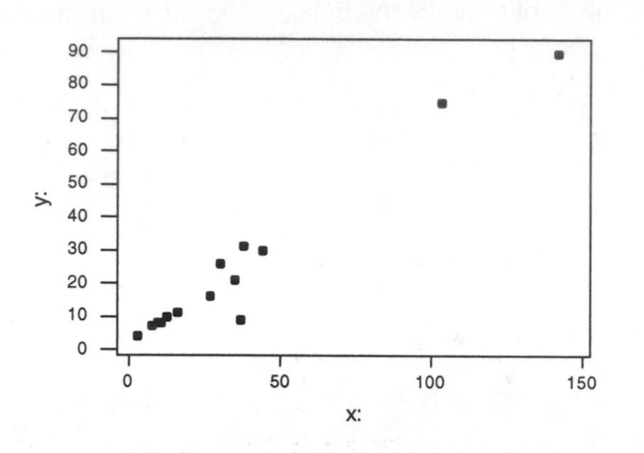

There is a strong linear relationship between BOD mass loading and BOD mass removal. As the loading increases, so does the removal. The two outliers seen on each of the boxplots are seen to be correlated here. There is one observation that appears not to match the liner pattern. This value is (37, 9). One might have expected a larger value for BOD mass removal.

5.

a The scatter plot with axes intersecting at (0,0) is shown below.

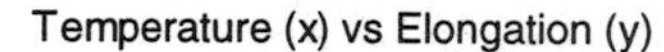

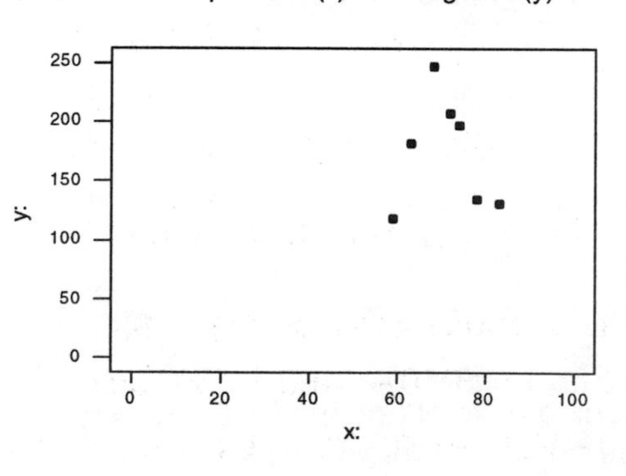

b The scatter plot with axes intersecting at (55, 100) is shown below.

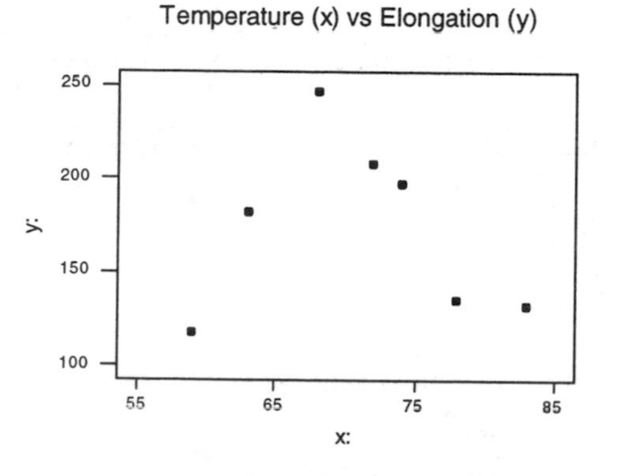

c A parabola appears to provide a good fit to both graphs.

6. There appears to be a linear relationship between racket resonance frequency and sum of peak-to-peak acceleration. As the resonance frequency increases the sum of peak-to-peak acceleration tends to decrease. However, there is not a perfect relationship. Variation does exist. One should also notice that there are two tennis rackets that appear to differ from the other 21 rackets. Both have very high resonance frequency values. One might investigate if these rackets differ in other ways as well.

7.

a $\mu_{Y \cdot 2500} = 1800 + 1.3(2500) = 5050$

b expected change = slope = $\beta_1 = 1.3$

c expected change = $100\beta_1 = 130$

d expected change = $-100\beta_1 = -130$

8.

a $\mu_{Y \cdot 2000} = 1800 + 1.3(2000) = 4400$, and $\sigma = 350$, so $P(Y > 5000)$

$$= P\left(Z > \frac{5000 - 4400}{350}\right) = P(Z > 1.71) = .0436$$

b Now E(Y) = 5050, so $P(Y > 5000) = P(Z > .14) = .4443$

c $E(Y_2 - Y_1) = E(Y_2) - E(Y_1) = 5050 - 4400 = 650$, and

$V(Y_2 - Y_1) = V(Y_2) + V(Y_1) = (350)^2 + (350)^2 = 245{,}000$, so the s.d. of $Y_2 - Y_1 = 494.97$.

Thus $P(Y_2 - Y_1 > 0) = P\left(z > \frac{100 - 650}{494.97}\right) = P(Z > .71) = .2389$

d The standard deviation of $Y_2 - Y_1 = 494.97$ (from **c**), and $E(Y_2 - Y_1) = 1800 + 1.3x_2 - (1800 + 1.3x_1) = 1.3(x_2 - x_1)$. Thus $P(Y_2 > Y_1) = P(Y_2 - Y_1 > 0) = P\left(z > \frac{-1.3(x_2 - x_1)}{494.97}\right) = .95$ implies that $-1.645 = \frac{-1.3(x_2 - x_1)}{494.97}$, so $x_2 - x_1 = 626.33$.

9.

a β_1 = expected change in flow rate (y) associated with a one inch increase in pressure drop (x) = .095.

b We expect flow rate to decrease by $5\beta_1 = .475$.

c $\mu_{Y \cdot 10} = -.12 + .095(10) = .83$, and $\mu_{Y \cdot 15} = -.12 + .095(15) = 1.305$.

d $P(Y > .835) = P\left(Z > \frac{.835 - .830}{.025}\right) = P(Z > .20) = .4207$

$P(Y > .840) = P\left(Z > \frac{.840 - .830}{.025}\right) = P(Z > .40) = .3446$

e Let Y_1 and Y_2 denote pressure drops for flow rates of 10 and 11, respectively. Then $\mu_{Y \cdot 11} = .925$, so $Y_1 - Y_2$ has expected value .830 - .925 = -.095, and s.d. $\sqrt{(.025)^2 + (.025)^2} = .035355$. Thus $P(Y_1 > Y_2) = P(Y_1 - Y_2 > 0) = P\left(z > \frac{+.095}{.035355}\right) = P(Z > 2.69) = .0036$

10. Y has expected value 14,000 when x = 1000 and 24,000 when x = 2000, so the two probabilities become $P\left(z > \frac{-8500}{\sigma}\right) = .05$ and $P\left(z > \frac{-17{,}500}{\sigma}\right) = .10$. Thus $\frac{-8500}{\sigma} = -1.645$ and $\frac{-17{,}500}{\sigma} = -1.28$. This gives two different values for σ, a contradiction, so the answer to the question posed is no.

11.

a β_1 = expected change for a one degree increase = -.01, and $10\beta_1 = -.1$ is the expected change for a 10 degree increase.

b $\mu_{Y \cdot 200} = 5.00 - .01(200) = 3$, and $\mu_{Y \cdot 250} = 2.5$.

c The probability that the first observation is between 2.4 and 2.6 is
$P(2.4 \le Y \le 2.6) = P\left(\frac{2.4 - 2.5}{.075} \le Z \le \frac{2.6 - 2.5}{.075}\right)$
$= P(-1.33 \le Z \le 1.33) = .8164$. The probability that any particular one of the other four observations is between 2.4 and 2.6 is also .8164, so the probability that all five are between 2.4 and 2.6 is $(.8164)^5 = .3627$.

d Let Y_1 and Y_2 denote the times at the higher and lower temperatures, respectively. Then Y_1 - Y_2 has expected value $5.00 - .01(x+1) - (5.00 - .01x) = -.01$. The standard deviation of Y_1 - Y_2 is $\sqrt{(.075)^2 + (.075)^2} = .10607$. Thus
$P(Y_1 - Y_2 > 0) = P\left(z > \frac{-(-.01)}{.10607}\right) = P(Z > .09) = .4641$.

Section 12.2

12.

a $S_{xx} = 39{,}095 - \frac{(517)^2}{14} = 20{,}002.929$, $S_{xy} = 25{,}825 - \frac{(517)(346)}{14} = 13047.714$;
$\hat{\beta}_1 = \frac{S_{xy}}{S_{xx}} = \frac{13{,}047.714}{20{,}002.929} = .652$; $\hat{\beta}_0 = \frac{\Sigma y - \hat{\beta}_1 \Sigma x}{n} = \frac{346 - (.652)(517)}{14} = .626$,
so the equation of the least squares regression line is $y = .626 + .652x$.

b $\hat{y}_{(35)} = .626 + .652(35) = 23.456$. The residual is $y - \hat{y} = 21 - 23.456 = -2.456$.

c $S_{yy} = 17{,}454 - \frac{(346)^2}{14} = 8902.857$, so
$SSE = 8902.857 - (.652)(1307.714) = 8049.848$.
$\hat{\sigma} = \sqrt{\frac{SSE}{n-2}} = \sqrt{\frac{8049.848}{12}} = 25.900$.

d $SST = S_{yy} = 8902.857$; $r^2 = 1 - \frac{SSE}{SST} = 1 - \frac{8049.848}{8902.857} = .096$.

e Without the two upper extreme observations, the new summary values are $n = 12, \Sigma x = 272, \Sigma x^2 = 8322, \Sigma y = 181, \Sigma y^2 = 3729, \Sigma xy = 5320$. The new $S_{xx} = 2156.667, S_{yy} = 998.917, S_{xy} = 1217.333$. New $\hat{\beta}_1 = .56445$ and $\hat{\beta}_0 = 2.2891$, which yields the new equation $y = 2.2891 + .56445x$. Removing the two values changes the position of the line considerably, and the slope slightly. The new $r^2 = 1 - \frac{311.79}{998.917} = .6879$, which is a great improvement over the original set of observations.

13. For this data, n = 4, $\Sigma x_i = 200$, $\Sigma y_i = 5.37$, $\Sigma x_i^2 = 12.000$, $\Sigma y_i^2 = 9.3501$, $\Sigma x_i y_i = 333$. $S_{xx} = 12{,}000 - \frac{(200)^2}{4} = 2000$, $S_{yy} = 9.3501 - \frac{(5.37)^2}{4} = 2.140875$, and $S_{xy} = 333 - \frac{(200)(5.37)}{4} = 64.5$. $\hat{\beta}_1 = \frac{S_{xy}}{S_{xx}} = \frac{64.5}{2000} = .03225$ and

$\hat{\beta}_0 = \frac{5.37}{4} - (.03225)\frac{200}{4} = -.27000$.

$SSE = S_{yy} - \hat{\beta}_1 S_{xy} = 2.14085 - (.03225)(64.5) = .060750$.

$r^2 = 1 - \frac{SSE}{SST} = 1 - \frac{.060750}{2.14085} = .972$. This is a very high value of r^2, which confirms the authors' claim that there is a strong linear relationship between the two variables.

14.

a n = 24, $\Sigma x_i = 4308$, $\Sigma y_i = 40.09$, $\Sigma x_i^2 = 773{,}790$, $\Sigma y_i^2 = 76.8823$,

$\Sigma x_i y_i = 7{,}243.65$. $S_{xx} = 773{,}790 - \frac{(4308)^2}{24} = 504.0$,

$S_{yy} = 76.8823 - \frac{(40.09)^2}{24} = 9.9153$, and

$S_{xy} = 7{,}243.65 - \frac{(4308)(40.09)}{24} = 45.8246$. $\hat{\beta}_1 = \frac{S_{xy}}{S_{xx}} = \frac{45.8246}{504} = .09092$ and

$\hat{\beta}_0 = \frac{40.09}{24} - (.09092)\frac{4308}{24} = -14.6497$. The equation of the estimated regression line is $\hat{y} = -14.6497 + .09092x$.

b When x = 182, $\hat{y} = -14.6497 + .09092(182) = 1.8997$. So when the tank temperature is 182, we would predict an efficiency ratio of 1.8997.

c The four observations for which temperature is 182 are: (182, .90), (182, 1.81), (182, 1.94), and (182, 2.68). Their corresponding residuals are: $.90 - 1.8997 = -0.9977$, $1.81 - 1.8997 = -0.0877$, $1.94 - 1.8997 = 0.0423$, $2.68 - 1.8997 = 0.7823$. These residuals do not all have the same sign because in the cases of the first two pairs of observations, the observed efficiency ratios were smaller than the predicted value of 1.8997. Whereas, in the cases of the last two pairs of observations, the observed efficiency ratios were larger than the predicted value.

d $SSE = S_{yy} - \hat{\beta}_1 S_{xy} = 9.9153 - (.09092)(45.8246) = 5.7489$.

$r^2 = 1 - \frac{SSE}{SST} = 1 - \frac{5.7489}{9.9153} = .4202$. (42.02% of the observed variation in efficiency ratio can be attributed to the approximate linear relationship between the efficiency ratio and the tank temperature.

15.

a The following stem and leaf display shows that: a typical value for this data is a number in the low 40's. there is some positive skew in the data. There are some potential outliers (79.5 and 80.0), and there is a reasonably large amount of variation in the data (e.g., the spread 80.0-29.8 = 50.2 is large compared with the typical values in the low 40's).

```
2|9
3|33            stem = tens
3|5566677889    leaf = ones
4|1223
4|56689
5|1
5|
6|2
6|9
7|
7|9
8|0
```

b No, the strength values are not uniquely determined by the MoE values. For example, note that the two pairs of observations having strength values of 42.8 have different MoE values.

c The least squares line is $\hat{y} = 3.2925 + .10748x$. For a beam whose modulus of elasticity is x = 40, the predicted strength would be $\hat{y} = 3.2925 + .10748(40) = 7.59$. The value x = 100 isfar beyond the range of the x values in the data, so it would be dangerous (i.e., potentially misleading) to extrapolated the linear relationship that far.

d From the output, SSE = 18.736, SST = 71.605, and the coefficient of determination is r^2 = .738 (or 73.8%). The r^2 value is large, which suggests that the linear relationship is a useful approximation to the true relationship between these two variables.

16.

a

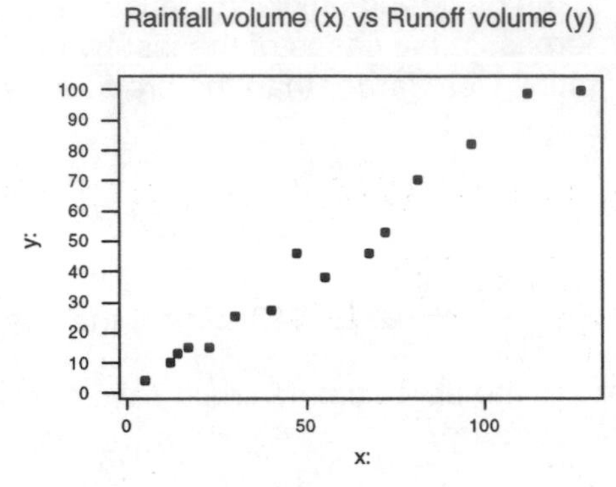

Yes, the scatterplot shows a strong linear relationship between rainfall volume and runoff volume, thus it supports the use of the simple linear regression model.

b $\bar{x} = 53.200$, $\bar{y} = 42.867$, $S_{xx} = 63040 - \dfrac{(798)^2}{15} = 20{,}586.4$,

$S_{yy} = 41{,}999 - \dfrac{(643)^2}{15} = 14{,}435.7$, and $S_{xy} = 51{,}232 - \dfrac{(798)(643)}{15} = 17{,}024.4$.

$\hat{\beta}_1 = \dfrac{S_{xy}}{S_{xx}} = \dfrac{17{,}024.4}{20{,}586.4} = .82697$ and $\hat{\beta}_0 = 42.867 - (.82697)53.2 = -1.1278$.

c $\mu_{y\cdot 50} = -1.1278 + .82697(50) = 40.2207$.

d $SSE = S_{yy} - \hat{\beta}_1 S_{xy} = 14{,}435.7 - (.82697)(17{,}324.4) = 357.07$.

$s = \hat{\sigma} = \sqrt{\dfrac{SSE}{n-2}} = \sqrt{\dfrac{357.07}{13}} = 5.24$.

e $r^2 = 1 - \dfrac{SSE}{SST} = 1 - \dfrac{357.07}{14{,}435.7} = .9753$. So 97.53% of the observed variation in runoff volume can be attributed to the simple linear regression relationship between runoff and rainfall.

17. Note: n = 23 in this study.

a For a one (mg/cm^2) increase in dissolved material, one would expect a .144 (g/l) increase in calcium content. Secondly, 86% of the observed variation in calcium content can be attributed to the simple linear regression relationship between calcium content and dissolved material.

b $\mu_{y\cdot 50} = 3.678 + .144(50) = 10.878$

c $r^2 = .86 = 1 - \frac{SSE}{SST}$, so $SSE = (SST)(1-.86) = (320.398)(.14) = 44.85572$.

Then $s = \sqrt{\frac{SSE}{n-2}} = \sqrt{\frac{44.85572}{21}} = 1.46$

18.

a $\hat{\beta}_1 = \frac{15(987.645)-(1425)(10.68)}{15(139{,}037.25)-(1425)^2} = \frac{-404.3250}{54{,}933.7500} = -.00736023$

$\hat{\beta}_0 = \frac{10.68-(-.00736023)(1425)}{15} = 1.41122185$, $y = 1.4112 - .007360x$.

b $\hat{\beta}_1 = -.00736023$

c With x now denoting temperature in $\circ C$, $y = \hat{\beta}_0 + \hat{\beta}_1\left(\frac{9}{5}x + 32\right)$

$= \left(\hat{\beta}_0 + 32\hat{\beta}_1\right) + \frac{9}{5}\hat{\beta}_1 x = 1.175695 - .0132484x$, so the new $\hat{\beta}_1$ is -.0132484 and the new $\hat{\beta}_0 = 1.175695$.

d Using the equation of **a**, predicted $y = \hat{\beta}_0 + \hat{\beta}_1(200) = -.0608$, but the deflection factor cannot be negative.

19. N = 14, $\Sigma x_i = 3300$, $\Sigma y_i = 5010$, $\Sigma x_i^2 = 913{,}750$, $\Sigma y_i^2 = 2{,}207{,}100$, $\Sigma x_i y_i = 1{,}413{,}500$

a $\hat{\beta}_1 = \frac{3{,}256{,}000}{1{,}902{,}500} = 1.71143233$, $\hat{\beta}_0 = -45.55190543$, so we use the equation $y = -45.5519 + 1.7114x$.

b $\hat{\mu}_{Y\cdot 225} = -45.5519 + 1.7114(225) = 339.51$

c Estimated expected change $= -50\hat{\beta}_1 = -85.57$

d No, the value 500 is outside the range of x values for which observations were available (the danger of extrapolation).

20.

a $\hat{\beta}_0 = .3651$, $\hat{\beta}_1 = .9668$

b .8485

c $\hat{\sigma} = .1932$

d SST = 1.4533, 71.7% of this variation can be explained by the model. Note: $\frac{SSR}{SST} = \frac{1.0427}{1.4533} = .717$ which matches R-squared on output.

21.

a Verification

b $\hat{\beta}_1 = \frac{491.4}{744.16} = .66034186$, $\hat{\beta}_0 = -2.18247148$

c predicted $y = \hat{\beta}_0 + \hat{\beta}_1(15) = 7.72$

d $\hat{\mu}_{Y \cdot 15} = \hat{\beta}_0 + \hat{\beta}_1(15) = 7.72$

22.

a $\hat{\beta}_1 = \frac{-404.325}{54.933.75} = -.00736023$, $\hat{\beta}_0 = 1.41122185$,
$SSE = 7.8518 - (1.41122185)(10.68) - (-.00736023)(987.654) = .049245$,
$s^2 = \frac{.049245}{13} = .003788$, and $\hat{\sigma} = s = .06155$

b $SST = 7.8518 - \frac{(10.68)^2}{15} = .24764$ so $r^2 = 1 - \frac{.049245}{.24764} = 1 - .199 = .801$

23.

a Using the $y_i's$ given to one decimal place accuracy is the answer to Exercise 19, $SSE = (150-125.6)^2 + ... + (670-639.0)^2 = 16,213.64$. The computation formula gives $SSE = 2,207,100 - (-45.55190543)(5010) - (1.71143233)(1,413,500) = 16,205.45$

b $SST = 2,207,100 - \frac{(5010)^2}{14} = 414,235.71$ so $r^2 = 1 - \frac{16,205.45}{414,235.71} = .961$.

24.

a

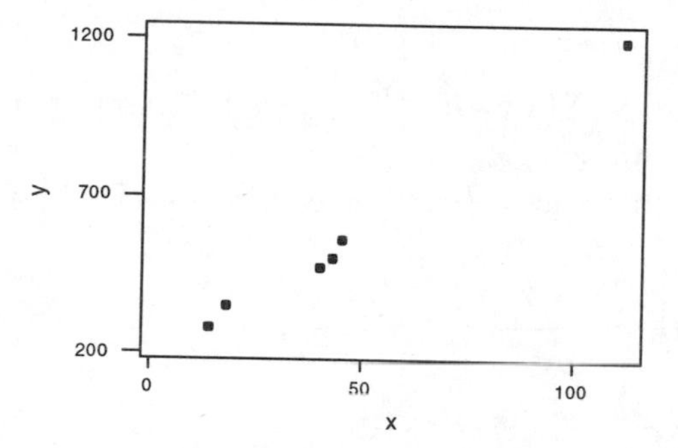

According to the scatter plot of the data, a simple linear regression model does appear to be plausible.

b The regression equation is `y = 138 + 9.31 x`

c The desired value is the coefficient of determination, $r^2 = 99.0\%$.

d The new equation is `y* = 190 + 7.55 x*`. This new equation appears to differ significantly. If we were to predict a value of y^* for $x^* = 50$, the value would be 567.9, where using the original data, the predicted value for x = 50 would be 603.5.

25. Substitution of $\hat{\beta}_0 = \frac{\Sigma y_i - \hat{\beta}_1 \Sigma x_i}{n}$ and $\hat{\beta}_1$ for b_o and b_1 on the left hand side of the normal equations yields $\frac{n(\Sigma y_i - \hat{\beta}_1 \Sigma x_i)}{n} + \hat{\beta}_1 \Sigma x_i = \Sigma y_i$ from the first equation and $\frac{\Sigma x_i(\Sigma y_i - \hat{\beta}_1 \Sigma x_i)}{n} + \hat{\beta}_1 \Sigma x_i^2 = \frac{\Sigma x_i \Sigma y_i}{n} + \frac{\hat{\beta}_1 (n \Sigma x_i^2 - (\Sigma x_i)^2)}{n}$

$\frac{\Sigma x_i \Sigma y_i}{n} + \frac{n \Sigma x_i y_i}{n} - \frac{\Sigma x_i \Sigma y_i}{n} = \Sigma x_i y_i$ from the second equation.

26. We show that when $\bar{x}$ is substituted for x in $\hat{\beta}_0 + \hat{\beta}_1 x$, $\bar{y}$ results, so that $(\bar{x}, \bar{y})$ is on the line $y = \hat{\beta}_0 + \hat{\beta}_1 x$: $\hat{\beta}_0 + \hat{\beta}_1 \bar{x} = \frac{\Sigma y_i - \hat{\beta}_1 \Sigma x_i}{n} + \hat{\beta}_1 \bar{x} = \bar{y} - \hat{\beta}_1 \bar{x} + \hat{\beta}_1 \bar{x} = \bar{y}$.

27. We wish to find b_1 to minimize $\Sigma(y_i - b_1 x_i)^2 = f(b_1)$. Equating $f'(b_1)$ to 0 yields $2\Sigma(y_i - b_1 x_i)(-x_i) = 0$ so $\Sigma x_i y_i = b_1 \Sigma x_i^2$ and $b_1 = \frac{\Sigma x_i y_i}{\Sigma x_i^2}$. The least squares estimator of $\hat{\beta}_1$ is thus $\hat{\beta}_1 = \frac{\Sigma x_i Y_i}{\Sigma x_i^2}$.

28.

a Subtracting $\bar{x}$ from each x_i shifts the plot in a rigid fashion $\bar{x}$ units to the left without otherwise altering its character. The last squares line for the new plot will thus have the same slope as the one for the old plot. Since the new line is $\bar{x}$ units to the left of the old one, the new y intercept (height at x = 0) is the height of the old line at x = $\bar{x}$, which is $\hat{\beta}_0 + \hat{\beta}_1 \bar{x} = \bar{y}$ (since from exercise 26, $(\bar{x}, \bar{y})$ is on the old line). Thus the new y intercept is $\bar{y}$.

b We wish b_0 and b_1 to minimize $f(b_0, b_1) = \Sigma[y_i - (b_0 + b_1(x_i - \bar{x}))]^2$. Equating $\frac{\partial f}{\partial b_0}$ to $\frac{\partial f}{\partial b_1}$ to 0 yields $nb_0 + b_1 \Sigma(x_i - \bar{x}) = \Sigma y_i$, $b_0 \Sigma(x_i - \bar{x}) + b_1 \Sigma(x_i - \bar{x})^2 = \Sigma(x_i - \bar{x})^2 = \Sigma(x_i - \bar{x}) y_i$. Since $\Sigma(x_i - \bar{x}) = 0$, $b_0 = \bar{y}$, and since $\Sigma(x_i - \bar{x}) y_i = \Sigma(x_i - \bar{x})(y_i - \bar{y})$ [because $\Sigma(x_i - \bar{x})\bar{y} = \bar{y}\Sigma(x_i - \bar{x})$], $b_1 = \hat{\beta}_1$. Thus $\hat{\beta}_0^* = \bar{Y}$ and $\hat{\beta}_1^* = \hat{\beta}_1$.

29. For data set #1, $r^2 = .43$ and $\hat{\sigma} = s = 4.03$; whereas these quantities are .99 and 4.03 for #2, and .99 and 1.90 for #3. In general, one hopes for both large r^2 (large % of variation explained) and small s (indicating that observations don't deviate much from the estimated line). Simple linear regression would thus seem to be most effective in the third situation.

Section 12.3

30.

a $\Sigma(x_i - \bar{x})^2 = 7{,}000{,}000$, so $V(\hat{\beta}_1) = \dfrac{(350)^2}{7{,}000{,}000} = .0175$ and the standard deviation of $\hat{\beta}_1$ is $\sqrt{.0175} = .1323$.

b $P(1.0 \le \hat{\beta}_1 \le 1.5) = P\left(\dfrac{1.0-1.25}{1.323} \le Z \le \dfrac{1.5-1.25}{1.323}\right)$
$= P(-1.89 \le Z \le 1.89) = .9412$.

c Although n = 11 here and n = 7 in **a**, $\Sigma(x_i - \bar{x})^2 = 1{,}100{,}000$ now, which is smaller than in **a**. Because this appears in the denominator of $V(\beta_1)$, the variance is smaller for the choice of x values in **a**.

31.

a $\hat{\beta}_1 = -.00736023$, $\hat{\beta}_0 = 1.41122185$, so
$SSE = 7.8518 - (1.41122185)(10.68) - (-.00736023)(987.654) = .04925$,
$s^2 = .003788$, $s = .06155$. $\hat{\sigma}^2_{\hat{\beta}_1} = \dfrac{s^2}{\Sigma x_i^2 - (\Sigma x_i)^2/n} = \dfrac{.003788}{3662.25} = .00000103$,
$\hat{\sigma}_{\hat{\beta}_1} = s_{\hat{\beta}_1}$ = estimated s.d. of $\hat{\beta}_1 = \sqrt{.00000103} = .001017$.

b $-.00736 \pm (2.160)(.001017) = -.00736 \pm .00220 = (-.00956, -.00516)$

32. Let β_1 denote the true average change in runoff for each 1 m^3 increase in rainfall. To test the hypotheses $H_o: \beta_1 = 0$ vs. $H_a: \beta_1 \ne 0$, the calculated t statistic is
$t = \dfrac{\hat{\beta}_1}{s_{\hat{\beta}_1}} = \dfrac{.82697}{.03652} = 22.64$ which (from the printout) has an associated p-value of P = 0.000. Therefore, since the p-value is so small, H_o is rejected and we conclude that there is a useful linear relationship between runoff and rainfall.

A confidence interval for β_1 is based on n – 2 = 15 – 2 = 13 degrees of freedom. $t_{.025,13} = 2.160$, so the interval estimate is

$\hat{\beta}_1 \pm t_{.025,13} \cdot s_{\hat{\beta}_1} = .82697 \pm (2.160)(.03652) = (.748,.906)$. Therefore, we can be confident that the true average change in runoff, for each 1 m^3 increase in rainfall, is somewhere between .748 m^3 and .906 m^3.

33.

a From the printout in Exercise 15, the error d.f. = n – 2 = 25, $t_{.025,25} = 2.060$. The confidence interval is then

$\hat{\beta}_1 \pm t_{.025,25} \cdot s_{\hat{\beta}_1} = .10748 \pm (2.060)(.01280) = (.081,.134)$. Therefore, we estimate with a high degree of confidence that the true average change in strength associated with a 1 Gpa increase in modulus of elasticity is between .081 MPa and .134 MPa.

b We wish to test $H_o : \beta_1 = .1$ vs. $H_a : \beta_1 > .1$. The calculated t statistic is

$t = \frac{\hat{\beta}_1 - .1}{s_{\hat{\beta}_1}} = \frac{.10748 - .1}{.01280} = .58$, which yields a p-value of .277. A large p-value such as this would not lead to rejecting H$_o$, so there is not enough evidence to contradict the prior belief.

34.

a $H_o : \beta_1 = 0$; $H_a : \beta_1 \neq 0$

RR: $|t| > t_{\alpha/2,n-2}$ or $|t| > 3.106$

$t = 5.29$: Reject H$_o$. The slope differs significantly from 0, and the model appears to be useful.

b At the level $\alpha = 0.01$, reject h$_o$ if the p-value is less than 0.01. In this case, the reported p-value was 0.000, therefore reject H$_o$. The conclusion is the same as that of part **a**.

c $H_o : \beta_1 = 1.5$; $H_a : \beta_1 < 1.5$

RR: $t < -t_{\alpha,n-2}$ or $t < -2.718$

$t = \frac{0.9668 - 1.5}{0.1829} = -2.92$: Reject H$_o$. The data contradict the prior belief.

35.

a $\hat{\beta}_1 \pm t_{\alpha/2,n-2} \cdot s_{\hat{\beta}_1}$: $0.5549 \pm 2.306(0.3101) = (-0.16, 1.27)$ The interval contains the value 0, indicating 0 is a possible value for β_1, and showing little usefulness of the model.

b The p-value associated with the model is 0.11, which exceeds standard levels of α. This indicates least squares is not a good way to predict age from transparent dentive content.

36.

a We reject H$_o$ if $t > t_{.01,13} = 2.650$. With $\Sigma x_i^2 - \frac{(\Sigma x_i)^2}{n} = 324.40$,

$t = \frac{1.7035 - 1}{3.725/\sqrt{324.40}} = \frac{.7035}{.2068} = 3.40$. Since $3.40 \geq 2.650$, H$_o$ is rejected in favor of H$_a$.

b $t_{.005,13} = 3.012$, so the C.I is $1.7035 \pm \frac{(3.012)(3.725)}{\sqrt{324.40}} = 1.7035 \pm .6229$ $= (1.08, 2.32)$

37.

a n = 10, $\Sigma x_i = 2615$, $\Sigma y_i = 39.20$, $\Sigma x_i^2 = 860{,}675$, $\Sigma y_i^2 = 161.94$, $\Sigma x_i y_i = 11{,}453.5$, so $\hat{\beta}_1 = \frac{12{,}027}{1{,}768{,}525} = .00680058$, $\hat{\beta}_0 = 2.14164770$, from which SSE = .09696713, s = .11009492 $s = .11009492 \doteq .110 = \hat{\sigma}$,

$\hat{\sigma}_{\hat{\beta}_1} = \frac{.110}{\sqrt{176{,}852}} = .000262$

b We wish to test $H_o: \beta_1 = .0060$ vs $H_a: \beta_1 \neq .0060$. At level .10, H$_o$ is rejected if either $t \geq t_{.05,8} = 1.860$ or $t \leq -t_{.05,8} = -1.860$. Since $t = \frac{.0068 - .0060}{.000262} = 3.06 \geq 1.1860$, H$_o$ is rejected.

38.

a From Exercise 23, SSE = 16.205.45, so $s^2 = 1350.454$, $s = 36.75$, and $s_{\hat{\beta}_1} = \frac{36.75}{368.636} = .0997$. Thus $t = \frac{1.711}{.0997} = 17.2 > 4.318 = t_{.0005,14}$, so p-value < .001. Because the p-value < .01, $H_o: \beta_1 = 0$ is rejected at level .01 in favor of the conclusion that the model is useful $(\beta_1 \neq 0)$.

b The C.I. for β_1 is $1.711 \pm (2.179)(.0997) = 1.711 \pm .217 = (1.494, 1.928)$. Thus the C.I. for $10\beta_1$ is $(14.94, 19.28)$.

39. SSE = 5,390,382 – (-12.84159)(7034) – (36.18385)(149,354.4) = 76,492.54, and SST = 892,458.73

Source	df	SS	MS	f
Regr	1	815,966.19	815,966.19	96.0
Error	9	76,492.54	8499.17	
Total	10	892,458.73		

Since no α is specified, let's use $\alpha = .01$. Then $F_{.01,1,9} = 10.56 < 96.0$, so $H_o : \beta_1 = 0$ is rejected and the model is judged useful. $s = \sqrt{8499.17} = 92.19$, $\Sigma(x_i - \bar{x})^2 = 623.2218$, so $t = \dfrac{36.184}{92.19 / \sqrt{623.2218}} = 9.80$, and $t^2 = (9.80)^2 = 96.0 = f$.

40. We use the fact that $\hat{\beta}_1$ is unbiased for β_1. $E(\hat{\beta}_0) = \dfrac{E(\Sigma y_i - \hat{\beta}_1 \Sigma x_i)}{n}$

$$= \frac{E(\Sigma y_i)}{n} - E(\hat{\beta}_1)\bar{x} = \frac{E(\Sigma Y_i)}{n} - \beta_1 \bar{x} = \frac{\Sigma(\beta_0 + \beta_1 x_i)}{n} - \beta_1 \bar{x} = \beta_0 + \beta_1 \bar{x} - \beta_1 \bar{x} = \beta_0 .$$

41.

a Let $c = n\Sigma x_i^2 - (\Sigma x_i)^2$. Then $E(\hat{\beta}_1) = \dfrac{1}{c} E[n\Sigma x_i Y_i ... Y_i - (\Sigma x_i)...(\Sigma x_i)(\Sigma Y_i)]$

$$= \frac{n}{c} \sum x_i E(Y_i) - \frac{\Sigma x_i}{c} \sum E(Y_i) = \frac{n}{c} \sum x_i (\beta_0 + \beta_1 x_i) - \frac{\Sigma x_i}{c} \sum (\beta_0 + \beta_1 x_i)$$

$$\frac{\beta_1}{c}\left[n\Sigma x_i^2 - (\Sigma x_i)^2\right] = \beta_1 .$$

b With $c = \Sigma(x_i - \bar{x})^2$, $\hat{\beta}_1 = \dfrac{1}{c}\Sigma(x_i - \bar{x})(Y_i - \bar{Y}) = \dfrac{1}{c}\Sigma(x_i - \bar{x})Y_i$ (since $\Sigma(x_i - \bar{x})\bar{Y} = \bar{Y}\Sigma(x_i - \bar{x}) = \bar{Y} \cdot 0 = 0$), so $V(\hat{\beta}_1) = \dfrac{1}{c^2}\Sigma(x_i - \bar{x})^2 Var(Y_i)$

$$= \frac{1}{c^2}\Sigma(x_i - \bar{x})^2 \cdot \sigma^2 = \frac{\sigma^2}{\Sigma(x_i - \bar{x})^2} = \frac{\sigma^2}{\Sigma x_i^2 - (\Sigma x_i)^2 / n}$$, as desired.

42. $t = \hat{\beta}_1 \frac{\sqrt{\Sigma x_i^2 - (\Sigma x_i)^2 / n}}{s}$. The numerator of $\hat{\beta}_1$ will be changed by the factor cd (since both $\Sigma x_i y_i$ and $(\Sigma x_i)(\Sigma y_i)$ appear) while the denominator of $\hat{\beta}_1$ will change by the factor c^2 (since both Σx_i^2 and $(\Sigma x_i)^2$ appear). Thus $\hat{\beta}_1$ will change by the factor d/c. Because $SSE = \Sigma(y_i - \hat{y}_i)^2$, SSE will change by the factor d^2, so s will change by the factor d. Since $\sqrt{\bullet}$ in t changes by the factor c, t itself will change by $\frac{d}{c} \cdot \frac{c}{d} = 1$, or not at all.

43. The numerator of d is |1 – 2| = 1, and the denominator is $\frac{4\sqrt{14}}{\sqrt{324.40}} = .831$, so $d = \frac{1}{.831} = 1.20$. The approximate power curve is for n – 2 df = 13, and β is read from Table A.17 as approximately .1.

Section 12.4

44.

a The mean of the x data in Exercise 12.15 is $\bar{x} = 45.11$. Since x = 40 is closer to 45.11 than is x = 60, the quantity $(40 - \bar{x})^2$ must be smaller than $(60 - \bar{x})^2$. Therefore, since these quantities are the only ones that are different in the two $s_{\hat{y}}$ values, the $s_{\hat{y}}$ value for x = 40 must necessarily be smaller than the $s_{\hat{y}}$ for x = 60. Said briefly, the closer x is to $\bar{x}$, the smaller the value of $s_{\hat{y}}$.

b From the printout in Exercise 12.15, the error degrees of freedom is df = 25. $t_{.025,25} = 2.060$, so the interval estimate when x = 40 is : $7.592 \pm (2.060)(.179)$ $7.592 \pm .369 = (7.223, 7.961)$. We estimate, with a high degree of confidence, that the true average strength for all beams whose MoE is 40 GPa is between 7.223 MPa and 7.961 MPa.

c From the printout in Exercise 12.15, s = .8657, so the 95% prediction interval is
$\hat{y} \pm t_{.025,25}\sqrt{s^2 + s_{\hat{y}}^2} = 7.592 \pm (2.060)\sqrt{(.8657)^2 + (.179)^2}$
$= 7.592 \pm 1.821 = (5.771, 9.413)$. Note that the prediction interval is almost 5 times as wide as the confidence interval.

d For two 95% intervals, the simultaneous confidence level is at least 100(1 – 2(.05)) = 90%

45.

a $\hat{\beta}_0 + 25\hat{\beta}_1 = 50.80$, $\bar{x} = 26.80$, $\sqrt{\frac{1}{15} + \frac{15(26.80-25)^2}{4866}} = .2769$, $s = 3.725$, and $t_{.05,13} = 1.771$, so the 90% C.I. is $50.80 \pm (1.771)(3.725)(.2769)$ $= 50.80 \pm 1.83 = (48.97, 52.63)$.

b Utilizing the computation of **a**, the denominator of t is (3.725)(.2769) = 1.031, so $t = \frac{50.80-50}{1.031} = .78$. H_o is rejected if $t \geq t_{.10,13} = 1.350$, but .78 <1.350 so H_o cannot be rejected.

46.

a A 95% CI for $\mu_{Y \cdot 500}$: $\hat{y}_{(500)} = -.311 + (.00143)(500) = .40$ and

$s_{\hat{y}(500)} = .131\sqrt{\frac{1}{13} + \frac{(500-471.54)^2}{131{,}519.23}} = .03775$, so the interval is

$\hat{y}_{(500)} \pm t_{.025,11} \cdot S_{\hat{y}(500)} = .40 \pm 2.210(.03775) = .40 \pm .08 = (.32, .48)$

b The width at x = 400 will be wider than that of x = 500 because x = 400 is farther away from the mean ($\bar{x} = 471.54$).

c A 95% CI for β_1:

$\hat{\beta}_1 \pm t_{.025,11} \cdot s_{\hat{\beta}_1} = .00143 \pm 2.201(.0003602) = (.000637, .002223)$

d We wish to test $H_0 : y_{(400)} = .25$ vs. $H_0 : y_{(400)} \neq .25$. The test statistic is

$t = \frac{\hat{y}_{(400)} - .25}{s_{\hat{y}(400)}}$, and we reject H_o if $|t| \geq t_{.025,11} = 2.201$.

$\hat{y}_{(400)} = -.311 + .00143(400) = .2614$ and

$s_{\hat{y}(400)} = .131\sqrt{\frac{1}{13} + \frac{(400-471.54)^2}{131{,}519.23}} = .0445$, so the calculated

$t = \frac{.2614-.25}{.0445} = .2561$, which is not ≥ 2.201, so we do not reject H_o. This sample data does not contradict the prior belief.

47.

a $\hat{y}_{(40)} = -1.128 + .82697(40) = 31.95$, $t_{.025,13} = 2.160$; a 95% PI for runoff is $31.95 \pm 2.160\sqrt{(5.24)^2 + (1.44)^2} = 31.95 \pm 11.74 = (20.21, 43.69)$. No, the resulting interval is very wide, therefore the available information is not very precise.

b $\Sigma x = 798, \Sigma x^2 = 63{,}040$ which gives $S_{xx} = 20{,}586.4$, which in turn gives $s_{\hat{y}(50)} = 5.24\sqrt{\frac{1}{15} + \frac{(50 - 53.20)^2}{20{,}586.4}} = 1.358$, so the PI for runoff when x = 50 is $40.22 \pm 2.160\sqrt{(5.24)^2 + (1.358)^2} = 40.22 \pm 11.69 = (28.53, 51.92)$. The simultaneous prediction level for the two intervals is at least $100(1 - 2\alpha)\% = 90\%$.

48.

a $S_{xx} = 18.24 - \frac{(12.6)^2}{9} = .60$, $S_{xy} = 40.968 - \frac{(12.6)(27.68)}{9} = 2.216$;

$S_{yy} = 93.3448 - \frac{(27.68)^2}{9} = 8.213$ $\hat{\beta}_1 = \frac{S_{xy}}{S_{xx}} = \frac{2.216}{.60} = 3.693$;

$\hat{\beta}_0 = \frac{\Sigma y - \hat{\beta}_1 \Sigma x}{n} = \frac{27.68 - (3.693)(12.6)}{9} = -2.095$, so the point estimate is $\hat{y}_{(1.5)} = -2.095 + 3.693(1.5) = 3.445$. $SSE = 8.213 - 3.693(2.216) = .0293$, which yields $s = \sqrt{\frac{SSE}{n-2}} = \sqrt{\frac{.0293}{7}} = .0647$. Thus

$s_{\hat{y}(1.5)} = .0647\sqrt{\frac{1}{9} + \frac{(1.5 - 1.4)^2}{.60}} = .0231$. The 95% CI for $\mu_{y \cdot 1.5}$ is $3.445 \pm 2.365(.0231) = 3.445 \pm .055 = (3.390, 3.50)$.

b A 95% PI for y when x = 1.5 is similar: $3.445 \pm 2.365\sqrt{(.0647)^2 + (.0231)^2} = 3.445 \pm .162 = (3.283, 3.607)$. The prediction interval for a future y value is wider than the confidence interval for an average value of y when x is 1.5.

c A new PI for y when x = 1.2 will be wider since x = 1.2 is farther away from the mean $\bar{x} = 1.4$.

49. 95% CI: (462.1, 597.7); midpoint = 529.9; $t_{.025,8} = 2.306$;

$$529.9 + (2.306)\left(\hat{s}_{\hat{\beta}_0+\hat{\beta}_1(15)}\right) = 597.7$$

$$\hat{s}_{\hat{\beta}_0+\hat{\beta}_1(15)} = 29.402$$

99% CI: $529.9 \pm (3.355)(29.402) = (431.3, 628.5)$

50. $\hat{\beta}_1 = 18.87349841$, $\hat{\beta}_0 = -8.77862227$, SSE = 2486.209, s = 16.6206

a $\hat{\beta}_0 + \hat{\beta}_1(18) = 330.94$, $\bar{x} = 20.2909$, $\sqrt{\dfrac{1}{11} + \dfrac{11(18 - 20.2909)^2}{3834.26}} = .3255$,

$t_{.025,9} = 2.262$, so the CI is $330.94 \pm (2.262)(16.6206)(.3255)$

$= 330.94 \pm 12.24 = (318.70, 343.18)$

b $\sqrt{1 + \dfrac{1}{11} + \dfrac{11(18 - 20.2909)^2}{3834.26}} = 1.0516$, so the P.I. is

$330.94 \pm (2.262)(16.6206)(1.0516) = 330.94 \pm 39.54 = (291.40, 370.48)$.

c To obtain simultaneous confidence of at least 97% for the three intervals, we compute each one using confidence level 99%, (with $t_{.005,9} = 3.250$). For x = 15, the interval is $274.32 \pm 22.35 = (251.97, 296.67)$. For x = 18, $330.94 \pm 17.58 = (313.36, 348.52)$. For x = 20, $368.69 \pm 0.84 = (367.85, 369.53)$.

51.

a 0.40 is closer to $\bar{x}$.

b $\hat{\beta}_0 + \hat{\beta}_1(0.40) \pm t_{\alpha/2,n-2} \cdot \left(\hat{s}_{\hat{\beta}_0+\hat{\beta}_1(0.40)}\right)$ or $0.8104 \pm (2.101)(0.0311)$

$= (0.745, 0.876)$

c $\hat{\beta}_0 + \hat{\beta}_1(1.20) \pm t_{\alpha/2,n-2} \cdot \sqrt{s^2 + s^2_{\hat{\beta}_0+\hat{\beta}_1(1.20)}}$ or

$0.2912 \pm (2.101) \cdot \sqrt{(0.1049)^2 + (0.0352)^2} = (.059, .523)$

52.

a We wish to test $H_o: \beta_1 = 0$ vs $H_a: \beta_1 \neq 0$. The test statistic $t = \frac{10.6026}{.9985} = 10.62$ leads to a p-value of < .006 (2P(t > 4.0) from the 7 df row of table A.8), and H_o is rejected since the p-value is smaller than any reasonable α. The data suggests that this model does specify a useful relationship between chlorine flow and etch rate.

b A 95% confidence interval for β_1: $10.6026 \pm (2.365)(.9985) = (8.24, 12.96)$. We can be highly confident that when the flow rate is increased by 1 SCCM, the associated expected change in etch rate will be between 824 and 1296 A/min.

c A 95% CI for $\mu_{Y \cdot 3.0}$: $38.256 \pm 2.365\left(2.546\sqrt{\frac{1}{9} + \frac{9(3.0 - 2.667)^2}{58.50}}\right)$

$= 38.256 \pm 2.365(2.546)(.35805) = 38.256 \pm 2.156 = (36.100, 40.412)$, or 3610.0 to 4041.2 A/min.

d The 95% PI is $38.256 \pm 2.365\left(2.546\sqrt{1 + \frac{1}{9} + \frac{9(3.0 - 2.667)^2}{58.50}}\right)$

$= 38.256 \pm 2.365(2.546)(1.06) = 38.256 \pm 6.398 = (31.859, 44.655)$, or 3185.9 to 4465.5 A/min.

e The intervals for x* = 2.5 will be narrower than those above because 2.5 is closer to the mean than is 3.0.

f No. a value of 6.0 is not in the range of observed x values, therefore predicting at that point is meaningless.

53. Choice **a** will be the smallest, with d being largest. **a** is less than **b** and **c** (obviously), and **b** and **c** are both smaller than **d**. Nothing can be said about the relationship between **b** and **c**.

54.

a There is a linear pattern in the scatter plot, although the pot also shows a reasonable amount of variation about any straight line fit to the data. The simple linear regression model provides a sensible starting point for a formal analysis.

b n = 141, $\Sigma x_i = 1185, \Sigma x_i^2 = 151{,}825, \Sigma y_i = 5960, \Sigma y_i^2 = 2{,}631{,}200$, and $\Sigma x_i y_i = 449{,}850$, from which
$\hat{\beta}_1 = -1.060132, \hat{\beta}_0 = 515.446887, SSE = 36{,}036.93,$
$r^2 = .616, s^2 = 3003.08, s = 54.80, s_{\hat{\beta}_1} = \dfrac{54.80}{\sqrt{51{,}523.21}} = .241$ $H_o: \beta_1 = 0$ vs $H_a: \beta_1 \neq 0$, $t = \dfrac{\hat{\beta}_1}{s_{\hat{\beta}_1}}$. Reject H_o at level .05 if either $t \geq t_{.025,12} = 2.179$ or $t \leq -2.179$. We calculate $t = \dfrac{-1.060}{.241} = -4.39$. Since $-4.39 \leq -2.179$ H_o is rejected. The simple linear regression model does appear to specify a useful relationship.

c A confidence interval for $\beta_0 + \beta_1(75)$ is requested. The interval is centered at $\hat{\beta}_0 + \hat{\beta}_1(75) = 435.9$. $s_{\hat{\beta}_0 + \hat{\beta}_1(75)} = s\sqrt{\dfrac{1}{n} + \dfrac{n(75-\bar{x})^2}{n\Sigma x_i^2 - (\Sigma x_i)^2}} = 14.83$ (using s = 54.803). Thus a 95% CI is $435.9 \pm (2.179)(14.83) = (403.6, 559.7)$.

55.

a $x_2 = x_3 = 12$, yet $y_2 \neq y_3$

b

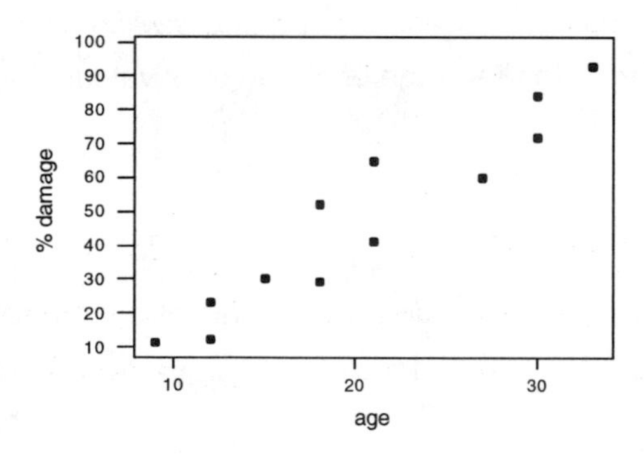

Based on a scatterplot of the data, a simple linear regression model does seem a reasonable way to describe the relationship between the two variables.

c $\hat{\beta}_1 = \frac{2296}{699} = 3.284692$, $\hat{\beta}_0 - 19.669528$, $y = -19.67 + 3.285x$

d $SSE = 35{,}634 - (-19.669528)(572) - (3.284692)(14{,}022) = 827.0188$,

$s^2 = 82.70188, s = 9.094$. $s_{\hat{\beta}_0 + \hat{\beta}_1(20)} = 9.094\sqrt{\frac{1}{12} + \frac{12(20-20.5)^2}{8388}} = 2.6308$,

$\hat{\beta}_0 + \hat{\beta}_1(20) = 46.03$, $t_{.025,10} = 2.228$. The PI is $46.03 \pm 2.228\sqrt{s^2 + s^2_{\hat{\beta}_0 + \hat{\beta}_1(20)}}$

$= 46.03 \pm 21.09 = (24.94, 67.12)$.

56. $$\hat{\beta}_0 + \hat{\beta}_1 x = \bar{Y} - \hat{\beta}_1 \bar{x} + \hat{\beta}_1 x = \bar{Y} + (x - \bar{x})\hat{\beta}_1 = \frac{1}{n}\sum Y_i + \frac{(x-\bar{x})\sum(x_i - \bar{x})Y_i}{n\Sigma x_i^2 - (\Sigma x_i)^2} = \Sigma d_i Y_i$$

where $d_i = \frac{1}{n} + \frac{(x-\bar{x})(x_i - \bar{x})}{n\Sigma x_i^2 - (\Sigma x_i)^2}$. Thus $Var(\hat{\beta}_0 + \hat{\beta}_1 x) = \sum d_i^2 Var(Y_i) = \sigma^2 \Sigma d_i^2$, which, after some algebra, yields the desired expression.

Section 12.5

57. Most people acquire a license as soon as they become eligible. If, for example, the minimum age for obtaining a license is 16, then the time since acquiring a license, y, is usually related to age by the equation $y \approx x - 16$, which is the equation of a straight line. In other words, the majority of people in a sample will have y values that closely follow the line $y = x - 16$.

58.

a Summary values: $\Sigma x = 44{,}615$, $\Sigma x^2 = 170{,}355{,}425$, $\Sigma y = 3{,}860$, $\Sigma y^2 = 1{,}284{,}450$, $\Sigma xy = 14{,}755{,}500$, $n = 12$. Using these values we calculate $S_{xx} = 4{,}480{,}572.92$, $S_{yy} = 42{,}816.67$, and $S_{xy} = 404{,}391.67$. So

$$r = \frac{S_{xy}}{\sqrt{S_{xx}}\sqrt{S_{yy}}} = .9233.$$

b The value of r does not depend on which of the two variables is labeled as the x variable. Thus, had we let x = RBOT time and y = TOST time, the value of r would have remained the same.

c The value of r does no depend on the unit of measure for either variable. Thus, had we expressed RBOT time in hours instead of minutes, the value of r would have remained the same.

d

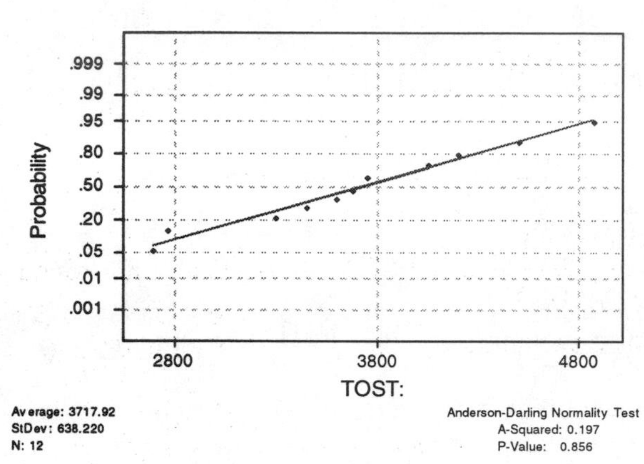

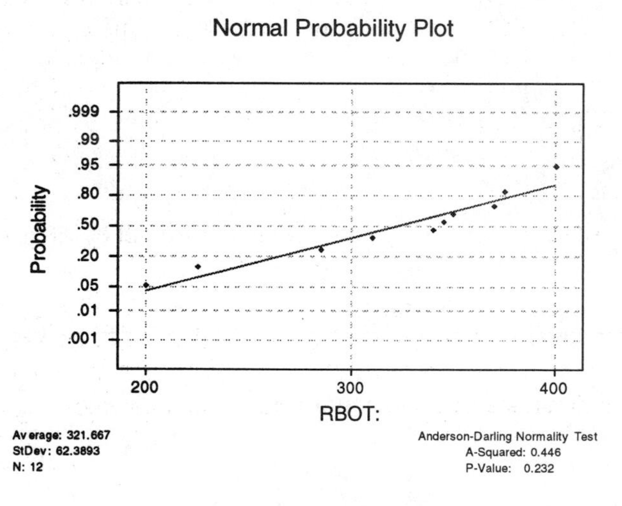

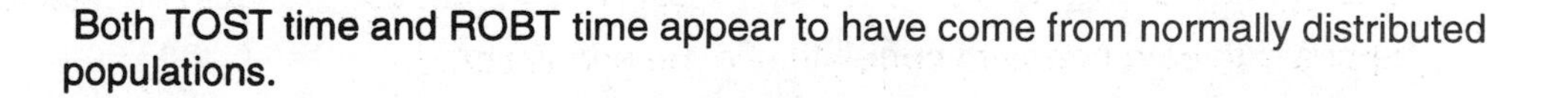
Both TOST time and ROBT time appear to have come from normally distributed populations.

e $H_o: \rho_1 = 0$ vs $H_a: \rho \neq 0$. $t = \dfrac{r\sqrt{n-2}}{\sqrt{1-r^2}}$; Reject H_o at level .05 if either $t \geq t_{.025,10} = 2.228$ or $t \leq -2.228$. r = .923, t = 7.58, so H_o should be rejected. The model is useful.

59.

a $S_{xx} = 251{,}970 - \dfrac{(1950)^2}{18} = 40{,}720$, $S_{yy} = 130.6074 - \dfrac{(47.92)^2}{18} = 3.033711$, and $S_{xy} = 5530.92 - \dfrac{(1950)(47.92)}{18} = 339.586667$, so

$r = \dfrac{339.586667}{\sqrt{40{,}720}\sqrt{3.033711}} = .9662$. There is a very strong positive correlation between the two variables.

b Because the association between the variables is positive, the specimen with the larger shear force will tend to have a larger percent dry fiber weight.

c Changing the units of measurement on either (or both) variables will have no effect on the calculated value of r, because any change in units will affect both the numerator and denominator of r by exactly the same multiplicative constant.

d $r^2 = (.966)^2 = .933$

e $H_o: \rho = 0$ vs $H_a: \rho > 0$. $t = \dfrac{r\sqrt{n-2}}{\sqrt{1-r^2}}$; Reject H_o at level .01 if

$t \geq t_{.01,16} = 2.583$. $t = \dfrac{.966\sqrt{16}}{\sqrt{1-.966^2}} = 14.94 \geq 2.583$, so H_o should be rejected .
The data indicates a positive linear relationship between the two variables.

60. $H_o: \rho = 0$ vs $H_a: \rho \neq 0$. $t = \dfrac{r\sqrt{n-2}}{\sqrt{1-r^2}}$; Reject H_o at level .01 if either $t \geq t_{.005,22} = 2.819$ or $t \leq -2.819$. r = .5778, t = 3.32, so H_o should be rejected. There appears to be a non-zero correlation in the population.

61.

a R = .9066, t = 7.75, $H_o : \rho_1 = 0$ vs $H_a : \rho \neq 0$, and the p-value for an upper-tailed test satisfies p-value < .0005, so H_o would be rejected at any reasonable significance level.

b $r^2 = (.9066)^2 = .82$

62.

a $H_o : \rho_1 = 0$ vs $H_a : \rho \neq 0$, Reject H_o if; Reject H_o at level .05 if either $t \geq t_{.025,12} = 2.179$ or $t \leq -2.179$. $t = \dfrac{r\sqrt{n-2}}{\sqrt{1-r^2}} = \dfrac{(.449)\sqrt{12}}{\sqrt{1-(.449)^2}} = 1.74$. Fail to reject H_o/ the data does not suggest that the population correlation coefficient differs from 0.

b $(.449)^2 = .20$ so 20 percent of the observed variation in gas porosity can be accounted or by variation in hydrogen content.

63. n = 6, $\Sigma x_i = 111.71, \Sigma x_i^2 = 2{,}724.7643, \Sigma y_i = 2.9, \Sigma y_i^2 = 1.6572$, and $\Sigma x_i y_i = 63.915$. $r = \dfrac{(6)(63.915)-(111.71)(2.9)}{\sqrt{(6)(2{,}724.7943)-(111.73)^2} \cdot \sqrt{(6)(1.6572)-(2.9)^2}} = .7729$.

$H_o : \rho_1 = 0$ vs $H_a ; \rho \neq 0$; Reject H_o at level .05 if $|t| \geq t_{.025,4} = 2.776$.

$t = \dfrac{(.7729)\sqrt{4}}{\sqrt{1-(.7729)^2}} = 2.436$. Fail to reject H_o. The data does not indicate that the population correlation coefficient differs from 0. This result may seem surprising due to the relatively large size of r (.77), however, it can be attributed to a small sample size (6).

64. $r = \dfrac{-757.6423}{\sqrt{(3756.96)(465.34)}} = -.5730$

a $v = .5\ln\left(\dfrac{.427}{1.573}\right) = -.652$, so (12.11) is $-.652 \pm \dfrac{(1.645)}{\sqrt{26}} = (-.976, -.3290)$, and the desired interval for ρ is $(-.751, -.318)$.

b $z = (-.652 + .549)\sqrt{23} = -.49$, so H_o cannot be rejected at any reasonable level.

c $r^2 = .328$

d Again, $r^2 = .328$

65.

a Although the normal probability plot of the x's appears somewhat curved, such a pattern Is not terribly unusual when n is small; the test of normality presented in section 14.2 (p. 625) does not reject the hypothesis of population normality. The normal probability plot of the y's is much straighter.

b $H_o : \rho_1 = 0$ will be rejected in favor of $H_a : \rho \neq 0$ at level .01 if $|t| \geq t_{.005,8} = 3.355$. $\Sigma x_i = 864, \Sigma x_i^2 = 78{,}142, \Sigma y_i = 138.0, \Sigma y_i^2 = 1959.1$ and $\Sigma x_i y_i = 12{,}322.4$, so $r = \dfrac{3992}{(186.8796)(23.3880)} = .913$ and $t = \dfrac{.913(2.8284)}{.4080} = 6.33 \geq 3.355$, so reject H_o. There does appear to be a linear relationship.

66.

a Because p-value = .00032 < α = .001, H_o should be rejected at this significance level.

b Not necessarily. For this n, the test statistic *t* has approximately a standard normal distribution when $H_o : \rho_1 = 0$ is true, and a p-value of .00032 corresponds to $z = 3.60$ (or –3.60). Solving $3.60 = \dfrac{r\sqrt{498}}{\sqrt{1}} - r^2$ for r yields r = .159. This r suggests only a weak linear relationship between x and y, one that would typically have little practical import.

67. $t = 2.20 \geq t_{.025,9998} = 1.96$, so H_o is rejected in favor of H_a. The value t = 2.20 is statistically significant -- it cannot be attributed just to sampling variability in the case $\rho = 0$. But with this n, r = .022 implies $\rho = .022$, which in turn shows an extremely weak linear relationship.

Supplementary Exercises

68.

a $n=8$, $\Sigma x_i = 207, \Sigma x_i^2 = 6799, \Sigma y_i = 621.8, \Sigma y_i^2 = 48{,}363.76$ and $\Sigma x_i y_i = 15{,}896.8$, which gives $\hat{\beta}_1 = \frac{-1538.20}{11{,}543} = -.133258$, $\hat{\beta}_0 = 81.173051$, and $y = 81.173 - .1333x$ as the equation of the estimated line.

b We wish to test $H_0 : \beta_1 = 0$ vs $H_0 : \beta_1 \neq 0$. At level .01, H_o will be rejected (and the model judged useful) if either $t \geq t_{.005,6} = 3.707$ or $t \leq -3.707$. SSE = 8.732664, s = 1.206, and $t = \frac{-.1333}{1.206/37.985} = \frac{-.1333}{.03175} = -4.2$, which is ≤ -3.707, so we do reject H_o and find the model useful.

c The larger the value of $\sum (x_i - \overline{x})^2$, the smaller will be $\hat{\sigma}_{\hat{\beta}_1}$ and the more accurate the estimate will tend to be. For the given x_i's, $\sum (x_i - \overline{x})^2 = 1442.88$, whereas the proposed x values $x_1 = ... = x_4 = 0$, $x_5 = ... = x_8 = 50$, $\sum (x_i - \overline{x})^2 = 5000$. Thus the second set of x values is preferable to the first set. With just 3 observations at x = 0 and 3 at x = 50, $\sum (x_i - \overline{x})^2 = 3750$, which is again preferable to the first set of x_i's.

d $\hat{\beta}_0 + \hat{\beta}_1(25) = 77.84$, and $s_{\hat{\beta}_0 + \hat{\beta}_1(25)} = s\sqrt{\frac{1}{n} + \frac{n(25-\overline{x})^2}{n\Sigma x_i^2 - (\Sigma x_i)^2}} = 14.83$

$= 1.206\sqrt{\frac{1}{8} + \frac{8(25-25.875)^2}{11.543}} = .426$, so the 95% CI is

$77.84 \pm (2.447)(.426) = 77.84 \pm 1.04 = (76.80, 78.88)$

69.

a The test statistic value is $t = \frac{\hat{\beta}_1 - 1}{s_{\hat{\beta}_1}}$, and H_o will be rejected if either

$t \geq t_{.025,11} = 2.201$ or $t \leq -2.201$. With

$\Sigma x_i = 243, \Sigma x_i^2 = 5965, \Sigma y_i = 241, \Sigma y_i^2 = 5731$ and $\Sigma x_i y_i = 5805$,

$\hat{\beta}_1 = .913819$, $\hat{\beta}_0 = 1.457072$, $SSE = 75.126$, $s = 2.613$, and $s_{\hat{\beta}_1} = .0693$,

$t = \frac{.9138 - 1}{.0693} = -1.24$. Because -1.24 is neither ≤ -2.201 nor ≥ 2.201, H_o cannot be rejected. It is plausible that $\beta_1 = 1$.

b $r = \frac{16{,}902}{(136)(128.15)} = .970$

70.

a sample size = 8

b $\hat{y} = 326.976038 - (8.403964)x$. When x = 35.5, $\hat{y} = 28.64$.

c Yes, the model utility test is statistically significant at the level .01.

d $r = \sqrt{r^2} = \sqrt{0.9134} = 0.9557$

e First check to see if the value x = 40 falls within the range of x values used to generate the least-squares regression equation. If it does not, this equation should not be used. Furthermore, for this particular model an x value of 40 yields a g value of –9.18, which is an impossible value for y.

71.

a $r^2 = .5073$

b $r = +\sqrt{r^2} = \sqrt{.5073} = .7122$ (positive because $\hat{\beta}_1$ is positive.)

c We test test $H_0 : \beta_1 = 0$ vs $H_0 : \beta_1 \neq 0$. The test statistic t = 3.93 gives p-value = .0013, which is < .01, the given level of significance, therefore we reject H_o and conclude that the model is useful.

d We use a 95% CI for $\mu_{Y \cdot 50}$. $\hat{y}_{(50)} = .787218 + .007570(50) = 1.165718$, $t_{.025,15} = 2.131$, s = "Root MSE" = .020308, so

$$s_{\hat{y}(50)} = .20308\sqrt{\frac{1}{17} + \frac{17(50-42.33)^2}{17(41{,}575)-(719.60)^2}} = .051422$$. The interval is , then,

$$1.165718 \pm 2.131(.051422) = 1.165718 \pm .109581 = (1.056137, 1.275299).$$

e $\hat{y}_{(30)} = .787218 + .007570(30) = 1.0143$. The residual is $y - \hat{y} = .80 - 1.0143 = -.2143$.

72.

a

Regression Plot

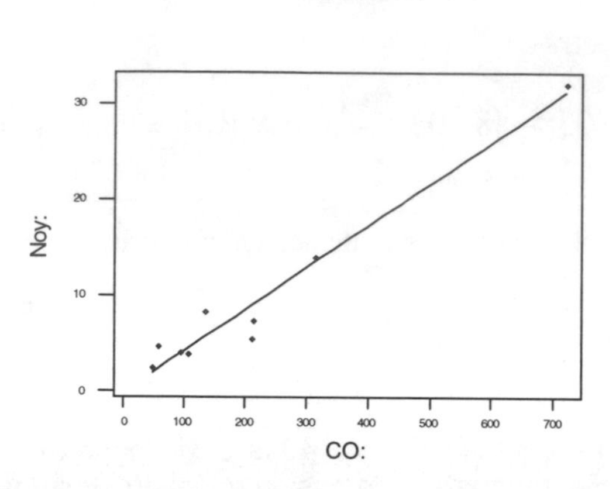

The above analysis was created in Minitab. A simple linear regression model seems to fit the data well. The least squares regression equation is $\hat{y} = -.220 + .0436x$. The model utility test obtained from Minitab produces a t test statistic equal to 12.72. The corresponding p-value is extremely small. So we have sufficient evidence to claim that ΔCO is a good predictor of ΔNO_y.

b $\hat{y} = -.220 + .0436(400) = 17.228$. A 95% prediction interval produced by Minitab is (11.953, 22.503). Since this interval is so wide, it does not appear that ΔNO_y is accurately predicted.

c While the large ΔCO value appears to be "near" the least squares regression line, the value has extremely high leverage. The least squares line that is obtained when excluding the value is $\hat{y} = 1.00 + .0346x$. The r^2 value with the value included is 96% and is reduced to 75% when the value is excluded. The value of s with the value included is 2.024, and with the value excluded is 1.96. So the large ΔCO value does appear to effect our analysis in a substantial way.

73.

a n = 9, $\Sigma x_i = 228, \Sigma x_i^2 = 5958, \Sigma y_i = 93.76, \Sigma y_i^2 = 982.2932$ and $\Sigma x_i y_i = 2348.15$, giving $\hat{\beta}_1 = \frac{-243.93}{1638} = -.148919$, $\hat{\beta}_0 = 14.190392$, and the equation $\hat{y} = 14.19 - (.1489)x$.

b β_1 is the expected increase in load associated with a one-day age increase (so a negative value of β_1 corresponds to a decrease). We wish to test $H_0 : \beta_1 = -.10$ vs. $H_0 : \beta_1 < -.10$ (the alternative contradicts prior belief). H_o will be rejected at level .05 if $t = \dfrac{\hat{\beta}_1 - (-.10)}{s_{\hat{\beta}_1}} \le -t_{.05,7} = -1.895$. With SSE = 1.4862, s = .4608, and

$s_{\hat{\beta}_1} = \dfrac{.4608}{\sqrt{182}} = .0342$. Thus $t = \dfrac{-.1489 + 1}{.0342} = -1.43$. Because –1.43 is not ≤ -1.895, do not reject H_o.

c $\Sigma x_i = 306, \Sigma x_i^2 = 7946$, so $\sum (x_i - \bar{x})^2 = 7946 - \dfrac{(306)^2}{12} = 143$ here, as contrasted with 182 for the given 9 x_i's. Even though the sample size for the proposed x values is larger, the original set of values is preferable.

d $(t_{.025,7})(s)\sqrt{\dfrac{1}{9} + \dfrac{9(28 - 25.33)^2}{1638}} = (2.365)(.4608)(.3877) = .42$, and $\hat{\beta}_0 + \hat{\beta}_1(28) = 10.02$, so the 95% CI is $10.02 \pm .42 = (9.60, 10.44)$.

74.

a $\hat{\beta}_1 = \dfrac{3.5979}{44.713} = .0805$, $\hat{\beta}_0 = 1.6939$, $\hat{y} = 1.69 + (.0805)x$.

b $\hat{\beta}_1 = \dfrac{3.5979}{.2943} = 12.2254$, $\hat{\beta}_0 = -20.4046$, $\hat{y} = -20.40 + (12.2254)x$.

c r = .992, so r^2 = .984 for either regression.

75.

a The plot suggests a strong linear relationship between x and y.

b n = 9, $\Sigma x_i = 1797, \Sigma x_i^2 = 4334.41, \Sigma y_i = 7.28, \Sigma y_i^2 = 7.4028$ and $\Sigma x_i y_i = 178.683$, so $\hat{\beta}_1 = \dfrac{299.931}{6717.6} = .04464854$, $\hat{\beta}_0 = -.08259353$, and the equation of the estimated line is $\hat{y} = -.08259 - (.044649)x$.

c $SSE = 7.4028 - (-601281) - 7.977935 = .026146,$

$SST = 7.4028 - \frac{(7.28)^2}{9} = .026146, = 1.5141$, and $r^2 = 1 - \frac{SSE}{SST} = .983$, so 93.8% of the observed variation is "explained."

d $\hat{y}_4 = -.08259 - (.044649)(19.1) = .7702$, and $y_4 - \hat{y}_4 = .68 - .7702 = -.0902$.

e s = .06112, and $s_{\hat{\beta}_1} = \frac{.06112}{\sqrt{746.4}} = .002237$, so the value of t for testing $H_0 : \beta_1 = 0$ vs $H_0 : \beta_1 \neq 0$ is $t = \frac{.044649}{.002237} = 19.96$. From Table A.5, $t_{.0005,7} = 5.408$, so $p - value < 2(.0005) = .001$. There is strong evidence for a useful relationship.

f A 95% CI for β_1 is $.044649 \pm (2.365)(.002237) = .044649 \pm .005291$ $= (.0394,.0499)$.

g A 95% CI for $\beta_0 + \beta_1(20)$ is $.810 \pm (2.365)(.002237)(.3333356)$ $= .810 \pm .048 = (.762,.858)$

76. Substituting x* = 0 gives the CI $\hat{\beta}_0 \pm t_{\alpha/2,n-2} \cdot s\sqrt{\frac{1}{n} + \frac{n\bar{x}^2}{n\Sigma x_i^2 - (\Sigma x_i)^2}}$. From Example 12.8, $\hat{\beta}_0 = 3.621$, SSE = .262453, n = 14, $\Sigma x_i = 890, \bar{x} = 63.5714, \Sigma x_i^2 = 67{,}182$, so with s = .1479, $t_{.025,12} = 2.179$, the CI is $3.621 \pm 2.179(.1479)\sqrt{\frac{1}{12} + \frac{56{,}578.52}{148{,}448}}$ $= 3.621 \pm 2.179(.1479)(.6815) = 3.62 \pm .22 = (3.40,3.84)$.

77. $SSE = \Sigma y^2 - \hat{\beta}_0 \Sigma y - \hat{\beta}_1 \Sigma xy$. Substituting $\hat{\beta}_0 = \frac{\Sigma y - \hat{\beta}_1 \Sigma x}{n}$, SSE becomes

$$SSE = \Sigma y^2 - \frac{\Sigma y(\Sigma y - \hat{\beta}_1 \Sigma x)}{n} - \hat{\beta}_1 \Sigma xy = \Sigma y^2 - \frac{(\Sigma y)^2}{n} + \frac{\hat{\beta}_1 \Sigma x \Sigma y}{n} - \hat{\beta}_1 \Sigma xy$$

$$= \left[\Sigma y^2 - \frac{(\Sigma y)^2}{n}\right] - \hat{\beta}_1\left[\Sigma xy - \frac{\Sigma x \Sigma y}{n}\right] = S_{yy} - \hat{\beta}_1 S_{xy}$$, as desired.

78. The value of the sample correlation coefficient using the squared y values would not necessarily be approximately 1. If the y values are greater than 1, then the squared y values would differ from each other by more than the y values differ from one another. Hence, the relationship between x and y^2 would be less like a straight line, and the resulting value of the correlation coefficient would decrease.

79.

a With $s_{xx} = \sum (x_i - \bar{x})^2$, $s_{yy} = \sum (y_i - \bar{y})^2$, note that $\frac{s_y}{s_x} = \sqrt{\frac{s_{yy}}{s_{xx}}}$ (since the factor n-1 appears in both the numerator and denominator, so cancels). Thus

$$y = \hat{\beta}_0 + \hat{\beta}_1 x = \bar{y} + \hat{\beta}_1 (x - \bar{x}) = \bar{y} + \frac{s_{xy}}{s_{xx}}(x - \bar{x}) = \bar{y} + \sqrt{\frac{s_{yy}}{s_{xx}}} \cdot \frac{s_{xy}}{\sqrt{s_{xx}s_{yy}}}(x - \bar{x})$$

$= \bar{y} + \frac{s_y}{s_x} \cdot r \cdot (x - \bar{x})$, as desired.

b By .573 s.d.'s above, (above, since r < 0) or (since s_y = 4.3143) an amount 2.4721 above.

80. With s_{xy} given in the text, $r = \frac{s_{xy}}{\sqrt{s_{xx}s_{yy}}}$ (where e.g. $s_{xx} = \sum (x_i - \bar{x})^2$), and $\hat{\beta}_1 = \frac{s_{xy}}{s_{xx}}$.

Also, $s = \sqrt{\frac{SSE}{n-2}}$ and $SSE = \Sigma y_i^2 - \hat{\beta}_0 \Sigma y_i - \hat{\beta}_1 \Sigma x_i y_i = s_{yy} - \hat{\beta}_1 s_{xy}$. Thus the t statistic for $H_o : \hat{\beta}_1 = 0$ is $t = \frac{\hat{\beta}_1}{s / \sqrt{\sum (x_i - \bar{x})^2}} = \frac{(s_{xy} / s_{xx}) \cdot \sqrt{s_{xx}}}{\sqrt{(s_{yy} - s_{xy}^2 / s_{xx})/(n-2)}}$

$= \frac{s_{xy} \cdot \sqrt{n-2}}{\sqrt{(s_{xx}s_{yy} - s_{xy}^2)}} = \frac{(s_{xy} / \sqrt{s_{xx}s_{yy}})\sqrt{n-2}}{\sqrt{1 - s_{xy}^2 / s_{xx}s_{yy}}} = \frac{r\sqrt{n-2}}{\sqrt{1-r^2}}$ as desired.

81. Using the notation of the exercise above, $SST = s_{yy}$ and $SSE = s_{yy} - \hat{\beta}_1 s_{xy}$

$= s_{yy} - \frac{s_{xy}^2}{s_{xx}}$, so $1 - \frac{SSE}{SST} = 1 - \frac{s_{yy} - \frac{s_{xy}^2}{s_{xx}}}{s_{yy}} = \frac{s_{xy}^2}{s_{xx}s_{yy}} = r^2$, as desired.

82.

a $\hat{\beta}_1 = \frac{20(71.51) - (63.5)(17.26)}{20(311.74) - (63.5)^2} = \frac{334.190}{2202.550} = .15172868$,

$\hat{\beta}_0 = \frac{17.26 - .15172868(63.5)}{20} = .38126144$, so the equation is

$y = .3813 + .1517x$.

b New $\Sigma x_i = 63.5 - 9.8 = 53.7$, new $\Sigma y_i = 17.26 - 1.9 = 15.36$, new $\Sigma x_i^2 = 311.74 - (9.8)^2 = 215.70$, new $\Sigma x_i y_i = 71.51 - (9.8)(1.9) = 52.89$, new $n = 19$, so $\hat{\beta}_1 = \dfrac{180.078}{1214.610} = .148260$, $\hat{\beta}_0 = .389391$, so now $y = .3894 + .1483x$. For x = 9.8, the predicted y = 1.84.

83. For the second boiler, $n = 19$, $\Sigma x_i = 125$, $\Sigma y_i = 472.0$, $\Sigma x_i^2 = 3625$, $\Sigma y_i^2 = 37{,}140.82$, and $\Sigma x_i y_i = 9749.5$, giving $\hat{\gamma}_1$ = estimated slope $= \dfrac{-503}{6125} = -.0821224$, $\hat{\gamma}_0 = 80.377551$, $SSE_2 = 3.26827$, $SSx_2 = 1020.833$. For boiler #1, n = 8, $\hat{\beta}_1 = -.1333$, $SSE_1 = 8.733$, and $SSx_1 = 1442.875$. Thus $\hat{\sigma}^2 = \dfrac{8.733 + 3.286}{10} = 1.2$, $\hat{\sigma} = 1.095$, and $t = \dfrac{-.1333 + .0821}{1.095\sqrt{\frac{1}{1442.875} + \frac{1}{1020.833}}}$ $= \dfrac{-.0512}{.0448} = -1.14$. $t_{.025,10} = 2.228$ and -1.14 is neither ≥ 2.228 nor ≤ -2.228, so H_o is not rejected. It is plausible that $\beta_1 = \gamma_1$.

Chapter 13

Section 13.1

1.

a $\bar{x} = 15$ and $\sum (x_j - \bar{x})^2 = 250$, so s.d. of $Y_i - \hat{Y}_i$ is $10\sqrt{1 - \frac{1}{5} - \frac{(x_i - 15)^2}{250}} =$ 6.32, 8.37, 8.94, 8.37, and 6.32 for i = 1, 2, 3, 4, 5.

b Now $\bar{x} = 20$ and $\sum (x_i - \bar{x})^2 = 1250$, giving standard deviations 7.87, 8.49, 8.83, 8.94, and 2.83 for i = 1, 2, 3, 4, 5.

c The deviation from the estimated line is likely to be much smaller for the observation made in the experiment of **b** for x = 50 than for the experiment of **a** when x = 25. That is, the observation (50, Y) is more likely to fall close to the least squares line than is (25, Y).

2. The pattern gives no cause for questioning the appropriateness of the simple linear regression model, and no observation appears unusual.

3.

a This plot indicates there are no outliers, the variance of ε is constant, and the ε are normally distributed. A straight-line regression function is a reasonable choice for a model.

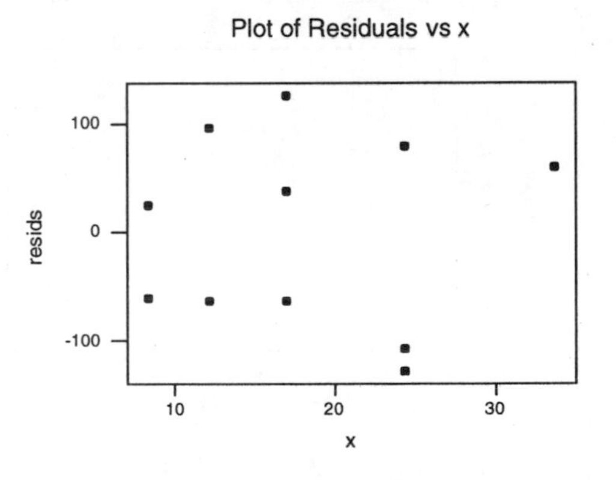

b With $\sum(x_j - \bar{x})^2 = \sum x_j^2 - \frac{(\Sigma x_j)^2}{n} = 4198.03 - 3574.81 = 623.22$

$e_i^* = \frac{e_i}{92.19}\sqrt{1 - \frac{1}{11} - \frac{(x_i - 18.03)^2}{623.22}}$. This gives e_i^* = -.75, .31, -.74, 1.13, .43, .72, 1.43, .93, -1.51, -1.27, .90 for i = 1, …, 11; $\frac{e_2}{s} = \frac{24.52}{92.19} = .27$ while $e_2^* = .31$, and $\frac{e_7}{s} = \frac{125.72}{92.19} = 1.36 \neq 1.43 = e_7^*$, so $\frac{e_i}{s}$ is not $\doteq e_i^*$.

c The plot of e_i^* vs x has almost exactly the same pattern as the plot of **a**; only the scale on the vertical axis is changed.

4. The (x, residual) pairs for the plot are (0, -.335), (7, -.508), (17. -.341), (114, .592), (133, .679), (142, .700), (190, .142), (218, 1.051), (237, -1.262), and (285, -.719). The plot shows substantial evidence of curvature.

5. The standardized residuals (in order corresponding to increasing x) are -.50, -.75, -.50, .79, .90, .93, .19, 1.46, -1.80, and -1.12. A standardized residual plot shows the same pattern as the residual plot discussed in the previous exercise. The z percentiles for the

Normal Probability Plot for the Standardized Residuals

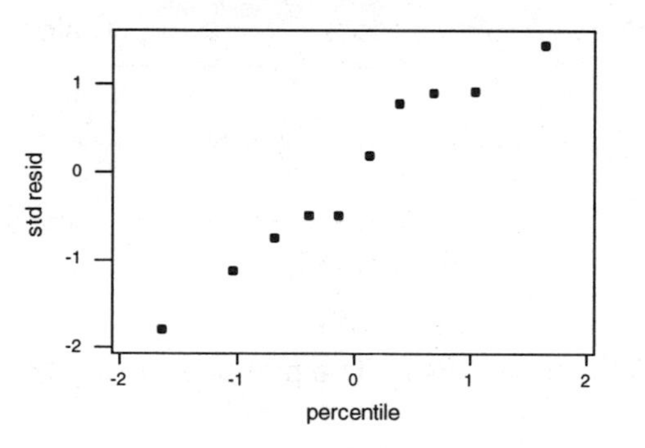

normal probability plot are –1.645, -1.04, -.68, -.39, -.13, .13, .39, .68, 1.04, 1.645. The plot follows. The points follow a linear pattern, so the standardized residuals appear to have a normal distribution.

6.

a $H_o: \beta_1 = 0$ vs. $H_a: \beta_1 \neq 0$. The test statistic is $t = \dfrac{\hat{\beta}_1}{s_{\hat{\beta}_1}}$, and we will reject H_o if $t \geq t_{.025,4} = 2.776$ or if $t \leq -2.776$. $s_{\hat{\beta}_1} = \dfrac{s}{\sqrt{S_{xx}}} = \dfrac{7.265}{12.869} = .565$, and $t = \dfrac{6.19268}{.565} = 10.97$. Since $10.97 \geq 2.776$, we reject H_o and conclude that the model is useful.

b $\hat{y}_{(7.0)} = 1008.14 + 6.19268(7.0) = 1051.49$, from which the residual is $y - \hat{y}_{(7.0)} = 1046 - 1051.49 = -5.49$. Similarly, the other residuals are -.73, 4.11, 7.91, 3.58, and –9.38. The plot of the residuals vs x follows:

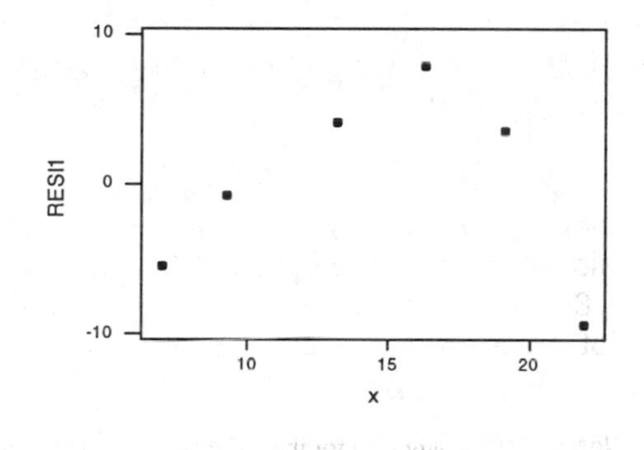

Because a curved pattern appears, a linear regression function may be inappropriate.

c The standardized residuals are calculated as

$$e_1^* = \frac{-5.49}{7.265\sqrt{1 + \frac{1}{6} + \frac{(7.0 - 14.48)^2}{165.5983}}} = -1.074$$

, and similarly the others are -.123, .624, 1.208, .587, and –1.841. The plot of e* vs x follows :

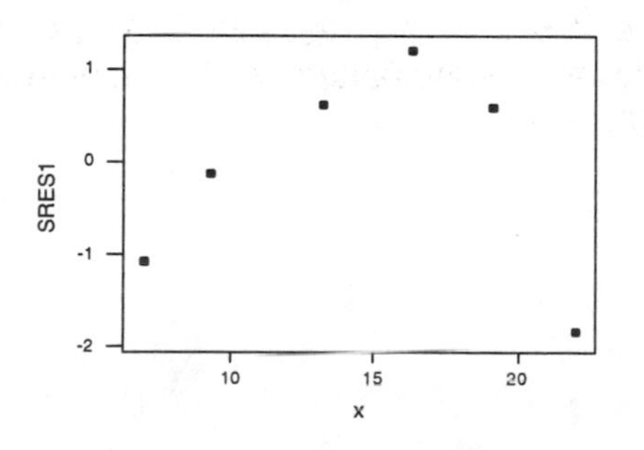

This plot (on the previous page) gives the same information as the previous plot. No values are exceptionally large, but the e* of –1.841 is close to 2 std deviations away from the expected value of 0.

7.

a

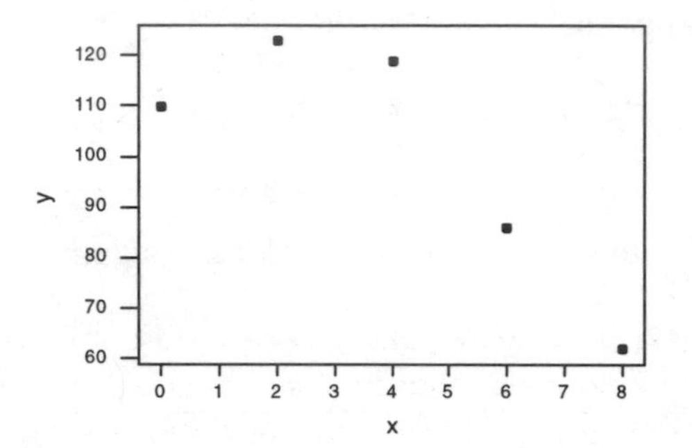

There is an obvious curved pattern in the scatter plot, which suggests that a simple linear model will not provide a good fit.

b The $\hat{y}'s$, e's, and e*'s are given below:

x	y	$\hat{y}$	e	e*
0	110	126.6	-16.6	-1.55
2	123	113.3	9.7	.68
4	119	100.0	19.0	1.25
6	86	86.7	-.7	-.05
8	62	73.4	-11.4	-1.06

8.

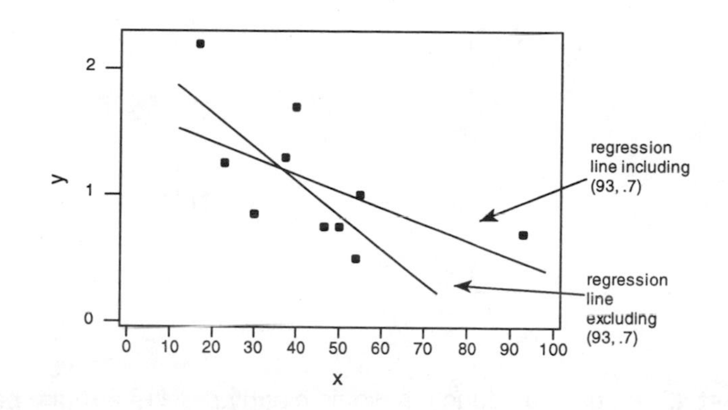

The picture clearly suggests that the point (93, .7) has been quite influential in obtaining the equation of the least squares line; the x value is quite far above the other x values and the point itself is somewhat inconsistent with the rest of the data, as is seen by deleting it form the sample and re-fitting the line (the result is y = 2.24 - .0278x). The standardized residuals for the original data are 1.84, -.47, -1.19, .22, 1.29, -.78, -.65, -1.12, .15 and 1.38. Neither a residual plot nor the magnitudes of any of the e*'s suggest that one point has been influential in determining the fit.

9. Both a scatter plot and residual plot (based on the simple linear regression model) for the first data set suggest that a simple linear regression model is reasonable, with no pattern or influential data points which would indicate that the model should be modified. However, scatter plots for the other three data sets reveal difficulties.

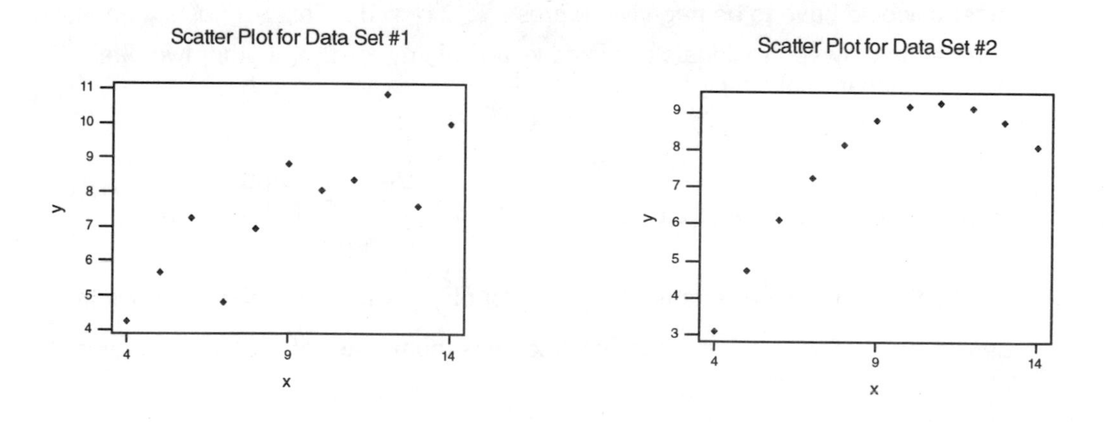

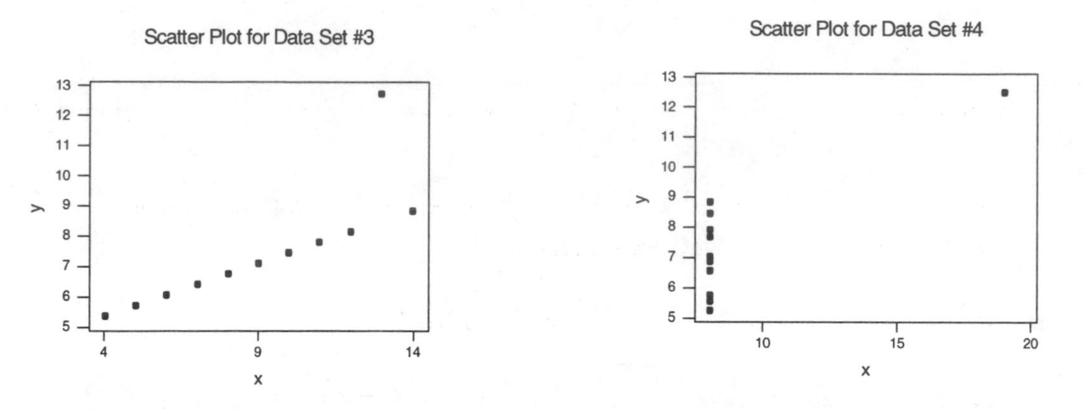

For data set #2, a quadratic function would clearly provide a much better fit. For data set #3, the relationship is perfectly linear except one outlier, which has obviously greatly influenced the fit even though its x value is not unusually large or small. The signs of the residuals here (corresponding to increasing x) are + + + + - - - - - + -, and a residual plot would reflect this pattern and suggest a careful look at the chosen model. For data set #4 it is clear that the slope of the least squares line has been determined entirely by the outlier, so this point is extremely influential (and its x value does lie far from the remaining ones).

10.

a $e_i = y_i - \left(\hat{\beta}_0 - \hat{\beta}_1 x_i\right) = y_i - \bar{y} - \hat{\beta}_1(x_i - \bar{x})$, so
$\Sigma e_i = \Sigma(y_i - \bar{y}) - \hat{\beta}_1 \Sigma(x_i - \bar{x}) = 0 + \hat{\beta}_1 \cdot 0 = 0$.

b Since $\Sigma e_i = 0$ always, the residuals cannot be independent. There is clearly a linear relationship between the residuals. If one e_l is large positive, then al least one other e_l would have to be negative to preserve $\Sigma e_i = 0$. This suggests a negative correlation between residuals (for fixed values of any n – 2, the other two obey a negative linear relationship).

c $\Sigma x_i e_i = \Sigma x_i y_i - \Sigma x_i \bar{y} - \hat{\beta}_1 \Sigma x_i (x_i - \bar{x}) = \left[\Sigma x_i y_i - \frac{(\Sigma x_i)(\Sigma y_i)}{n}\right] - \hat{\beta}_1 \left[\Sigma x_i^2 - \frac{(\Sigma x_i)^2}{n}\right]$,

but the first term in brackets is the numerator of $\hat{\beta}_1$, while the second term is the denominator of $\hat{\beta}_1$, so the difference becomes (numerator of $\hat{\beta}_1$) – (numerator of $\hat{\beta}_1$) = 0.

d The five e_i^*'s from Exercise 7 above are –1.55, .68, 1.25, -.05, and –1.06, which sum to -.73. This sum differs too much from 0 to be explained by rounding. In general it is not true that $\Sigma e_i^* = 0$.

Chapter 13

11.

a $Y_i - \hat{Y}_i = Y_i - \bar{Y} - \hat{\beta}_1(x_i - \bar{x}) = Y_i - \frac{1}{n}\sum_j Y_j - \frac{(x_i - \bar{x})\sum_j (x_j - \bar{x})Y_j}{\sum_j (x_j - \bar{x})^2} = \sum_j c_j Y_j$,

where $c_j = 1 - \frac{1}{n} - \frac{(x_i - \bar{x})^2}{n\Sigma(x_j - \bar{x})^2}$ for j = i and $c_j = 1 - \frac{1}{n} - \frac{(x_i - \bar{x})(x_j - \bar{x})}{\Sigma(x_j - \bar{x})^2}$ for $j \neq i$. Thus $Var(Y_i - \hat{Y}_i) = \Sigma Var(c_j Y_j)$ (since the Y_j's are independent) = $\sigma^2 \Sigma c_j^2$ which, after some algebra, gives equation (13.2).

b $\sigma^2 = Var(Y_i) = Var(\hat{Y}_i + (Y_i - \hat{Y}_i)) = Var(\hat{Y}_i) + Var(Y_i - \hat{Y}_i)$, so

$Var(Y_i - \hat{Y}_i) = \sigma^2 - Var(\hat{Y}_i) = \sigma^2 - \sigma^2\left[\frac{1}{n} + \frac{(x_i - \bar{x})^2}{n\Sigma(x_j - \bar{x})^2}\right]$, which is exactly (13.2).

c As x_i moves further from $\bar{x}$, $(x_i - \bar{x})^2$ grows larger, so $Var(\hat{Y}_i)$ increases (since $(x_i - \bar{x})^2$ has a positive sign in $Var(\hat{Y}_i)$), but $Var(Y_i - \hat{Y}_i)$ decreases (since $(x_i - \bar{x})^2$ has a negative sign).

12.

a $\Sigma e_i = 34$, which is not = 0, so these cannot be the residuals.

b Each $x_i e_i$ is positive (since x_i and e_i have the same sign) so $\Sigma x_i e_i > 0$, which contradicts the result of exercise 10**c**, so these cannot be the residuals for the given x values.

13. The distribution of any particular standardized residual is also a t distribution with n – 2 d.f., since e_i^* is obtained by taking standard normal variable $\frac{(Y_i - \hat{Y}_i)}{(\sigma_{Y_i - \hat{Y}})}$ and substituting the estimate of σ in the denominator (exactly as in the predicted value case). With E_i^* denoting the ith standardized residual as a random variable, when n = 25 E_i^* has a t distribution with 23 d.f. and $t_{.01,23} = 2.50$, so P(E_i^* outside (-2.50, 2.50)) = $P(E_i^* \geq 2.50) + P(E_i^* \leq -2.50) = .01 + .01 = .02$.

14. $n_1 = n_2 = 3$ (3 observations at 110 and 3 at 230), $n_3 = n_4 = 4$, $\bar{y}_{1.} = 202.0$, $\bar{y}_{2.} = 149.0$, $\bar{y}_{3.} = 110.5$, $\bar{y}_{4.} = 107.0$, $\Sigma\Sigma y_{ij}^2 = 288{,}013$, so $SSPE = 288{,}013 - \left[3(202.0)^2 + 3(149.0)^2 + 4(110.5)^2 + 4(107.0)^2\right] = 4361$. With $\Sigma x_i = 4480$, $\Sigma y_i = 1923$, $\Sigma x_i^2 = 1{,}733{,}500$, $\Sigma y_i^2 = 288{,}013$ (as above), and $\Sigma x_i\, y_i = 544{,}730$, SSE = 7241 so SSLF = 7241-4361=2880. With c – 2 = 2 and n – c = 10, $F_{.05,2,10} = 4.10$. $MSLF = \frac{2880}{2} = 1440$ and $SSPE = \frac{4361}{10} = 436.1$, so the computed value of F is $\frac{1440}{436.1} = 3.30$. Since 3.30 is not ≥ 4.10, we do not reject H_o.

This formal test procedure does not suggest that a linear model is inappropriate. However, the scatter plot clearly reveals a curved pattern which suggests that a nonlinear model would be more reasonable and provide a better fit than a linear model.

Section 13.2

15.

a

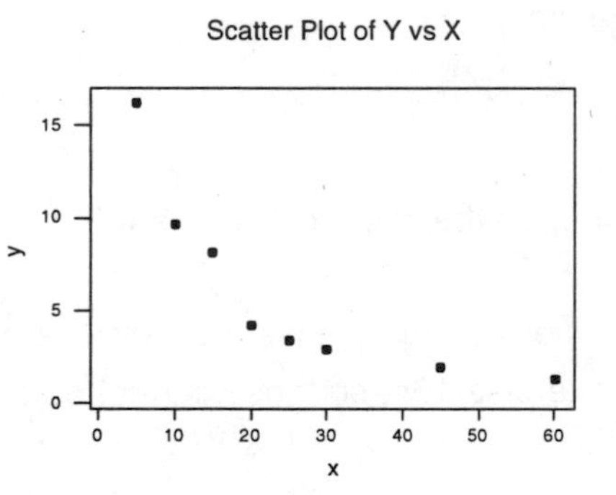

The points have a definite curved pattern. A linear model would not be appropriate.

b In this plot we have a strong linear pattern.

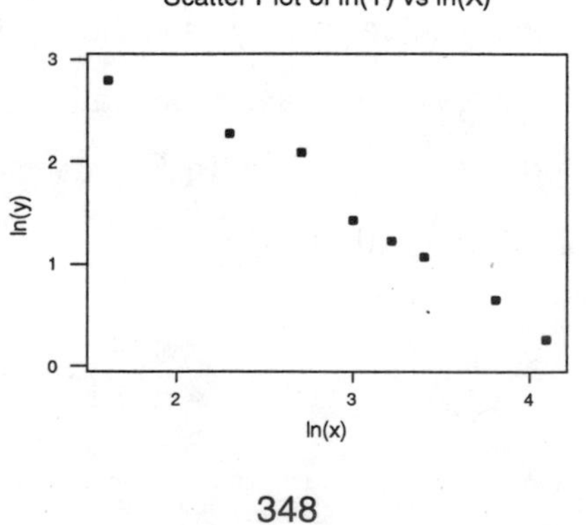

c The linear pattern in **b** above would indicate that a transformed regression using the natural log of both x and y would be appropriate. The probabilistic model is then $y = \alpha x^{\beta} \cdot \varepsilon$. (The power function with an error term!)

d A regression of ln(y) on ln(x) yields the equation $\ln(y) = 4.6384 - 1.04920\ln(x)$. Using Minitab we can get a P.I. for y when x = 20 by first transforming the x value: ln(20) = 2.996. The computer generated 95% P.I. for ln(y) when ln(x) = 2.996 is (1.1188,1.8712). We must now take the antilog to return to the original units of Y: $\left(e^{1.1188}, e^{1.8712}\right) = (3.06, 6.50)$.

e A computer generated residual analysis:

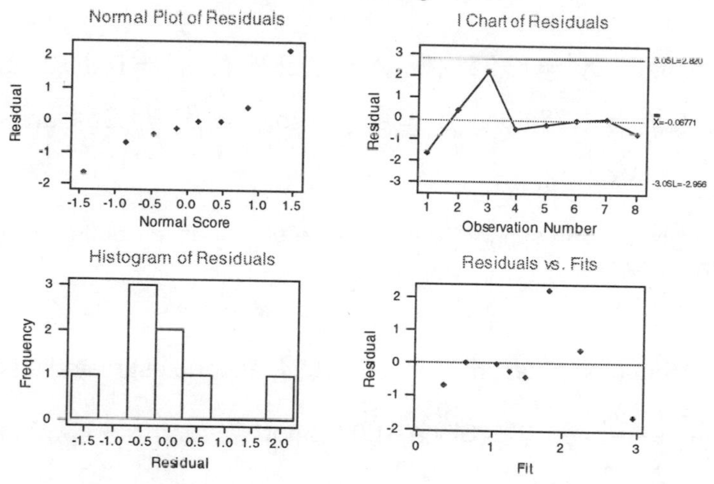

Looking at the residual vs. fits (bottom right), one standardized residual, corresponding to the third observation, is a bit large. There are only two positive standardized residuals, but two others are essentially 0. The patterns in the residual plot and the normal probability plot (upper left) are marginally acceptable.

16.

a $\Sigma x_i = 9.72$, $\Sigma y_i' = 313.10$, $\Sigma x_i^2 = 8.0976$, $\Sigma y_i'^2 = 288{,}013$, $\Sigma x_i\, y_i' = 255.11$, (all from computer printout, where $y_i' = \ln(L_{178})$), from which $\hat{\beta}_1 = 6.6667$ and $\hat{\beta}_0 = 20.6917$ (again from computer output). Thus $\hat{\beta} = \hat{\beta}_1 = 6.6667$ and $\hat{\alpha} = e^{\hat{\beta}_0} = 968{,}927{,}163$.

b We first predict y' using the linear model and then exponentiate:

$y' = 20.6917 + 6.6667(.75) = 25.6917$, so

$\hat{y} = \hat{L}_{178} = e^{25.6917} = 1.438051363 \times 10^{11}$.

c We first compute a prediction interval for the transformed data and then exponentiate. With $t_{.025,10} = 2.228$, s = .5946, and

$\sqrt{1 + \frac{1}{12} + \frac{(.95 - \bar{x})^2}{\Sigma x^2 - (\Sigma x)^2 / 12}} = 1.082$, the prediction interval for y' is

$27.0251 \pm (2.228)(.5496)(1.082) = 27.0251 \pm 1.4334 = (25.5917, 28.4585)$. The P.I. for y is then $\left(e^{25.5917}, e^{28.4585}\right)$.

17.

a $\Sigma x_i' = 15.501$, $\Sigma y_i' = 13.352$, $\Sigma x_i'^2 = 20.228$, $\Sigma y_i'^2 = 16.572$, $\Sigma x_i' \, y_i' = 18.109$, from which $\hat{\beta}_1 = 1.254$ and $\hat{\beta}_0 = -.468$ so $\hat{\beta} = \hat{\beta}_1 = 1.254$ and $\hat{\alpha} = e^{-.468} = .626$.

b The plots give strong support to this choice of model; in addition, $r^2 = .960$ for the transformed data.

c SSE = .11536 (computer printout), s = .1024, and the estimated sd of $\hat{\beta}_1$ is .0775, so $t = \frac{1.25 - 1.33}{.0775} = -1.07$. Since –1.07 is not $\le -t_{.05,11} = -1.796$, H_o cannot be rejected in favor of H_a.

d The claim that $\mu_{Y \cdot 5} = 2\mu_{Y \cdot 2.5}$ is equivalent to $\alpha \cdot 5^{\beta} = 2\alpha(2.5)^{\beta}$, or that $\beta = 1$. Thus we wish test $H_o : \beta_1 = 1$ vs. $H_a : \beta_1 \ne 1$. With $t = \frac{1 - 1.33}{.0775} = -4.30$ and RR $-t_{.005,11} \le -3.106$, H_o is rejected at level .01 since $-4.30 \le -3.106$.

18.

a $Y = \beta_0 + \beta_1 \ln(x) + \varepsilon$ where x = concentration and y = critical minimum pH. With $x' = \ln(x)$, $\Sigma x_i' = -10.648$, $\Sigma y_i = 24.0$, $\Sigma x_i'^2 = 40.120$, $\Sigma y_i^2 = 146.76$, $\Sigma x_i' \, y_i = -58.195$, the estimated model parameters are $\hat{\beta}_1 = .483$ and $\hat{\beta}_0 = 7.29$.

b For x = 1, $x' = \ln(1) = 0$, so $\hat{y} = \hat{\beta}_0 + \hat{\beta}_1(0) = 7.29$. With SSE = .008440, s = .06496, $t_{.025,2} = 4.303$ and $\sqrt{1+\frac{1}{4}+\frac{(0-(-2.66))^2}{40.120-(-10.648)^2/4}} = 1.360$, the P.I. is $7.29 \pm (4.303)(.06496)(1.360) = 7.29 \pm .38 = (6.91, 7.67)$.

19.

a No, there is definite curvature in the plot.

b $Y' = \beta_0 + \beta_1(x') + \varepsilon$ where $x' = \frac{1}{temp}$ and $y' = \ln(lifetime)$. Plotting y' vs. x' gives a plot which has a pronounced linear appearance (and in fact r^2 = .954 for the straight line fit).

c $\Sigma x_i' = .082273$, $\Sigma y_i' = 123.64$, $\Sigma x_i'^2 = .00037813$, $\Sigma y_i'^2 = 879.88$, $\Sigma x_i' y_i' = .57295$, from which $\hat{\beta}_1 = 3735.4485$ and $\hat{\beta}_0 = -10.2045$ (values read from computer output). With x = 220, $x' = .00445$ so $\hat{y}' = -10.2045 + 3735.4485(.00445) = 6.7748$ and thus $\hat{y} = e^{\hat{y}'} = 875.50$.

d For the transformed data, SSE = 1.39857, and $n_1 = n_2 = n_3 = 6$, $\bar{y}_{1.}' = 8.44695$, $\bar{y}_{2.}' = 6.83157$, $\bar{y}_{3.}' = 5.32891$, from which SSPE = 1.36594, SSLF = .02993, $f = \frac{.02993/1}{1.36594/15} = .33$. Comparing this to $F_{.01,1,15} = 8.68$, it is clear that H_o cannot be rejected.

20. After examining a scatter plot and a residual plot for each of the five suggested models as well as for y vs. x, I felt that the power model $Y = \alpha x^{\beta} \cdot \varepsilon$ ($y' = \ln(y)$ vs. $x' = \ln(x)$) provided the bet fit. The transformation seemed to remove most of the curvature from the scatter plot, the residual plot appeared quite random, $|e_i'^*| < 1.65$ for every i, there was no indication of any influential observations, and r^2 = .785 for the transformed data.

21.

a The suggested model is $Y = \beta_0 + \beta_1(x') + \varepsilon$ where $x' = \frac{10^4}{x}$. The summary quantities are $\Sigma x_i' = 159.01$, $\Sigma y_i = 121.50$, $\Sigma x_i'^2 = 4058.8$, $\Sigma y_i^2 = 1865.2$, $\Sigma x_i' y_i = 2281.6$, from which $\hat{\beta}_1 = -.1485$ and $\hat{\beta}_0 = 18.1391$, and the estimated regression function is $y = 18.1391 - \frac{1485}{x}$.

b $x = 500 \Rightarrow \hat{y} = 18.1391 - \frac{1485}{500} = 15.17$.

22.

a $\frac{1}{y} = \alpha + \beta x$, so with $y' = \frac{1}{y}$, $y' = \alpha + \beta x$. The corresponding probabilistic model is $\frac{1}{y} = \alpha + \beta x + \varepsilon$.

b $\frac{1}{y} - 1 = e^{\alpha + \beta x}$, so $\ln\left(\frac{1}{y} - 1\right) = \alpha + \beta x$. Thus with $y' = \ln\left(\frac{1}{y} - 1\right)$, $y' = \alpha + \beta x$. The corresponding probabilistic model is $Y' = \alpha + \beta x + \varepsilon'$, or equivalently $Y = \frac{1}{1 + e^{\alpha + \beta x} \cdot \varepsilon}$ where $\varepsilon = e^{\varepsilon'}$.

c $\ln(y) = e^{\alpha + \beta x} = \ln(\ln(y)) = \alpha + \beta x$. Thus with $y' = \ln(\ln(y))$, $y' = \alpha + \beta x$. The probabilistic model is $Y' = \alpha + \beta x + \varepsilon'$, or equivalently, $Y = e^{e^{\alpha + \beta x}} \cdot \varepsilon$ where $\varepsilon = e^{\varepsilon'}$.

d This function cannot be linearized.

23. $Var(Y) = Var(\alpha e^{\beta x} \cdot \varepsilon) = \left[\alpha e^{\beta x}\right]^2 \cdot Var(\varepsilon) = \alpha^2 e^{2\beta x} \cdot \tau^2$ where we have set $Var(\varepsilon) = \tau^2$. If $\beta > 0$, this is an increasing function of x so we expect more spread in y for large x than for small x, while the situation is reversed if $\beta < 0$. It is important to realize that a scatter plot of data generated from this model will not spread out uniformly about the exponential regression function throughout the range of x values; the spread will only be uniform on the transformed scale. Similar results hold for the multiplicative power model.

24. $H_0 : \beta_1 = 0$ vs $H_a : \beta_1 \neq 0$. The value of the test statistic is z = .73, with a corresponding p-value of .463. Since the p-value is greater than any sensible choice of alpha we do not reject H_o. There is insufficient evidence to claim that age has a significant impact on the presence of kyphosis.

25. The point estimate of β_1 is $\hat{\beta}_1 = .17772$, so the estimate of the odds ratio is $e^{\hat{\beta}_1} = e^{.17772} \approx 1.194$. That is , when the amount of experience increases by one year (i.e. a one unit increase in x), we estimate that the odds ratio increase by about 1.194. The z value of 2.70 and its corresponding p-value of .007 imply that the null hypothesis $H_0 : \beta_1 = 0$ can be rejected at any of the usual significance levels (e.g., .10, .05, .025, .01). Therefore, there is clear evidence that β_1 is not zero, which means that experience does appear to affect the likelihood of successfully performing the task. This is consistent with the confidence interval (1.05, 1.36) for the odds ratio given in the printout, since this interval does not contain the value 1. A graph of $\hat{\pi}$ appears below.

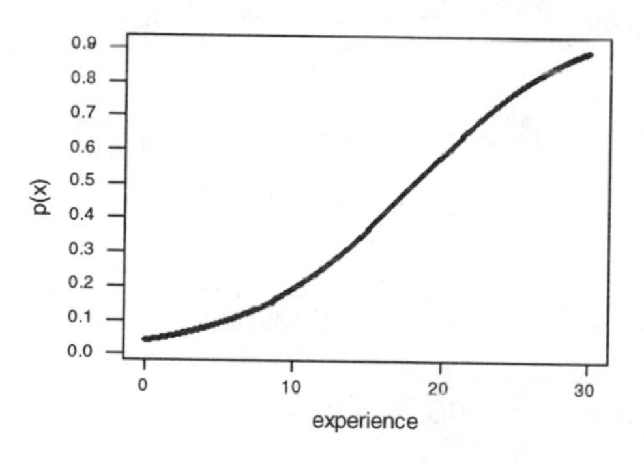

26.

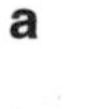
a

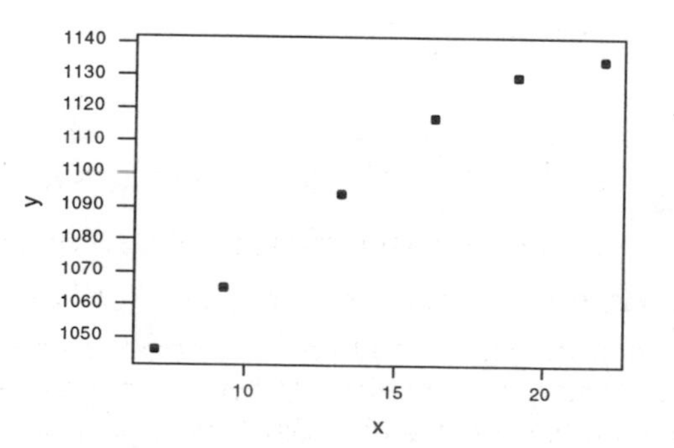

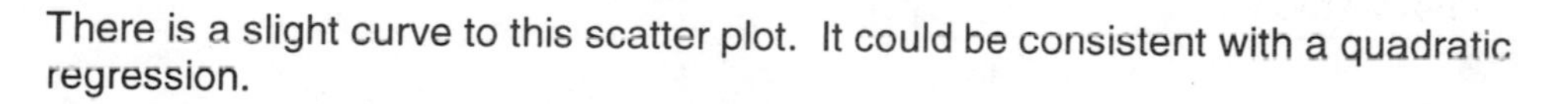
There is a slight curve to this scatter plot. It could be consistent with a quadratic regression.

b We desire R^2, which we find in the output: $R^2 = 93.8\%$

c $H_0 : \beta_1 = \beta_2 = 0$ vs H_a : at least one $\beta_i \neq 0$. The test statistic is

$f = \dfrac{MSR}{MSE} = 22.51$, and the corresponding p-value is .016. Since the p-value < .05,

we reject H_o and conclude that the model is useful.

d We want a 99% confidence interval, but the output gives us a 95% confidence

interval of (452.71, 529.48), which can be rewritten as 491.10 ± 38.38;

$t_{.025,3} = 3.182$, so $s_{\hat{y}\cdot 14} = \dfrac{38.38}{3.182} = 12.06$; Now, $t_{.005,3} = 5.841$, so the 99% C.I. is

$491.10 \pm 5.841(12.06) = 491.10 \pm 70.45 = (420.65, 561.55)$.

e $H_0 : \beta_2 = 0$ vs $H_a : \beta_2 \neq 0$. The test statistic is t = -3.81, with a corresponding p-

value of .032, which is < .05, so we reject H_o. the quadratic term appears to be

useful in this model.

27.

a A scatter plot of the data indicated a quadratic regression model might be appropriate.

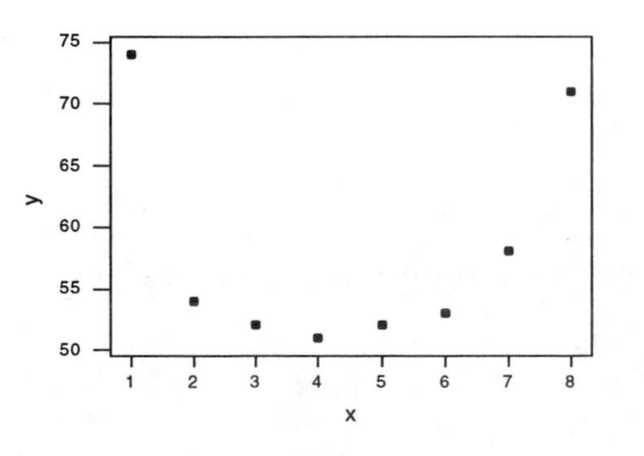

b $\hat{y} = 84.482 - 15.875(6) + 1.7679(6)^2 = 52.88$; residual = $y_6 - \hat{y}_6 = 53 - 52.88 = .12$;

c $SST = \Sigma y_i^2 - \frac{(\Sigma y_i)^2}{n} = 586.88$, so $R^2 = 1 - \frac{61.77}{586.88} = .895$.

d The first two residuals are the largest, but they are both within the interval (-2, 2). Otherwise, the standardized residual plot does not exhibit any troublesome features. For the Normal Probability Plot:

Residual	Zth percentile
-1.95	-1.53
-.66	-.89
-.25	-.49
.04	-.16
.20	.16
.58	.49
.90	.89
1.91	1.53

The normal probability plot does not exhibit any troublesome features.

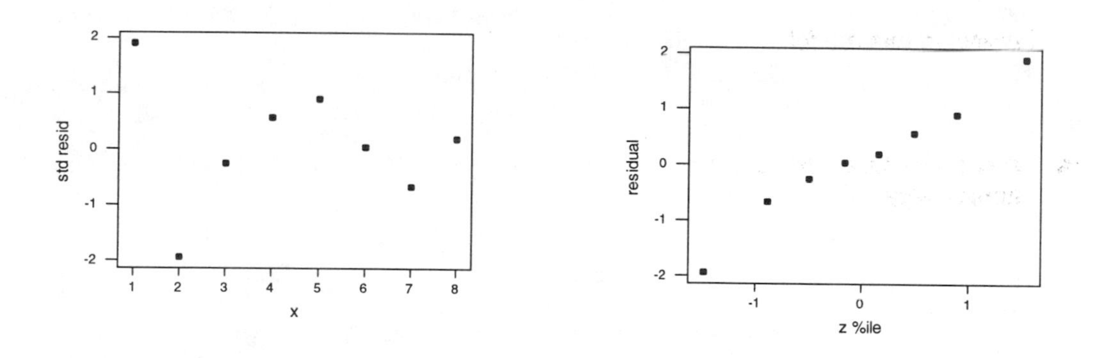

e $\hat{\mu}_{Y \cdot 6} = 52.88$ (from **b**) and $t_{.025,n-3} = t_{.025,5} = 2.571$, so the C.I. is $52.88 \pm (2.571)(1.69) = 52.88 \pm 4.34 = (48.54, 57.22)$.

f SSE = 61.77 so $s^2 = \frac{61.77}{5} = 12.35$ and $\sqrt{12.35 + (1.69)^2} = 3.90$. From section 13.2, the P.I. is $52.88 \pm (2.571)(3.90) = 52.88 \pm 10.03 = (42.85, 62.91)$.

Section 13.3

28.

a $\hat{\mu}_{Y\cdot 75} = \hat{\beta}_0 + \hat{\beta}_1(75) + \hat{\beta}_2(75)^2 = -113.0937 + 3.36684(75) - .01780(75)^2 = 39.41$

b $\hat{y} = \hat{\beta}_0 + \hat{\beta}_1(60) + \hat{\beta}_2(60)^2 = 24.93$.

c $SSE = \Sigma y_i^2 - \hat{\beta}_0 \Sigma y_i - \hat{\beta}_1 \Sigma x_i y_i - \hat{\beta}_2 \Sigma x_i^2 y_i = 8386.43 - (-113.0937)(210.70)$ $-(3.3684)(17{,}002) - (-.0178)(1{,}419{,}780) = 217.82$, $s^2 = \frac{SSE}{n-3} = \frac{217.82}{3} = 72.61$, s = 8.52

d $R^2 = 1 - \frac{217.82}{987.35} = .779$

e H_o will be rejected in favor of H_a if either $t \ge t_{.005,3} = 5.841$ or if $t \le -5.841$. The computed value of t is $t = \frac{-.01780}{.00226} = -7.88$, and since $-7.88 \le -5.841$, we reject H_o.

29.

a From computer output:

$\hat{y}$:	111.89	120.66	114.71	94.06	58.69
$y - \hat{y}$:	-1.89	2.34	4.29	-8.06	3.31

Thus $SSE = (-1.89)^2 + \ldots + (3.31)^2 = 103.37$, $s^2 = \frac{103.37}{2} = 51.69$, $s = 7.19$.

b $SST = \Sigma y_i^2 - \frac{(\Sigma y_i)^2}{n} = 2630$, so $R^2 = 1 - \frac{103.37}{2630} = .961$.

c $H_0 : \beta_2 = 0$ will be rejected in favor of $H_a : \beta_2 \neq 0$ if either $t \geq t_{.025,2} = 4.303$ or if $t \leq -4.303$. With $t = \frac{-1.84}{.480} = -3.83$, H_o cannot be rejected; the data does not argue strongly for the inclusion of the quadratic term.

d To obtain joint confidence of at least 95%, we compute a 98% C.I. for each coefficient using $t_{.01,2} = 6.965$. For β_1 the C.I. is $8.06 \pm (6.965)(4.01)$ $= (-19.87, 35.99)$ (an extremely wide interval), and for β_2 the C.I. is $-1.84 \pm (6.965)(.480) = (-5.18, 1.50)$.

e $t_{.05,2} = 2.920$ and $\hat{\beta}_0 + 4\hat{\beta}_1 + 16\hat{\beta}_2 = 114.71$, so the C.I. is $114.71 \pm (2.920)(5.01)$ $= 114.71 \pm 14.63 = (100.08, 129.34)$.

f If we knew $\hat{\beta}_0, \hat{\beta}_1, \hat{\beta}_2$, the value of x which maximizes $\hat{\beta}_0 + \hat{\beta}_1 x + \hat{\beta}_2 x^2$ would be obtained by setting the derivative of this to 0 and solving:

$\beta_1 + 2\beta_2 x = 0 \Rightarrow x = -\frac{\beta_1}{2\beta_2}$. The estimate of this is $x = -\frac{\hat{\beta}_1}{2\hat{\beta}_2} = 2.19$.

30.

a $R^2 = 0.853$. This means 85.3% of the variation in wheat yield is accounted for by the model.

b $-135.44 \pm (2.201)(41.97) = (-227.82, -43.06)$

c $H_0 : \mu_{y \cdot 2.5} = 1500$

$H_a : \mu_{y \cdot 2.5} < 1500$

$RR: t \leq -t_{.01,11} = -2.718$

When x = 2.5, $\hat{y} = 1402.15$

$$t = \frac{1{,}402.15 - 1500}{53.5} = -1.83$$

Fail to reject H_o. The data does not indicate $\mu_{y \cdot 2.5}$ is less than 1500.

d $1402.15 \pm (2.201)\sqrt{(136.5)^2 + (53.5)^2} = (1081.3, 1725.0)$

31.

a Using Minitab, the regression equation is $y = 13.6 + 11.4x - 1.72x^2$.

b Again, using Minitab, the predicted and residual values are:

$\hat{y}$:	23.327	23.327	29.587	31.814	31.814	31.814	20.317
$y-\hat{y}$:	-.327	1.173	1.587	.914	.186	1.786	-.317

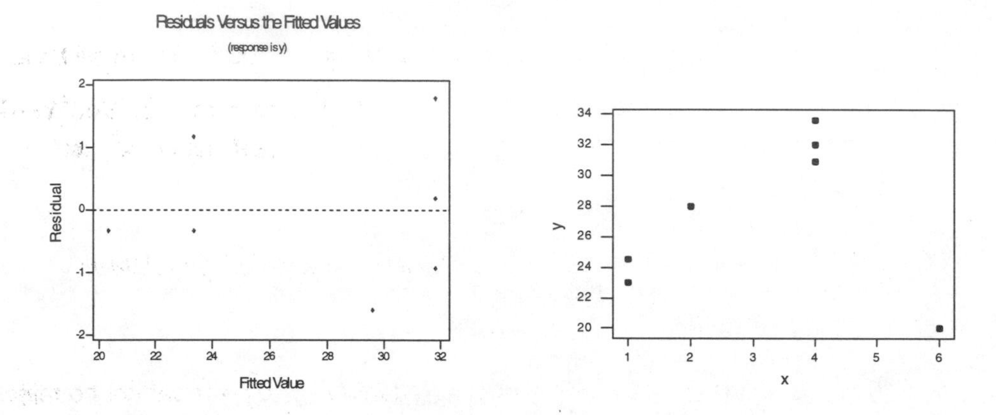

The residual plot is consistent with a quadratic model (no pattern which would suggest modification), but it is clear from the scatter plot that the point (6, 20) has had a great influence on the fit – it is the point which forced the fitted quadratic to have a maximum between 3 and 4 rather than, for example, continuing to curve slowly upward to a maximum someplace to the right of x = 6.

c From Minitab output, s^2 = MSE = 2.040, and R^2 = 94.7%. The quadratic model thus explains 94.7% of the variation in the observed y's , which suggests that the model fits the data quite well.

d $\sigma^2 = Var(\hat{Y}_i) + Var\left(Y_i - \hat{Y}_i\right)$ suggests that we can estimate $Var\left(Y_i - \hat{Y}_i\right)$ by $s^2 - s_{\hat{y}}^2$ and then take the square root to obtain the estimated standard deviation of each residual. This gives $\sqrt{2.040 - (.955)^2}$ = 1.059, (and similarly for all points) 10.59, 1.236, 1.196, 1.196, 1.196, and .233 as the estimated std dev's of the residuals. The standardized residuals are then computed as $\dfrac{-.327}{1.059} = -.31$, (and similarly) 1.10, -1.28, -.76, .16, 1.49, and –1.28, none of which are unusually large. (Note: Minitab regression output can produce these values.) The resulting residual plot is virtually identical to the plot of **b**. $\dfrac{y-\hat{y}}{s} = \dfrac{-.327}{1.426} = -.229 \neq -.31$, so standardizing using just s would not yield the correct standardized residuals.

e $Var(Y_f)+Var(\hat{Y}_f)$ is estimated by $2.040+(.777)^2=2.638$, so $s_{y_f+\hat{y}_f}=\sqrt{2.638}=1.624$. With $\hat{y}=31.81$ and $t_{.05,4}=2.132$, the desired P.I. is $31.81\pm(2.132)(1.624)=(28.35,35.27)$.

32.

a $.3463-1.2933(x-\bar{x})+2.3964(x-\bar{x})^2-2.3968(x-\bar{x})^3$.

b From **a**, the coefficient of x^3 is -2.3968, so $\hat{\beta}_3=-2.3968$. There sill be a contribution to x^2 both from $2.3964(x-4.3456)^2$ and from $-2.3968(x-4.3456)^3$. Expanding these and adding yields 33.6430 as the coefficient of x^2, so $\hat{\beta}_2=33.6430$.

c $x=4.5\Rightarrow x'=x-\bar{x}=.1544$; substituting into **a** yields $\hat{y}=.1949$.

d $t=\dfrac{-2.3968}{2.4590}=-.97$, which is not significant ($H_0:\beta_3=0$ cannot be rejected), so the inclusion of the cubic term is not justified.

33.

a $\bar{x}=20$ and s_x = 10.8012 so $x'=\dfrac{x-20}{10.8012}$. For x = 20, $x'=0$, and $\hat{y}=\hat{\beta}_0^*=.9671$. For x = 25, $x'=.4629$, so $\hat{y}=.9671-.0502(.4629)-.0176(.4629)^2+.0062(.4629)^3=.9407$.

b $\hat{y}=.9671-.0502\left(\dfrac{x-20}{10.8012}\right)-.0176\left(\dfrac{x-20}{10.8012}\right)^2+.0062\left(\dfrac{x-20}{10.8012}\right)^3$
$.00000492x^3-.000446058x^2+.007290688x+.96034944$.

c $t=\dfrac{.0062}{.0031}=2.00$. We reject H_o if either $t\ge t_{.025,n-4}=t_{.025,3}=3.182$ or if $t\le-3.182$. Since 2.00 is neither ≥ 3.182 nor ≤ -3.182, we cannot reject H_o; the cubic term should be deleted.

d $SSE=\Sigma(y_i-\hat{y}_i)$ and the $\hat{y}_i$'s are the same from the standardized as from the unstandardized model, so SSE, SST, and R^2 will be identical for the two models.

e $\Sigma y_i^2 = 6.355538$, $\Sigma y_i = 6.664$, so SST = .011410. For the quadratic model R^2 = .987 and for the cubic model, R^2 = .994; The two R^2 values are very close, suggesting intuitively that the cubic term is relatively unimportant.

34.

a $\bar{x} = 49.9231$ and s_x = 41.3652 so for x = 50, $x' = \dfrac{x - 49.9231}{41.3652} = .001859$ and

$\hat{\mu}_{Y \cdot 50} = .8733 - .3255(.001859) + .0448(.001859)^2 = .873$.

b SST = 1.456923 and SSE = .117521, so R^2 = .919.

c $.8733 - .3255\left(\dfrac{x - 49.9231}{41.3652}\right) + .0448\left(\dfrac{x - 49.9231}{41.3652}\right)^2$

$1.200887 - .01048314x + .00002618x^2$.

d $\hat{\beta}_2 = \dfrac{\hat{\beta}_2^*}{s_x^2}$ so the estimated sd of $\hat{\beta}_2$ is the estimated sd of $\hat{\beta}_2^*$ multiplied by $\dfrac{1}{s_x}$:

$s_{\hat{\beta}_2} = (.0319)\left(\dfrac{1}{41.3652}\right) = .00077118$.

e $t = \dfrac{.0448}{.0319} = 1.40$ which is not significant (compared to $\pm t_{.025,9}$ at level .05), so the quadratic term should not be retained.

35. $Y' = \ln(Y) = \ln\alpha + \beta x + \gamma x^2 + \ln(\varepsilon) = \beta_0 + \beta_1 x + \beta_2 x^2 + \varepsilon'$ where $\varepsilon' = \ln(\varepsilon)$, $\beta_0 = \ln(\alpha)$, $\beta_1 = \beta$, and $\beta_2 = \gamma$. That is, we should fit a quadratic to $(x, \ln(y))$. The resulting estimated quadratic (from computer output) is $2.00397 + .1799x - .0022x^2$, so $\hat{\beta} = .1799$, $\hat{\gamma} = -.0022$, and $\hat{\alpha} = e^{2.0397} = 7.6883$. (The ln(y)'s are 3.6136, 4.2499, 4.6977, 5.1773, and 5.4189, and the summary quantities can then be computed as before.)

Section 13.4

36.

a Holding age, time, and heart rate constant, maximum oxygen uptake will increase by .01 L/min for each 1 kg increase in weight. Similarly, holding weight, age, and heart rate constant, the maximum oxygen uptake decreases by .13 L/min with every 1 minute increase in the time necessary to walk 1 mile.

b $\hat{\mu}_{y\cdot 76,20,12,140} = 5.0 + .01(76) - .05(20) - .13(12) - .01(140) = 1.8$ L/min.

c $\hat{\mu} = 1.8$ from **b**, and $\sigma = .4$, so, assuming y follows a normal distribution,

$$P(1.00 < Y < 2.60) = P\left(\frac{1.00 - 1.8}{.4} < Z < \frac{2.6 - 1.8}{.4}\right) = P(-2.0 < Z < 2.0) = .9544$$

37.

a The mean value of y when $x_1 = 50$ and $x_2 = 3$ is
$\mu_{y\cdot 50,3} = -.800 + .060(50) + .900(3) = 4.9$ hours.

b When the number of deliveries (x_2) is held fixed, then average change in travel time associated with a one-mile (i.e. one unit) increase in distance traveled (x_1) is .060 hours. Similarly, when distance traveled (x_1) is held fixed, then the average change in travel time associated with on extra delivery (i.e., a one unit increase in x_2) is .900 hours.

c Under the assumption that y follows a normal distribution, the mean and standard deviation of this distribution are 4.9 (because $x_1 = 50$ and $x_2 = 3$) and $\sigma = .5$ (since the standard deviation is assumed to be constant regardless of the values of x_1 and x_2). Therefore $P(y \le 6) = P\left(z \le \frac{6 - 4.9}{.5}\right) = P(z \le 2.20) = .9861$. That is, in the long run, about 98.6% of all days will result in a travel time of at most 6 hours.

38.

a mean life $= 125 + 7.75(40) + .0950(1100) - .009(40)(1100) = 143.50$

b First, the mean life when $x_1 = 30$ is equal to
$125 + 7.75(30) + .0950x_2 - .009(30)x_2 = 357.50 - .175x_2$. So when the load increases by 1, the mean life decreases by .175. Second, the mean life when $x_1 = 40$ is equal to $125 + 7.75(40) + .0950x_2 - .009(40)x_2 = 435 - .265x_2$. So when the load increases by 1, the mean life decreases by .265.

39.

a For $x_1 = 2$, $x_2 = 8$ (remember the units of x_2 are in 1000,s) and $x_3 = 1$ (since the outlet has a drive-up window) the average sales are
$\hat{y} = 10.00 - 1.2(2) + 6.8(8) + 15.3(1) = 77.3$ (i.e., $77,300).

b For $x_1 = 3$, $x_2 = 5$, and $x_3 = 0$ the average sales are
$\hat{y} = 10.00 - 1.2(3) + 6.8(5) + 15.3(0) = 40.4$ (i.e., $40,400).

c When the number of competing outlets (x_1) and the number of people within a 1-mile radius (x_2) remain fixed, the sales will increase by $15,300 when an outlet has a drive-up window.

40.

a $\hat{\mu}_{Y\cdot 10,.5,50,100} = 1.52 + .02(10) - 1.40(.5) + .02(50) - .0006(100) = 1.96$

b $\hat{\mu}_{Y\cdot 20,.5,50,30} = 1.52 + .02(20) - 1.40(.5) + .02(50) - .0006(30) = 1.40$

c $\hat{\beta}_4 = -.0006;\ 100\hat{\beta}_4 = -.06$.

d There are no interaction predictors – e.g., $x_5 = x_1 x_4$ -- in the model. There would be dependence if interaction predictors involving x_4 had been included.

e $R^2 = 1 - \dfrac{20.0}{39.2} = .490$. For testing $H_0 : \beta_1 = \beta_2 = \beta_3 = \beta_4 = 0$ vs. H_a: at least one among $\beta_1, \ldots, \beta_4$ is not zero, the test statistic is $F = \dfrac{R^2/k}{(1-R^2)/(n-k-1)}$. H_o will be rejected if $f \ge F_{.05,4,25} = 2.76$. $f = \dfrac{.490/4}{.510/25} = 6.0$. Because $6.0 \ge 2.76$, H_o is rejected and the model is judged useful (this even though the value of R^2 is not all that impressive).

41. $H_0 : \beta_1 = \beta_2 = \ldots = \beta_6 = 0$ vs. H_a: at least one among $\beta_1, \ldots, \beta_6$ is not zero. The test statistic is $F = \dfrac{R^2/k}{(1-R^2)/(n-k-1)}$. H_o will be rejected if $f \ge F_{.05,6,30} = 2.42$.

$f = \dfrac{.83/6}{(1-.83)/30} = 24.41$. Because $24.41 \ge 2.42$, H_o is rejected and the model is judged useful.

42.

a To test $H_0 : \beta_1 = \beta_2 = 0$ vs. H_a : at least one $\beta_i \neq 0$, the test statistic is $f = \frac{MSR}{MSE} = 319.31$ (from output). The associated p-value is 0, so at any reasonable level of significance, H_o should be rejected. There does appear to be a useful linear relationship between temperature difference and at leas one of the two predictors.

b The degrees of freedom for SSE = n – (k + 1) = 9 – (2 – 1) = 6 (which you could simply read in the DF column of the printout), and $t_{.025,6} = 2.447$, so the desired confidence interval is $3.000 \pm (2.447)(.4321) = 3.000 \pm 1.0573$, or about $(1.943, 4.057)$. Holding furnace temperature fixed, we estimate that the average change in temperature difference on the die surface will be somewhere between 1.943 and 4.057.

c When x_1 = 1300 and x_2 = 7, the estimated average temperature difference is $\hat{y} = -199.56 + .2100x_1 + 3.000x_2 = -199.56 + .2100(1300) + 3.000(7) = 94.44$. The desired confidence interval is then $94.44 \pm (2.447)(.353) = 94.44 \pm .864$, or $(93.58, 95.30)$.

d From the printout, s = 1.058, so the prediction interval is

$$94.44 \pm (2.447)\sqrt{(1.058)^2 + (.353)^2} = 94.44 \pm 2.729 = (91.71, 97.17).$$

43.

a The coefficient o multiple determination is $R^2 = 78\%$. So 78% of the observed variation in surface area can be attributed to the stated approximate relationship between surface area and the predictors.

b x_1 = 2.6, x_2 = 250, and x_1x_2 = (2.6)(250) = 650, so

$$\hat{y} = 185.49 - 45.97(2.6) - 0.3015(250) + 0.0888(650) = 48.313$$

c No, it is not legitimate to interpret β_1 in this way. It is not possible to increase by 1 unit the cobalt content, x_1, while keeping the interaction predictor, x_3, fixed. When x_1 changes, so does x_3, since $x_3 = x_1x_2$.

d Yes, there appears to be a useful linear relationship between y and the predictors. We determine this by observing that the p-value corresponding to the model utility test is < .0001 (F test statistic = 18.924).

e We wish to test $H_0 : \beta_3 = 0$ vs. $H_a : \beta_3 \neq 0$. The test statistic is t=3.496, with a corresponding p-value of .0030. Since the p-value is < alpha = .01, we reject H_o and conclude that the interaction predictor does provide useful information about y.

f A 95% C.I. for the mean value of surface area under the stated circumstances requires the following quantities:
$\hat{y} = 185.49 - 45.97(2) - 0.3015(500) + 0.0888(2)(500) = 31.598$. Next,
$t_{.025,16} = 2.120$, so the 95% confidence interval is
$31.598 \pm (2.120)(4.69) = 31.598 \pm 9.9428 = (21.6552, 41.5408)$

44.

a Holding starch damage constant, for every 1% increase in flour protein, the absorption rate will increase by 1.44%. Similarly, holding flour protein percentage constant, the absorption rate will increase by .336% for every 1-unit increase in starch damage.

b R^2 = .96447, so 96.447% of the observed variation in absorption can be explained by the model relationship.

c To answer the question, we test $H_0 : \beta_1 = \beta_2 = 0$ vs H_a : at least one $\beta_i \neq 0$. The test statistic is $f = 339.31092$, and has a corresponding p-value of zero, so at any significance level we will reject H_o. There is a useful relationship between absorption and at least one of the two predictor variables.

45.

a The appropriate hypotheses are $H_0 : \beta_1 = \beta_2 = \beta_3 = \beta_4 = 0$ vs. H_a : at least one $\beta_i \neq 0$. The test statistic is $f = \dfrac{R^2/k}{(1-R^2)/(n-k-1)} = \dfrac{.946/4}{(1-.946)/20} = 87.6 \geq 7.10 = F_{.001,4,20}$ (the smallest available significance level from Table A.9), so we can reject H_o at any significance level. We conclude that at least one of the four predictor variables appears to provide useful information about tenacity.

b The adjusted R^2 value is $1 - \dfrac{n-1}{n-(k+1)}\left(\dfrac{SSE}{SST}\right) = 1 - \dfrac{n-1}{n-(k+1)}(1-R^2)$
$= 1 - \dfrac{24}{20}(1 - .946) = .935$, which does not differ much from R^2 = .946.

c The estimated average tenacity when $x_1 = 16.5$, $x_2 = 50$, $x_3 = 3$, and $x_4 = 5$ is
$\hat{y} = 6.121 - .082x + .113x + .256x - .219x$
$\hat{y} = 6.121 - .082(16.5) + .113(50) + .256(3) - .219(5) = 10.091$. For a 99% C.I., $t_{.005,20} = 2.845$, so the interval is $10.091 \pm 2.845(.350) = (9.095, 11.087)$.
Therefore, when the four predictors are as specified in this problem, the true average tenacity is estimated to be between 9.095 and 11.087.

46.

a Yes, there does appear to be a useful linear relationship between repair time and the two model predictors. We determine this by conducting a model utility test: $H_0 : \beta_1 = \beta_2 = 0$ vs. H_a : at least one $\beta_i \neq 0$. We reject H$_o$ if $f \geq F_{.05,2,9} = 4.26$. The calculated statistic is

$$f = \frac{SSR/k}{SSE/(n-k-1)} = \frac{MSR}{MSE} = \frac{10.63/2}{(20.9)/9} = \frac{5.315}{.232} = 22.91$$

. Since $22.91 \geq 4.26$, we reject H$_o$ and conclude that at least one of the two predictor variables is useful.

b We will reject $H_0 : \beta_2 = 0$ in favor of $H_a : \beta_2 \neq 0$ if $|t| \geq t_{.005,9} = 3.25$. The test statistic is $t = \frac{1.250}{.312} = 4.01$ which is ≥ 3.25, so we reject H$_o$ and conclude that the "type of repair" variable does provide useful information about repair time, given that the "elapsed time since the last service" variable remains in the model.

c A 95% confidence interval for β_3 is: $1.250 \pm (2.262)(.312) = (.5443, 1.9557)$. We estimate, with a high degree of confidence, that when an electrical repair is required the repair time will be between .54 and 1.96 hours longer than when a mechanical repair is required, while the "elapsed time" predictor remains fixed.

d $\hat{y} = .950 + .400(6) + 1.250(1) = 4.6$, $s^2 = MSE = .23222$, and $t_{.005,9} = 3.25$, so the 99% P.I. is $4.6 \pm (3.25)\sqrt{(.23222) + (.192)^2} = 4.6 \pm 1.69 = (2.91, 6.29)$. The prediction interval is quite wide, suggesting a variable estimate for repair time under these conditions.

47.

a For a 1% increase in the percentage plastics, we would expect a 28.9 kcal/kg increase in energy content. Also, for a 1% increase in the moisture percentage, we would expect a 37.4 kcal/kg decrease in energy content.

b The appropriate hypotheses are $H_0 : \beta_1 = \beta_2 = \beta_3 = \beta_4 = 0$ vs. H_a : at least one $\beta_i \neq 0$. The value of the F test statistic is 167.71, with a corresponding p-value that is extremely small. So, we reject H$_o$ and conclude that at least one of the four predictors is useful in predicting energy content, using a linear model.

c $H_0 : \beta_3 = 0$ vs. $H_a : \beta_3 \neq 0$. The value of the t test statistic is t = 2.24, with a corresponding p-value of .034, which is less than the significance level of .05. So we can reject H$_o$ and conclude that percentage garbage provides useful information about energy consumption, given that the other three predictors remain in the model.

d $\hat{y} = 2244.9 + 28.925(20) + 7.644(25) + 4.297(40) - 37.354(45) = 1505.5$, and $t_{.025,25} = 2.060$. (Note an error in the text: $s_{\hat{y}} = 12.47$, not 7.46). So a 95% C.I for the true average energy content under these circumstances is $1505.5 \pm (2.060)(12.47) = 1505.5 \pm 25.69 = (1479.8, 1531.1)$. Because the interval is reasonably narrow, we would conclude that the mean energy content has been precisely estimated.

e A 95% prediction interval for the energy content of a waste sample having the specified characteristics is $1505.5 \pm (2.060)\sqrt{(31.48)^2 + (12.47)^2}$ $= 1505.5 \pm 69.75 = (1435.7, 1575.2)$.

48.

a $H_0 : \beta_1 = \beta_2 = ... = \beta_9 = 0$

H_a : at least one $\beta_i \neq 0$

RR: $f \geq F_{.01,9,5} = 10.16$

$$f = \frac{R^2/k}{(1-R^2)/(n-k-1)} = \frac{.938/9}{(1-.938)/(5)} = 8.41$$

Fail to reject H_o. The model does not appear to specify a useful relationship.

b $\hat{\mu}_y = 21.967$, $t_{\alpha/2,n-(k+1)} = t_{.025,5} = 2.571$, so the C.I. is $21.967 \pm (2.571)(1.248) = (18.76, 25.18)$.

c $s^2 = \dfrac{SSE}{n-(k+1)} = \dfrac{23.379}{5} = 4.6758$, and the C.I. is

$21.967 \pm (2.571)\sqrt{4.6758 + (1.248)^2} = (15.55, 28.39)$.

d $SSE_k = 23.379,\ SSE_l = 203.82,$

$H_0 : \beta_4 = \beta_5 = ... = \beta_9 = 0$

H_a : at least one of the above $\beta_i \neq 0$

RR: $f \geq F_{\alpha,k-l,n-(k+1)} = F_{.05,6,5} = 4.95$

$$f = \frac{(203.82-23.379)/(9-3)}{(23.379)/(5)} = 6.43.$$

Reject H_o. At least one of the second order predictors appears useful.

49.

a $\hat{\mu}_{y\cdot 18.9,43} = 21.967 = 96.8303$; Residual = 91 – 96.8303 = -5.8303.

b $H_0 : \beta_1 = \beta_2 = 0$

H_a : at least one $\beta_i \neq 0$

RR: $f \geq F_{.05,2,9} = 8.02$

$f = \dfrac{R^2/k}{(1-R^2)/(n-k-1)} = \dfrac{.768/2}{(1-.768)/9} = 14.90$. Reject H_o. The model appears useful.

c $96.8303 \pm (2.262)(8.20) = (78.28, 115.38)$

d $96.8303 \pm (2.262)\sqrt{24.45^2 + 8.20^2} = (38.50, 155.16)$

50.

a Here k = 5, n – (k+1) = 6, so H_o will be rejected in favor of H_a at level .05 if either $t \geq t_{.025,6} = 2.447$ or $t \leq -2.447$. The computed value of t is $t = \dfrac{.557}{.94} = .59$, so H_o cannot be rejected and inclusion of x_1x_2 as a carrier in the model is not justified.

b No, in the presence of the other four carriers, any particular carrier is relatively unimportant, but this is not equivalent to the statement that all carriers are unimportant.

c $SSE_k = SST(1 - R^2) = 3224.65$, so $f = \dfrac{(5384.18 - 3224.65)/3}{(3224.65)/6} = 1.34$, and since 1.34 is not $\geq F_{.05,3,6} = 4.76$, H_o cannot be rejected; the data does not argue for the inclusion of any second order terms.

51.

a No, there is no pattern in the plots which would indicate that a transformation or the inclusion of other terms in the model would produce a substantially better fit.

b k = 5, n – (k+1) = 8, so $H_0 : \beta_1 = ... = \beta_5 = 0$ will be rejected if $f \geq F_{.05,5,8} = 3.69$; $f = \dfrac{(.759)/5}{(.241)/8} = 5.04 \geq 3.69$, so we reject H_o. At least one of the coefficients is not equal to zero.

c When $x_1 = 8.0$ and $x_2 = 33.1$ the residual is e = 2.71 and the standardized residual is e* = .44; since e* = e/(sd of the residual), sd of residual = e/e* = 6.16. Thus the estimated variance of $\hat{Y}$ is $(6.99)^2 - (6.16)^2 = 10.915$, so the estimated sd is 3.304. Since $\hat{y} = 24.29$ and $t_{.025,8} = 2.306$, the desired C.I. is $24.29 \pm 2.306(3.304) = (16.67, 31.91)$.

d $F_{.05,3,8} = 4.07$, so $H_0 : \beta_3 = \beta_4 = \beta_5 = 0$ will be rejected if $f \geq 4.07$. With $SSE_k = 8, s^2 = 390.88$, and $f = \frac{(894.95-390.88)/3}{(390.88)/8} = 3.44$, and since 3.44 is not ≥ 4.07, H_o cannot be rejected and the quadratic terms should all be deleted. (n.b.: this is not a modification which would be suggested by a residual plot.

52.

a For $x_1 = x_2 = x_3 = x_4 = +1$, $\hat{y} = 84.67 + .650 - .258 + ... + .050 = 85.390$. The single y corresponding to these x_i values is 85.4, so $y - \hat{y} = 85.4 - 85.390 = .010$.

b Letting $x_1', ..., x_4'$ denote the uncoded variables, $x_1' = .1x_1 + .3$, $x_2' = .1x_2 + .3$, $x_3' = x_3 + 2.5$, and $x_4' = 15x_4 + 160$; Substitution of $x_1 = 10x_1' - 3$, $x_2 = 10x_2' - 3$, $x_3 = x_3' - 2.5$, and $x_4 = \frac{x_4' + 160}{15}$ yields the uncoded function.

c For the full model k = 14 and for the reduced model l – 4, while n – (k + 1) = 16. Thus $H_0 : \beta_5 = ... = \beta_{14} = 0$ will be rejected if $f \geq F_{.05,10,16} = 2.49$. $SSE = (1 - R^2)SST$ so $SSE_k = 1.9845$ and $SSE_l = 4.8146$, giving $f = \frac{(4.8146-1.9845)/10}{(1.9845)/16} = 2.28$. Since 2.28 is not ≥ 2.49, H_o cannot be rejected, so all higher order terms should be deleted.

d $H_0 : \mu_{Y\cdot 0,0,0,0} = 85.0$ will be rejected in favor of $H_a : \mu_{Y\cdot 0,0,0,0} < 85.0$ if $t \leq -t_{.05,26} = -1.706$. With $\hat{\mu} = \hat{\beta}_0 = 85.5548$, $t = \frac{85.5548 - 85}{.0772} = 7.19$, which is certainly not ≤ -1.706, so H_o is not rejected and prior belief is not contradicted by the data.

Section 13.5

53.

a $\ln(Q) = Y = \ln(\alpha) + \beta\ln(a) + \gamma\ln(b) + \ln(\varepsilon) = \beta_0 + \beta_1 x_1 + \beta_2 x_2 + \varepsilon'$ where $x_1 = \ln(a), x_2 = \ln(b), \beta_0 = \ln(\alpha), \beta_1 = \beta, \beta_2 = \gamma$ and $\varepsilon' = \ln(\varepsilon)$. Thus we transform to $(y, x_1, x_2) = (\ln(Q), \ln(a), \ln(b))$ (take the natural log of the values of each variable) and do a multiple linear regression. A computer analysis gave $\hat{\beta}_0 = 1.5652$, $\hat{\beta}_1 = .9450$, and $\hat{\beta}_2 = .1815$. For a = 10 and b = .01, x_1 = ln(10) = 2.3026 and x_2 = ln(.01) = -4.6052, from which $\hat{y} = 2.9053$ and $\hat{Q} = e^{2.9053} = 18.27$.

b Again taking the natural log, $Y = \ln(Q) = \ln(\alpha) + \beta a + \gamma b + \ln(\varepsilon)$, so to fit this model it is necessary to take the natural log of each Q value (and not transform a or b) before using multiple regression analysis.

c We simply exponentiate each endpoint: $(e^{.217}, e^{1.755}) = (1.24, 5.78)$.

54.

a $n = 20, k = 5, n - (k+1) = 14$, so $H_0 : \beta_1 = ... = \beta_5 = 0$ will be rejected in favor of H_a :at least one among $\beta_1, ..., \beta_5 \neq 0$, if $f \geq F_{.01,5,14} = 4.69$. With $f = \frac{(.769)/5}{(.231)/14} = 9.32 \geq 4.69$, so H_o is rejected. Wood specific gravity appears to be linearly related to at lest one of the five carriers.

b For the full model, adjusted $R^2 = \frac{(19)(.769) - 5}{14} = .687$, while for the reduced model, the adjusted $R^2 = \frac{(19)(.769) - 4}{15} = .707$.

c From **a**, $SSE_k = (.231)(.0196610) = .004542$, and $SSE_l = (.346)(.0196610) = .006803$, so $f = \frac{(.002261)/3}{(.004542)/14} = 2.32$. Since $F_{.05,3,14} = 3.34$ and 2.32 is not ≥ 3.34, we conclude that $\beta_1 = \beta_2 = \beta_4 = 0$.

d $x_3' = \frac{x_3 - 52.540}{5.4447} = -.4665$ and $x_5' = \frac{x_5 - 89.195}{3.6660} = .2196$, so $\hat{y} = .5255 - (.0236)(-.4665) + (.0097)(.2196) = .5386$.

e $t_{.025,17} = 2.110$ (error df = n – (k+1) = 20 – (2+1) = 17 for the two carrier model), so the desired C.I. is $-.0236 \pm 2.110(.0046) = (-.0333, -.0139)$.

f $y = .5255 - .0236\left(\dfrac{x_3 - 52.540}{5.4447}\right) + .0097\left(\dfrac{x_5 - 89.195}{3.6660}\right)$, so $\hat{\beta}_3$ for the unstandardized model $= \dfrac{-.0236}{5.447} = -.004334$. The estimated sd of the unstandardized $\hat{\beta}_3$ is $= \dfrac{.0046}{5.447} = -.000845$.

g $\hat{y} = .532$ and $\sqrt{s^2 + s_{\hat{\beta}_0 + \hat{\beta}_3 x_3' + \hat{\beta}_5 x_5'}} = .02058$, so the P.I. is $.532 \pm (2.110)(.02058) = .532 \pm .043 = (.489,.575)$.

55.

k	R^2	Adj. R^2	$C_k = \dfrac{SSE_k}{s^2} + 2(k+1) - n$
1	.676	.647	138.2
2	.979	.975	2.7
3	.9819	.976	3.2
4	.9824		4

Where $s^2 = 5.9825$

a Clearly the model with k = 2 is recommended on all counts.

b No. Forward selection would let x_4 enter first and would not delete it at the next stage.

56. At step #1 (in which the model with all 4 predictors was fit), t = .83 was the t ratio smallest in absolute magnitude. The corresponding predictor x_3 was then dropped from the model, and a model with predictors x_1, x_2, and x_4 was fit. The t ratio for x_4 , -1.53, was the smallest in absolute magnitude and 1.53 < 2.00, so the predictor x_4 was deleted. When the model with predictors x_1 and x_2 only was fit, both t ratios considerably exceeded 2 in absolute value, so no further deletion is necessary.

57. The choice of a "best" model seems reasonably clear–cut. The model with 4 variables including all but the summerwood fiber variable would seem bests. R^2 is as large as any of the models, including the 5 variable model. R^2 adjusted is at its maximum and CP is at its minimum . As a second choice, one might consider the model with k = 3 which excludes the summerwood fiber and springwood % variables.

58. Backwards Stepping:

Step 1: A model with all 5 variables is fit; the smallest t-ratio is t = .12, associated with variable x_2 (summerwood fiber %). Since t = .12 < 2, the variable x_2 was eliminated.
Step 2: A model with all variables except x_2 was fit. Variable x_4 (springwood light absorption) has the smallest t-ratio (t = -1.76), whose magnitude is smaller than 2. Therefore, x_4 is the next variable to be eliminated.
Step 3: A model with variables x_3 and x_5 is fit. Both t-ratios have magnitudes that exceed 2, so both variables are kept and the backwards stepping procedure stops at this step. The final model identified by the backwards stepping method is the one containing x_3 and x_5.

Forward Stepping:

Step 1: After fitting all 5 one-variable models, the model with x_3 had the t-ratio with the largest magnitude (t = -4.82). Because the absolute value of this t-ratio exceeds 2, x_3 was the first variable to enter the model.
Step 2: All 4 two-variable models that include x_3 were fit. That is, the models $\{x_3, x_1\}$, $\{x_3, x_2\}$, $\{x_3, x_4\}$, $\{x_3, x_5\}$ were all fit. Of all 4 models, the t-ratio 2.12 (for variable x_5) was largest in absolute value. Because this t-ratio exceeds 2, x_5 is the next variable to enter the model.
Step 3: (not printed): All possible tree-variable models involving x_3 and x_5 and another predictor, None of the t-ratios for the added variables have absolute values that exceed 2, so no more variables are added. There is no need to print anything in this case, so the results of these tests are not shown.
Note; Both the forwards and backwards stepping methods arrived at the same final model, $\{x_3, x_5\}$, in this problem. This often happens, but not always. There are cases when the different stepwise methods will arrive at slightly different collections of predictor variables.

59. If multicollinearity were present, at least one of the four R^2 values would be very close to 1, which is not the case. Therefore, we conclude that multicollinearity is not a problem in this data.

60. Looking at the h_{ii} column and using $\frac{2(k+1)}{n} = \frac{8}{19} = .421$ as the criteria, three observations appear to have large influence. With h_{ii} values of .712933, .516298, and .513214, observations 14, 15, 16, correspond to response (y) values 22.8, 41.8, and 48.6.

61.

a $\frac{2(k+1)}{n} = \frac{6}{10} = .6$; since $h_{44} > .6$, data point #4 would appear to have large influence. (Note: Formulas involving matrix algebra appear in the first edition.)

b For data point #2, $x'_{(2)} = (1 \quad 3.453 \quad -4.920)$, so $\hat{\beta} - \hat{\beta}_{(2)} =$

$$\frac{-.766}{1-.302}(X'X)^{-1}\begin{pmatrix}1\\3.453\\-4.920\end{pmatrix} = -1.0974\begin{pmatrix}.3032\\.1644\\.1156\end{pmatrix} = \begin{pmatrix}-.333\\-.180\\-.127\end{pmatrix}$$ and similar calculations

yield $\hat{\beta} - \hat{\beta}_{(4)} = \begin{pmatrix}.106\\-.040\\.030\end{pmatrix}$.

c Comparing the changes in the $\hat{\beta}_i$'s to the $s_{\hat{\beta}_i}$'s, none of the changes is all that substantial (the largest is 1.2sd's for the change in $\hat{\beta}_1$ when point #2 is deleted). Thus although h_{44} is large, indicating a potential high influence of point #4 on the fit, the actual influence does not appear to be great.

Supplementary Exercises

62.

a For every 1 cm^{-1} increase in inverse foil thickness (x), we estimate that we would expect steady-state permeation flux to increase by $.26042 \mu A / cm^2$. Also, 98% of the observed variation in steady-state permeation flux can be explained by its relationship to inverse foil thickness.

b A point estimate of flux when inverse foil thickness is 23.5 can be found in the Observation 3 row of the Minitab output: $\hat{y} = 5.722 \mu A / cm^2$.

c To test model usefulness, we test the hypotheses $H_0 : \beta_1 = 0$ vs. $H_a : \beta_1 \neq 0$. The test statistic is t = 17034, with associated p-value of .000, which is less than any significance level, so we reject H_o and conclude that the model is useful.

d With $t_{.025,6} = 2.447$, a 95% Prediction interval for $Y_{(45)}$ is

$11.321 \pm 2.447\sqrt{.203 + (.253)^2} = 11.321 \pm 1.264 = (10.057, 12.585)$. That is, we are confident that when inverse foil thickness is 45 cm^{-1}, a predicted value of steady-state flux will be between 10.057 and 12.585 $\mu A / cm^2$.

e

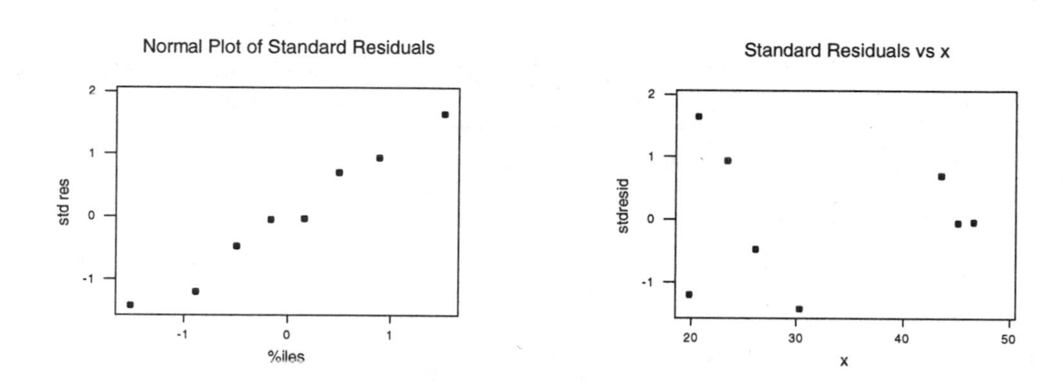

The normal plot gives no indication to question the normality assumption, and the residual plots against both x and y (only vs x shown) show no detectable pattern, so we judge the model adequate.

63.

a For a one-minute increase in the 1-mile walk time, we would expect the VO_2max to decrease by .0996, while keeping the other predictor variables fixed.

b We would expect male to have an increase of .6566 in VO_2max over females, while keeping the other predictor variables fixed.

c $\hat{y} = 3.5959 + .6566(1) + .0096(170) - .0996(11) - .0880(140) = 3.67$. The residual is $\hat{y} = (3.15 - 3.67) = -.52$.

d $R^2 = 1 - \frac{SSE}{SST} = 1 - \frac{30.1033}{102.3922} = .706$, or 70.6% of the observed variations in VO_2max can be attributed to the model relationship.

e $H_0 : \beta_1 = \beta_2 = \beta_3 = \beta_4 = 0$ will be rejected in favor of H_a :at least one among $\beta_1, \ldots, \beta_4 \neq 0$, if $f \geq F_{.05,4,15} = 8.25$. With $f = \frac{(.706)/4}{(1-.706)/15} = 9.005 \geq 8.25$, so H_o is rejected. It appears that the model specifies a useful relationship between VO_2max and at least one of the other predictors.

64.

a

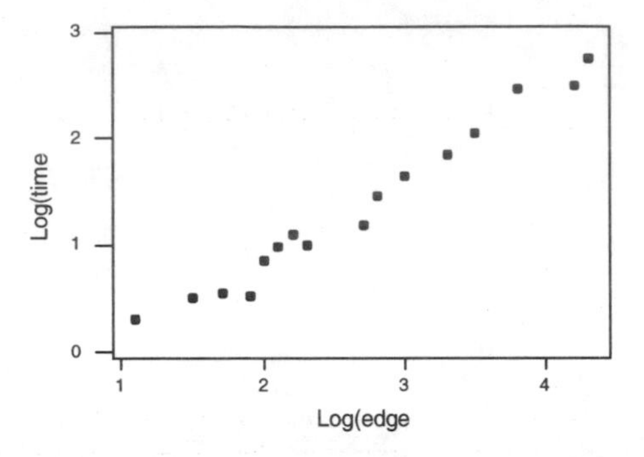

Yes, the scatter plot of the two transformed variables appears quite linear, and thus suggests a linear relationship between the two.

b Letting y denote the variable 'time', the regression model for the variables y' and x' is $\log_{10}(y) = y' = \alpha + \beta x' + \varepsilon'$. Exponentiating (taking the antilogs of) both sides gives $y = 10^{\alpha + \beta \log(x) + \varepsilon'} = (10^{\alpha})(x^{\beta})10^{\varepsilon'} = \gamma_0 x^{\gamma_1} \cdot \varepsilon$; i.e., the model is $y = \gamma_0 x^{\gamma_1} \cdot \varepsilon$ where $\gamma_0 = \alpha$ and $\gamma_1 = \beta$. This model is often called a "power function" regression model.

c Using the transformed variables y' and x', the necessary sums of squares are $S_{x'y'} = 68.640 - \frac{(42.4)(21.69)}{16} = 11.1615$ and $S_{x'x'} = 126.34 - \frac{(42.4)^2}{16} = 13.98$.

Therefore $\hat{\beta}_1 = \frac{S_{x'y'}}{S_{x'x'}} = \frac{11.1615}{13.98} = .79839$ and

$\hat{\beta}_0 = \frac{21.69}{16} - (.79839)\left(\frac{42.4}{16}\right) = -.76011$. The estimate of γ_1 is $\hat{\gamma}_1 = .7984$ and $\gamma_0 = 10^{\alpha} = 10^{-.76011} = .1737$. The estimated power function model is then $y = .1737x^{.7984}$. For x = 300, the predicted value of y is $\hat{y} = .1737(300)^{.7984} 16.502$, or about 16.5 seconds.

65.

a Based on a scatter plot (below), a simple linear regression model would not be appropriate. Because of the slight, but obvious curvature, a quadratic model would probably be more appropriate.

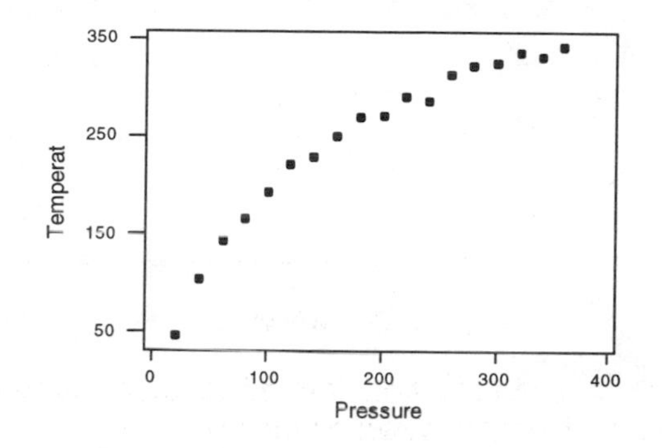

b Using a quadratic model, a Minilab generated regression equation is $\hat{y} = 35.423 + 1.7191x - .0024753x^2$, and a point estimate of temperature when pressure is 200 is $\hat{y} = 280.23$. Minitab will also generate a 95% prediction interval of (256.25, 304.22). That is, we are confident that when pressure is 200 psi, a single value of temperature will be between 256.25 and 304.22 $^\circ F$.

66.

a For the model excluding the interaction term, $R^2 = 1 - \frac{5.18}{8.55} = .394$, or 39.4% of the observed variation in lift/drag ratio can be explained by the model without the interaction accounted for. However, including the interaction term increases the amount of variation in lift/drag ratio that can be explained by the model to $R^2 = 1 - \frac{3.07}{8.55} = .641$, or 64.1%.

b Without interaction, we are testing $H_0 : \beta_1 = \beta_2 = 0$ vs. H_a : either β_1 or $\beta_2 \neq 0$. The test statistic is $f = \frac{R^2/k}{(1-R^2)/(n-k-1)}$, The rejection region is $f \geq F_{.05,2,6} = 5.14$, and the calculated statistic is $f = \frac{.394/2}{(1-.394)/6} = 1.95$, which does not fall in the rejection region, so we fail to reject H_0. This model is not useful. With the interaction term, we are testing $H_0 : \beta_1 = \beta_2 = \beta_3 = 0$ vs. H_a : at least one of the β_i's $\neq 0$. With rejection region $f \geq F_{.05,3,5} = 5.41$ and calculated statistic $f = \frac{.641/3}{(1-.641)/5} = 2.98$, we still fail to reject the null hypothesis. Even with the interaction term, there is not enough of a significant relationship between lift/drag ratio and the two predictor variables to make the model useful (a bit of a surprise!)

67.

a Using Minitab to generate the first order regression model, we test the model utility (to see if any of the predictors are useful), and with $f = 21.03$ and a p-value of .000, we determine that at least one of the predictors is useful in predicting palladium content. Looking at the individual predictors, the p-value associated with the pH predictor has value .169, which would indicate that this predictor is unimportant in the presence of the others.

b Testing $H_0 : \beta_1 = ... = \beta_{20} = 0$ vs. H_a : at least one of the $\beta_i 's \neq 0$. With calculated statistic $f = 6.29$, and p-value .002, this model is also useful at any reasonable significance level.

c Testing $H_0 : \beta_6 = ... = \beta_{20} = 0$ vs. H_a : at least one of the listed $\beta_i 's \neq 0$, the test statistic is $f = \frac{(SSE_l - SSE_k)/_{k-l}}{(SSE_k)/_{n-k-1}} = \frac{(716.10-290.27)/_{(20-5)}}{290.27/_{(32-20-1)}} = 1.07$. Using significance level .05, the rejection region would be $f \geq F_{.05,15,11} = 2.72$. Since 1.07 < 2.72, we fail to reject H_o and conclude that all the quadratic and interaction terms should not be included in the model. They do not add enough information to make this model significantly better than the simple first order model.

68.

a $R^2 = 1 - \frac{SSE}{SST} = 1 - \frac{.80017}{16.18555} = .9506$, or 95.06% of the observed variation in weld strength can be attributed to the given model.

b The complete second order model consists of nine predictors and nine corresponding coefficients. The hypotheses are $H_0 : \beta_1 = ... = \beta_9 = 0$ vs. H_a : at least one of the $\beta_i 's \neq 0$. The test statistic is $f = \frac{R^2/_k}{(1-R^2)/_{(n-k-1)}}$, where k = 9, and n = 37.The rejection region is $f \geq F_{.05,9,27} = 2.25$. The calculated statistic is $f = \frac{.9506/_9}{(1-.9506)/_{27}} = 57.68$ which is ≥ 2.25, so we reject the null hypothesis. The complete second order model is useful.

c To test $H_0 : \beta_7 = 0$ vs $H_a : \beta_7 \neq 0$ (the coefficient corresponding to the wc*wt predictor), $t = \sqrt{f} = \sqrt{2.32} = 1.52$. With df = 27, the p-value $\approx 2(.073) = .146$ (from Table A.8). With such a large p-value, this predictor is not useful in the presence of all the others, so it can be eliminated.

d The point estimate is $\hat{y} = 3.352 + .098(10) + .222(12) + .297(6) - .0102(10^2)$ $- .037(6^2) + .0128(10)(12) = 7.962$. With $t_{.025,27} = 2.052$, the 95% P.I. would be $7.962 \pm 2.052(.0750) = 7.962 \pm .154 = (7.808, 8.116)$. Because of the narrowness of the interval, it appears that the value of strength can be accurately predicted.

69.

a We wish to test $H_0 : \beta_1 = \beta_2 = 0$ vs. H_a : either β_1 or $\beta_2 \neq 0$. The test statistic is $f = \dfrac{R^2/k}{(1-R^2)/(n-k-1)}$, where k = 2 for the quadratic model. The rejection region is $f \geq F_{\alpha,k,n-k-1} = F_{.01,2,5} = 13.27$. $R^2 = 1 - \dfrac{.29}{202.88} = .9986$, giving f = 1783. No doubt about it, folks – the quadratic model is useful!

b The relevant hypotheses are $H_0 : \beta_2 = 0$ vs. $H_a : \beta_2 \neq 0$. The test statistic value is $t = \dfrac{\hat{\beta}_2}{s_{\hat{\beta}_2}}$, and H_o will be rejected at level .001 if either $t \geq 6.869$ or $t \leq -6.869$ (df = n – 3 = 5). Since $t = \dfrac{-.00163141}{.00003391} = -48.1 \leq -6.869$, H_o is rejected. The quadratic predictor should be retained.

c No. R^2 is extremely high for the quadratic model, so the marginal benefit of including the cubic predictor would be essentially nil – and a scatter plot doesn't show the type of curvature associated with a cubic model.

d $t_{.025,5} = 2.571$, and $\hat{\beta}_0 + \hat{\beta}_1(100) + \hat{\beta}_2(100)^2 = 21.36$, so the C.I. is $21.36 \pm (2.571)(.1141) = 21.36 \pm .69 = (20.67, 22.05)$

70. A scatter plot of $y' = \log_{10}(y)$ vs. x shows a substantial linear pattern, suggesting the model $Y = \alpha \cdot (10)^{\beta x} \cdot \varepsilon$, i.e. $Y' = \log(\alpha) + \beta x + \log(\varepsilon) = \beta_0 + \beta_1 x + \varepsilon'$. The necessary summary quantities are

$\Sigma x_i = 397,\ \Sigma x_i^2 = 14{,}263,\ \Sigma y_i' = -74.3,\ \Sigma y_i'^2 = 47{,}081,$ and $\Sigma x_i y_i' = -2358.1$, giving

$\hat{\beta}_1 = \dfrac{12(-2358.1) - (397)(-74.3)}{12(14{,}263) - (397)^2} = .08857312$ and $\hat{\beta}_0 = -9.12196058$. Thus $\hat{\beta} = .08857312$ and $\alpha = 10^{-9.12196058}$. The predicted value of y' when x = 35 is $-9.12196058 + .08857312(35) = -6.0219$, so $\hat{y} = 10^{-6.0219}$.

71.

a $H_0: \beta_1 = \beta_2 = 0$ will be rejected in favor of H_a : either β_1 or $\beta_2 \neq 0$ if $f = \frac{R^2/k}{(1-R^2)/(n-k-1)} \geq F_{\alpha,k,n-k-1} = F_{.01,2,7} = 9.55$. $SST = \Sigma y^2 - \frac{(\Sigma y)}{n} = 264.5$, so $R^2 = 1 - \frac{26.98}{264.5} = .898$, and $f = \frac{.898/2}{(.102)/7} = 30.8$. Because $30.8 \geq 9.55$ H_o is rejected at significance level .01 and the quadratic model is judged useful.

b The hypotheses are $H_0: \beta_2 = 0$ vs. $H_a: \beta_2 \neq 0$. The test statistic value is $t = \frac{\hat{\beta}_2}{s_{\hat{\beta}_2}} = \frac{-2.3621}{.3073} = -7.69$, and $t_{.0005,7} = 5.408$, so H_o is rejected at level .001 and p-value < .001. The quadratic predictor should not be eliminated.

c x = 1 here, and $\hat{\mu}_{Y \cdot 1} = \hat{\beta}_0 + \hat{\beta}_1(1) + \hat{\beta}_2(1)^2 = 45.96$. $t_{.025,7} = 1.895$, giving the C.I. $45.96 \pm (1.895)(1.031) = (44.01, 47.91)$.

72.

a 80.79

b Yes, p-value = .007 which is less than .01.

c No, p-value = .043 which is less than .05.

d $.14167 \pm (2.447)(.03301) = (.0609, .2224)$

e $\hat{\mu}_{y \cdot 9,66} = 6.3067$, using $\alpha = .05$, the interval is
$6.3067 \pm (2.447)\sqrt{(.4851)^2 + (.162)^2} = (5.06, 7.56)$

73.

a Estimate = $\hat{\beta}_0 + \hat{\beta}_1(15) + \hat{\beta}_2(3.5)^2 = 180 + (1)(15) + (10.5)(3.5) = 231.75$

b $R^2 = 1 - \frac{117.4}{1210.30} = .903$

c $H_0: \beta_1 = \beta_2 = 0$ vs. H_a : either β_1 or $\beta_2 \neq 0$ (or both) . $f = \frac{.903/2}{.097/9} = 41.9$, which greatly exceeds $F_{.01,2,9}$ so there appears to be a useful linear relationship.

d $s^2 = \frac{117.40}{12-3} = 13.044$, $\sqrt{s^2 + (est.st.dev)^2} = 3.806$, $t_{.025,9} = 2.262$. The P.I. is $229.5 \pm (2.262)(3.806) = (220.9, 238.1)$

74. The second order model has predictors $x_1, x_2, x_3, x_1^2, x_2^2, x_3^2, x_1x_2, x_1x_3, x_2x_3$ with corresponding coefficients $\beta_1, \beta_2, \beta_3, \beta_4, \beta_5, \beta_6, \beta_7, \beta_8, \beta_9$. We wish to test $H_0 : \beta_4 = \beta_5 = \beta_6 = \beta_7 = \beta_8 = \beta_9 = 0$ vs. the alternative that at least one of these six β_i's is not zero. The test statistic value is $f = \frac{(821.5-5027.1)/(9-3)}{(5027.1)/(20-10)} = \frac{530.9}{502.71} = 1.1$. Since 1.1 < $F_{.05,6,10} = 3.22$, H_o cannot be rejected. It doesn't appear as though any of the quadratic or interaction carriers should be included in the model.

75. There are obviously several reasonable choices in each case. In **a**, the model with 6 carriers is a defensible choice on all three grounds, as are those with 7 and 8 carriers. The models with 7, 8, or 9 carriers in **b** merit serious consideration. These models merit consideration because R_k^2, MSE_k, and CK meet the variable selection criteria given in Section 13.5.

76.

a $f = \frac{(.90)/(15)}{(.10)/(4)} = 2.4$. Because 2.4 < 5.86, $H_0 : \beta_1 = ... = \beta_{15} = 0$ cannot be rejected. There does not appear to be a useful linear relationship.

b The high R^2 value resulted from saturating the model with predictors. In general, one would be suspicious of a model yielding a high R^2 value when K is large relative to n.

c $\frac{(R^2)/(15)}{(1-R^2)/(4)} \geq 5.86$ iff $\frac{R^2}{1-R^2} \geq 21.975$ iff $R^2 \geq \frac{21.975}{22.975} = .9565$

77.

a The relevant hypotheses are $H_0 : \beta_1 = ... = \beta_5 = 0$ vs. H_a: at least one among $\beta_1, ..., \beta_5$ is not 0. $F_{.05,5,111} = 2.29$ and $f = \frac{(.827)/(5)}{(.173)/(111)} = 106.1$. Because $106.1 \geq 2.29$, H_o is rejected in favor of the conclusion that there is a useful linear relationship between Y and at least one of the predictors.

b $t_{.05,111} = 1.66$, so the C.I. is $.041 \pm (1.66)(.016) = .041 \pm .027 = (.014, .068)$. β_1 is the expected change in mortality rate associated with a one-unit increase in the particle reading when the other four predictors are held fixed; we cab be 90% confident that .014 < β_1 < .068.

c $H_0: \beta_4 = 0$ will be rejected in favor of $H_a: \beta_4 \neq 0$ if $t = \frac{\hat{\beta}_4}{s_{\hat{\beta}_4}}$ is either ≥ 2.62 or ≤ -2.62. $t = \frac{.014}{.007} = 5.9 \geq 2.62$, so H_o is rejected and this predictor is judged important.

d $\hat{y} = 19.607 + .041(166) + .071(60) + .001(788) + .041(68) + .687(.95) = 99.514$, and the corresponding residual is 103 – 99.514 = 3.486.

78.

a The set $x_1, x_3, x_4, x_5, x_6, x_8$ includes both x_1, x_4, x_5, x_8 and x_1, x_3, x_5, x_6, so $R^2_{1,3,4,5,6,8} \geq \max\left(R^2_{1,4,5,8}, R^2_{1,3,5,6}\right) = .723$.

b $R^2_{1,4} \leq R^2_{1,4,5,8} = .723$, but it is not necessarily $\leq .689$ since x_1, x_4 is not a subset of x_1, x_3, x_5, x_6.

Chapter 14

Section 14.1

1.

a We reject H_o if the calculated χ^2 value is greater than or equal to the tabled value of $\chi^2_{\alpha,k-1}$ from Table A.7. Since $12.25 \ge \chi^2_{.05,4} = 9.488$, we would reject H_o.

b Since 8.54 is not $\ge \chi^2_{.01,3} = 11.344$, we would fail to reject H_o.

c Since 4.36 is not $\ge \chi^2_{.10,2} = 4.605$, we would fail to reject H_o.

d Since 10.20 is not $\ge \chi^2_{.01,5} = 15.085$, we would fail to reject H_o.

2.

a In the d.f. = 2 row of Table A.7, our χ^2 value of 7.5 falls between $\chi^2_{.025,2} = 7.378$ and $\chi^2_{.01,2} = 9.210$, so the p-value is between .01 and .025, or .01 < p-value < .025.

b With d.f. = 6, our χ^2 value of 13.00 falls between $\chi^2_{.05,6} = 12.592$ and $\chi^2_{.025,6} = 14.440$, so .025 < p-value < .05.

c With d.f. = 9, our χ^2 value of 18.00 falls between $\chi^2_{.05,9} = 16.919$ and $\chi^2_{.025,9} = 19.022$, so .025 < p-value < .05.

d With k = 5, d.f. = k – 1 = 4, and our χ^2 value of 21.3 exceeds $\chi^2_{.005,4} = 14.860$, so the p-value < .005.

e The d.f. = k – 1 = 4 – 1 = 3; $\chi^2 = 5.0$ is less than $\chi^2_{.10,3} = 6.251$, so p-value > .10.

3. Using the number 1 for business, 2 for engineering, 3 for social science, and 4 for agriculture, let p_i = the true proportion of all clients from discipline i. If the Statistics department's expectations are correct, then the relevant null hypothesis is $H_o: p_1 = .40, p_2 = .30, p_3 = .20, p_4 = .10$, versus H_a : The Statistics department's expectations are not correct. With d.f = k – 1 = 4 – 1 = 3, we reject H_o if $\chi^2 \ge \chi^2_{.05,3} = 7.815$. Using the proportions in H_o, the expected number of clients are :

Client's Discipline	Expected Number
Business	(120)(.40) = 48
Engineering	(120)(.30) = 36
Social Science	(120)(.20) = 24
Agriculture	(120)(.10) = 12

Since all the expected counts are at least 5, the chi-squared test can be used. The value of the test statistic is $\chi^2 = \sum_{i=1}^{k} \frac{(n_i - np_i)^2}{np_i} = \sum_{allcells} \frac{(observed - expected)^2}{expected}$

$$= \left[\frac{(52-48)^2}{48} + \frac{(38-36)^2}{36} + \frac{(21-24)^2}{24} + \frac{(9-12)^2}{12} \right] = 1.57$$, which is not ≥ 7.815,

so we fail to reject H_o. (Alternatively, p-value = $P(\chi^2 \ge 1.57)$ which is > .10, and since the p-value is not < .05, we reject H_o). Thus we have no evidence to suggest that the statistics department's expectations are incorrect.

4. The uniform hypothesis implies that $p_{i0} = \frac{1}{8} = .125$ for I = 1, …, 8, so $H_o: p_{10} = p_{20} = \ldots = p_{80} = .125$ will be rejected in favor of H_a if $\chi^2 \ge \chi^2_{.10,7} = 12.017$. Each expected count is np_{i0} = 120(.125) = 15, so

$$\chi^2 = \left[\frac{(12-15)^2}{15} + \ldots + \frac{(10-15)^2}{15} \right] = 4.80$$. Because 4.80 is not ≥ 12.017, we fail to reject H_o. There is not enough evidence to disprove the claim.

5. We will reject H_o if the p-value < .10. The observed values, expected values, and corresponding χ^2 terms are :

Obs	4	15	23	25	38	21	32	14	10	8
Exp	6.67	13.33	20	26.67	33.33	33.33	26.67	20	13.33	6.67
χ^2	1.069	.209	.450	.105	.654	.163	1.065	1.800	.832	.265

(continued)

$\chi^2 = 1.069 + ... + .265 = 6.612$. With d.f. = 10 – 1 = 9, our χ^2 value of 6.612 is less than $\chi^2_{.10,9} = 14.684$, so the p value > .10, which is not < .10, so we cannot reject H_o. There is no evidence that the data is not consistent with the previously determined proportions.

6. A 9:3:4 ratio implies that $p_{10} = \frac{9}{16} = .5625$, $p_{20} = \frac{3}{16} = .1875$, and $p_{30} = \frac{4}{16} = .2500$. With n = 195 + 73 + 100 = 368, the expected counts are 207.000, 69.000, and 92.000, so $\chi^2 = \left[\frac{(195-207)^2}{207} + \frac{(73-69)^2}{69} + \frac{(100-92)^2}{92}\right] = 1.623$. With d.f. = 3 – 1 = 2, our χ^2 value of 1.623 is less than $\chi^2_{.10,2} = 4.605$, so the p-value > .10, which is not < .05, so we cannot reject H_o. the data does confirm the 9:3:4 theory.

7. We test $H_o: p_1 = p_2 = p_3 = p_4 = .25$ vs. H_a :at least one proportion $\neq .25$, and d.f. = 3. We will reject H_o if the p-value < .01.

Cell	1	2	3	4
Observed	328	334	372	327
Expected	340.25	340.25	340.25	34.025
χ^2 term	.4410	.1148	2.9627	.5160

$\chi^2 = 4.0345$, and with 3 d.f., p-value > .10, so we fail to reject H_o. The data fails to indicate a seasonal relationship with incidence of violent crime.

8. $H_o: p_1 = \frac{15}{365}, p_2 = \frac{46}{365}, p_3 = \frac{120}{365}, p_4 = \frac{184}{365}$, versus H_a :at least one proportion is not a stated in H_o. The degrees of freedom = 3, and the rejection region is $\chi^2 \geq \chi_{.01,3} = 11.344$.

Cell	1	2	3	4
Observed	11	24	69	96
Expected	8.22	25.21	65.75	100.82
χ^2 term	.9402	.0581	.1606	.2304

$\chi^2 = \sum \frac{(obs - \exp)^2}{\exp} = 1.3893$, which is not ≥ 11.344, so H_o is not rejected. The data does not indicate a relationship between patients' admission date and birthday.

9.

a Denoting the 5 intervals by [0, c_1), [c_1, c_2), …, [c_4, ∞), we wish c_1 for which $.2 = P(0 \le X \le c_1) = \int_0^{c_1} e^{-x}dx = 1 - e^{-c_1}$, so c_1 = -ln(.8) = .2231. Then $.2 = P(c_1 \le X \le c_2) \Rightarrow .4 = P(0 \le X_1 \le c_2) = 1 - e^{-c_2}$, so c_2 = -ln(.6) = .5108. Similarly, c_3 = -ln(.4) = .0163 and c_4 = -ln(.2) = 1.6094. the resulting intervals are [0, .2231), [.2231, .5108), [.5108, .9163), [.9163, 1.6094), and [1.6094, ∞).

b Each expected cell count is 40(.2) = 8, and the observed cell counts are 6, 8, 10, 7, and 9, so $\chi^2 = \left[\frac{(6-8)^2}{8} + \ldots + \frac{(9-8)^2}{8}\right] = 1.25$. Because 1.25 is not $\ge \chi^2_{.10,4} = 7.779$, even at level .10 H_o cannot be rejected; the data is quite consistent with the specified exponential distribution.

10.

a $\chi^2 = \sum_{i=1}^{k} \frac{(n_i - np_{i0})^2}{np_{i0}} = \sum_i \frac{N_i^2 - 2np_{i0}N_i + n^2p_{i0}^2}{np_{i0}} = \sum_i \frac{N_i^2}{np_{i0}} - 2\sum_i N_i + n\sum_i p_{i0}$

$= \sum_i \frac{N_i^2}{np_{i0}} - 2n + n(1) = \sum_i \frac{N_i^2}{np_{i0}} - n$ as desired. This formula involves only one subtraction, and that at the end of the calculation, so it is analogous to the shortcut formula for s^2.

b $\chi^2 = \frac{k}{n}\sum_i N_i^2 - n$. For the pigeon data, k = 8, n = 120, and $\Sigma N_i^2 = 1872$, so $\chi^2 = \frac{8(1872)}{120} - 120 = 124.8 - 120 = 4.8$ as before.

11.

a The six intervals must be symmetric about 0, so denote the 4th, 5th and 6th intervals by [0, a0, [a, b), [b, ∞). a must be such that $\Phi(a) = .6667(\frac{1}{2} + \frac{1}{6})$, which from Table A.3 gives $a \approx .43$. Similarly $\Phi(b) = .8333$ implies $b \approx .97$, so the six intervals are ($-\infty$, -.97), [-.97, -.43), [-.43, 0), [0, .43), [.43, .97), and [.97, ∞).

b The six intervals are symmetric about the mean of .5. From **a**, the fourth interval should extend from the mean to .43 standard deviations above the mean, i.e., from .5 to .5 + .43(.002), which gives [.5, .50086). Thus the third interval is [.5 - .00086, .5) = [.49914, .5). Similarly, the upper endpoint of the fifth interval is .5 + .97(.002) = .50194, and the lower endpoint of the second interval is .5 - .00194 = .49806. The resulting intervals are ($-\infty$, .49806), [.49806, .49914), [.49914, .5), [.5, .50086), [.50086, .50194), and [.50194, ∞).

c Each expected count is $45\left(\frac{1}{6}\right) = 7.5$, and the observed counts are 13, 6, 6, 8, 7, and 5, so $\chi^2 = 5.53$. With 5 d.f., the p-value > .10, so we would fail to reject H_o at any of the usual levels of significance. There is no evidence to suggest that the bolt diameters are not normally distributed.

Section 14.2

12.

a Let θ denote the probability of a male (as opposed to female) birth under the binomial model. The four cell probabilities (corresponding to x = 0, 1, 2, 3) are $\pi_1(\theta) = (1-\theta)^3$, $\pi_2(\theta) = 3\theta(1-\theta)^2$, $\pi_3(\theta) = 3\theta^2(1-\theta)$, and $\pi_4(\theta) = \theta^3$. The likelihood is $3^{n_2+n_3} \cdot (1-\theta)^{3n_1+2n_2+n_3} \cdot \theta^{n_2+2n_3+3n_4}$. Forming the log likelihood, taking the derivative with respect to θ, equating to 0, and solving yields $\hat{\theta} = \frac{n_2+2n_3+3n_4}{3n} = \frac{66+128+48}{480} = .504$. The estimated expected counts are $160(1-.504)^3 = 19.52$, $480(.504)(.496)^2 = 59.52$, 60.48, and 20.48, so $\chi^2 = \left[\frac{(14-19.52)^2}{19.52} + \ldots + \frac{(16-20.48)^2}{20.48}\right] = 1.56 + .71 + .20 + .98 = 3.45$. The number of degrees of freedom for the test is 4 – 1 – 1 = 2. H_o of a binomial distribution will be rejected using significance level .05 if $\chi^2 \geq \chi^2_{.05,2} = 5.992$. Because 3.45 < 5.992, H_o is not rejected, and the binomial model is judged to be quite plausible.

b Now $\hat{\theta} = \frac{53}{150} = .353$ and the estimated expected counts are 13.54, 22.17, 12.09, and 2.20. The last estimated expected count is much less than 5, so the chi-squared test based on 2 d.f. should not be used.

13. According to the stated model, the three cell probabilities are $(1-p)^2$, $2p(1-p)$, and p^2, so we wish the value of p which maximizes $(1-p)^{2n_1}[2p(1-p)]^{n_2} p^{2n_3}$. Proceeding as in example 14.6 gives $\hat{p} = \frac{n_2+2n_3}{2n} = \frac{234}{2776} = .0843$. The estimated expected cell counts are then $n(1-\hat{p})^2 = 1163.85$, $n[2\hat{p}(1-\hat{p})]^2 = 214.29$, $n\hat{p}^2 = 9.86$. This gives $\chi^2 = \left[\frac{(1212-1163.85)^2}{1163.85} + \frac{(118-214.29)^2}{214.29} + \frac{(58-9.86)^2}{9.86}\right] = 280.3$.
According to (14.15), H_o will be rejected if $\chi^2 \geq \chi^2_{\alpha,2}$, and since $\chi^2_{.01,2} = 9.210$, H_o is soundly rejected; the stated model is strongly contradicted by the data.

14.

a We wish to maximize $p^{\Sigma x_i - n}(1-p)^n$, or equivalently $(\Sigma x_i - n)\ln p + n\ln(1-p)$. Equating $\frac{d}{dp}$ to 0 yields $\frac{(\Sigma x_i - n)}{p} = \frac{n}{(1-p)}$, whence $p = \frac{(\Sigma x_i - n)}{\Sigma x_i}$. For the given data, $\Sigma x_i = (1)(1) + (2)(31) + ... + (12)(1) = 363$, so $\hat{p} = \frac{(363-130)}{363} = .642$, and $\hat{q} = .358$.

b Each estimated expected cell count is $\hat{p}$ times the previous count, giving $n\hat{q} = 130(.358) = 46.54$, $n\hat{q}\hat{p} = 46.54(.642) = 29.88$, 19.18, 12.31, 17.91, 5.08, 3.26, … . Grouping all values ≥ 7 into a single category gives 7 cells with estimated expected counts 46.54, 29.88, 19.18, 12.31, 7.91, 5.08 (sum = 120.9), and 130 – 120.9 = 9.1. The corresponding observed counts are 48, 31, 20, 9, 6, 5, and 11, giving $\chi^2 = 1.87$. With k = 7 and m = 1 (p was estimated), from (14.15) we need $\chi^2_{.10,5} = 9.236$. Since 1.87 is not ≥ 9.236, we don't reject H$_o$.

15. The part of the likelihood involving θ is $\left[(1-\theta)^4\right]^{n_1} \cdot \left[\theta(1-\theta)^3\right]^{n_2} \cdot \left[\theta^2(1-\theta)^2\right]^{n_3} \cdot \left[\theta^3(1-\theta)\right]^{n_4} \cdot \left[\theta^4\right]^{n_5} = \theta^{n_2+2n_3+3n_4+4n_5}(1-\theta)^{4n_1+3n_2+2n_3+n_4} = \theta^{233}(1-\theta)^{367}$, so $\ln(likelihood) = 233\ln\theta + 367\ln(1-\theta)$. Differentiating and equating to 0 yields $\hat{\theta} = \frac{233}{600} = .3883$, and $(1-\hat{\theta}) = .6117$ [note that the exponent on θ is simply the total # of successes (defectives here) in the n = 4(150) = 600 trials.] Substituting this θ' into the formula for p_i yields estimated cell probabilities .1400, .3555, .3385, .1433, and .0227. Multiplication by 150 yields the estimated expected cell counts are 21.00, 53.33, 50.78, 21.50, and 3.41. the last estimated expected cell count is less than 5, so we combine the last two categories into a single one (≥ 3 defectives), yielding estimated counts 21.00, 53.33, 50.78, 24.91, observed counts 26, 51, 47, 26, and $\chi^2 = 1.62$. With d.f. = 4 – 1 – 1 = 2, since $1.62 < \chi^2_{.10,2} = 4.605$, the p-value > .10, and we do not reject H$_o$. The data suggests that the stated binomial distribution is plausible.

16. $\hat{\lambda} = \bar{x} = \dfrac{(0)(6)+(1)(24)+(2)(42)+\ldots+(8)(6)+(9)(2)}{300} = \dfrac{1163}{300} = 3.88$, so the

estimated cell probabilities are computed from $\hat{p} = e^{-3.88}\dfrac{(3.88)^x}{x!}$. (on following page)

x	0	1	2	3	4	5	6	7	≥ 8
np(x)	6.2	24.0	46.6	60.3	58.5	45.4	29.4	16.3	13.3
obs	6	24	42	59	62	44	41	14	8

This gives $\chi^2 = 7.789$. To see whether the Poisson model provides a good fit, we need $\chi^2_{.10,9-1-1} = \chi^2_{.10,7} = 12.017$. Since $7.789 < 12.017$, the Poisson model does provide a good fit.

17. $\hat{\lambda} = \dfrac{380}{120} = 3.167$, so $\hat{p} = e^{-3.167}\dfrac{(3.167)^x}{x!}$.

x	0	1	2	3	4	5	6	≥ 7
$\hat{p}$	.0421	.1334	.2113	.2230	.1766	.1119	.0590	.0427
$n\hat{p}$	5.05	16.00	25.36	26.76	21.19	13.43	7.08	5.12
obs	24	16	16	18	15	9	6	16

The resulting value of $\chi^2 = 103.98$, and when compared to $\chi^2_{.01,7} = 18.474$, it is obvious that the Poisson model fits very poorly.

18. $\hat{p}_1 = P(X < .100) = P\left(Z < \dfrac{.100 - .173}{.066}\right) = \Phi(-1.11) = .1335$,
$\hat{p}_2 = P(.100 \leq X \leq .150) = P(-1.11 \leq Z \leq -.35) = .2297$,
$\hat{p}_3 = P(-.35 \leq Z \leq .41) = .2959$, $\hat{p}_4 = P(.41 \leq Z \leq 1.17) = .2199$, and $\hat{p}_5 = .1210$.
The estimated expected counts are then (multiply $\hat{p}_i$ by n = 83) 11.08, 19.07, 24.56, 18.25, and 10.04, from which $\chi^2 = 1.67$. Comparing this with $\chi^2_{.05,5-1-2} = \chi^2_{.05,2} = 5.992$, the hypothesis of normality cannot be rejected.

19. With A = $2n_1 + n_4 + n_5$, B = $2n_2 + n_4 + n_6$, and C = $2n_3 + n_5 + n_6$, the likelihood is proportional to $\theta_1^A \theta_2^B (1-\theta_1-\theta_2)^C$, where A + B + C = 2n. Taking the natural log and equating both $\frac{\partial}{\partial\theta_1}$ and $\frac{\partial}{\partial\theta_2}$ to zero gives $\frac{A}{\theta_1} = \frac{C}{1-\theta_1-\theta_2}$ and $\frac{B}{\theta_2} = \frac{C}{1-\theta_1-\theta_2}$, whence $\theta_2 = \frac{B\theta_1}{A}$. Substituting this into the first equation gives $\theta_1 = \frac{A}{A+B+C}$, and then $\theta_2 = \frac{B}{A+B+C}$. Thus $\hat{\theta}_1 = \frac{2n_1+n_4+n_5}{2n}$, $\hat{\theta}_2 = \frac{2n_2+n_4+n_6}{2n}$, and $\left(1-\hat{\theta}_1-\hat{\theta}_2\right) = \frac{2n_3+n_5+n_6}{2n}$. Substituting the observed n_i's yields $\hat{\theta}_1 = \frac{2(49)+20+53}{400} = .4275$, $\hat{\theta}_2 = \frac{110}{400} = .2750$, and $\left(1-\hat{\theta}_1-\hat{\theta}_2\right) = .2975$, from which $\hat{p}_1 = (.4275)^2 = .183$, $\hat{p}_2 = .076$, $\hat{p}_3 = .089$, $\hat{p}_4 = 2(.4275)(.275) = .235$, $\hat{p}_5 = .254$, $\hat{p}_6 = .164$.

Category	1	2	3	4	5	6
np	36.6	15.2	17.8	47.0	50.8	32.8
observed	49	26	14	20	53	38

This gives $\chi^2 = 29.1$. With $\chi^2_{.01,6-1-2} = \chi^2_{.01,3} = 11.344$, and $\chi^2_{.01,6-1} = \chi^2_{.01,5} = 15.085$, according to (14.15) H_o must be rejected since $29.1 \geq 15.085$.

20. The pattern of points in the plot appear to deviate from a straight line, a conclusion that is also supported by the small p-value (< .01000) of the Ryan-Joiner test. Therefore, it is implausible that this data came from a normal population. In particular, the observation 116.7 is a clear outlier. It would be dangerous to use the one-sample t interval as a basis for inference.

21. The Ryan-Joiner test p-value is larger than .10, so we conclude that the null hypothesis of normality cannot be rejected. This data could reasonably have come from a normal population. This means that it would be legitimate to use a one-sample t test to test hypotheses about the true average ratio.

22.

x_i	y_i	x_i	y_i	x_i	y_i
69.5	-1.967	75.5	-.301	79.6	.634
71.9	-1.520	75.7	-.199	79.7	.761
72.6	-1.259	75.8	-.099	79.9	.901
73.1	-1.063	76.1	.000	80.1	1.063
73.3	-.901	76.2	.099	82.2	1.259
73.5	-.761	76.9	.199	83.7	1.520
74.1	-.634	77.0	.301	93.7	1.967
74.2	-.517	77.9	.407		
75.3	-.407	78.1	.517		

n.b.: Minitab was used to calculate the y_i's. $\Sigma x_{(i)} = 1925.6$, $\Sigma x_{(i)}^2 = 148{,}871$, $\Sigma y_i = 0$, $\Sigma y_i^2 = 22.523$, $\Sigma x_{(i)} y_i = 103.03$, so $r = \dfrac{25(103.03)}{\sqrt{25(148{,}871) - (1925.6)^2}\sqrt{25(25.523)}} = .923$.

Since $c_{.01} = .9408$, and $.923 < .9408$, even at the very smallest significance level of .01, the null hypothesis of population normality must be rejected (the largest observation appears to be the primary culprit).

23. Minitab gives r = .967, though the hand calculated value may be slightly different because when there are ties among the $x_{(i)}$'s, Minitab uses the same y_i for each $x_{(i)}$ in a group of tied values. $C_{10} = .9707$, and $c_{.05} = 9639$, so .05 < p-value < .10. At the 5% significance level, one would have to consider population normality plausible.

Section 14.3

24. H_o: TV watching and physical fitness are independent of each other
H_a: the two variables are not independent
Df = (4 – 1)(2 – 1) = 3
With $\alpha = .05$, RR: $\chi^2 \geq 7.815$
Computed $\chi^2 = 6.161$
Fail to reject H_o. The data fail to indicate an association between daily TV viewing habits and physical fitness.

25. Let P_{ij} = the proportion of white clover in area of type i which has a type j mark (i = 1, 2; j = 1, 2, 3, 4, 5). The hypothesis H_o: $p_{1j} = p_{2j}$ for j = 1, ..., 5 will be rejected at level .01 if $\chi^2 \ge \chi^2_{.01,(2-1)(5-1)} = \chi^2_{.01,4} = 13.277$.

$\hat{E}_{ij}$	1	2	3	4	5		
1	449.66	7.32	17.58	8.79	242.65	726	$\chi^2 = 23.18$
2	471.34	7.68	18.42	9.21	254.35	761	
	921	15	36	18	497	1487	

Since $23.18 \ge 13.277$, H_o is rejected.

26. Let p_{i1} = the probability that a fruit given treatment 1 matures and p_{i2} = the probability that a fruit given treatment 1 aborts. Then H_o: $p_{i1} = p_{i2}$ for I = 1, 2, 3, 4, 5 will be rejected if $\chi^2 \ge \chi^2_{.01,4} = 13.277$.

Observed		Estimated Expected		
Matured	Aborted	Matured	Aborted	n_i
141	206	110.7	236.3	347
28	69	30.9	66.1	97
25	73	31.3	66.7	98
24	78	32.5	69.5	102
20	82	32.5	69.5	102
		238	508	746

Thus $\chi^2 = \frac{(141-110.7)^2}{110.7} + \ldots + \frac{(82-69.5)^2}{69.5} = 24.82$, which is ≥ 13.277, so H_o is rejected at level .01.

27. With i = 1 identified with men and i = 2 identified with women, and j = 1, 2, 3 denoting the 3 categories L>R, L=R, L<R, we wish to test H_o: $p_{1j} = p_{2j}$ for j = 1, 2, 3 vs. H_a: p_{1j} not equal to p_{2j} for at least one j. The estimated cell counts for men are 17.95, 8.82, and 13.23 and for women are 39.05, 19.18, 28.77, resulting in $\chi^2 = 44.98$. With (2 – 1)(3 – 1) = 2 degrees of freedom, since $44.98 > \chi^2_{.005,2} = 10.597$, p-value < .005, which strongly suggests that H_o should be rejected.

28. With p_{ij} denoting the probability of a type j response when treatment i is applied, H_o: $p_{1j} = p_{2j} = p_{3j} = p_{4j}$ for j = 1, 2, 3, 4 will be rejected at level .005 if $\chi^2 \ge \chi^2_{.005,9} = 23.587$.

$\hat{E}_{ij}$	1	2	3	4
1	24.1	10.0	21.6	40.4
2	25.8	10.7	23.1	43.3
3	26.1	10.8	23.4	43.8
4	30.1	12.5	27.0	50.5

$\chi^2 = 27.66 \ge 23.587$, so reject H_o at level .005

29. H_o: $p_{1j} = \ldots = p_{6j}$ for j = 1, 2, 3 is the hypothesis of interest, where p_{ij} is the proportion of the jth sex combination resulting from the ith genotype. H_o will be rejected at level .10 if $\chi^2 \ge \chi^2_{.10,10} = 15.987$.

$\hat{E}_{ij}$	1	2	3	
1	35.8	83.1	35.1	154
2	39.5	91.8	38.7	170
3	35.1	81.5	34.4	151
4	9.8	22.7	9.6	42
5	5.1	11.9	5.0	22
6	26.7	62.1	26.2	115
	152	353	149	654

χ^2	1	2	3	
	.02	.12	.44	
	.06	.66	1.01	
	.13	.37	.34	
	.32	.49	.26	
	.00	.06	.19	
	.40	.14	1.47	
				6.46

(carrying 2 decimal places in $\hat{E}_{ij}$ yields $\chi^2 = 6.49$). Since 6.46 < 15.987, H_o cannot be rejected at level .10.

30. H_o: the design configurations are homogeneous with respect to type of failure vs. H_a: the design configurations are not homogeneous with respect to type of failure.

$\hat{E}_{ij}$	1	2	3	4	
1	16.11	43.58	18.00	12.32	90
2	7.16	19.37	8.00	5.47	40
3	10.74	29.05	12.00	8.21	60
	34	92	38	26	190

$\chi^2 = \frac{(20-16.11)^2}{16.11} + \ldots + \frac{(5-8.21)^2}{8.21} = 13.253$. With 6 df,
$\chi^2_{.05,6} = 12.592 < 13.253 < \chi^2_{.025,6} = 14.440$, so .025 < p-value < .05. Since the p-value is < .05, we reject H_o. (If a smaller significance level were chosen, a different conclusion would be reached.) Configuration appears to have an effect on type of failure.

31. With I denoting the Ith type of car (I = 1, 2, 3, 4) and j the jth category of commuting distance, H_o: $p_{ij} = p_{i.}\, p_{.j}$ (type of car and commuting distance are independent) will be rejected at level .05 if $\chi^2 \geq \chi^2_{.05,6} = 12.592$.

$\hat{E}_{ij}$	1	2	3	
1	10.19	26.21	15.60	52
2	11.96	30.74	18.30	61
3	19.40	49.90	29.70	99
4	7.45	19.15	11.40	38
	49	126	75	250

$\chi^2 = 14.15 \geq 12.592$, so the independence hypothesis H_o is rejected at level .05 (but not at level .025!)

32.
$$\chi^2 = \frac{(479-494.4)^2}{494.4} + \frac{(173-151.5)^2}{151.5} + \frac{(119-125.2)^2}{125.2} + \frac{(214-177.0)^2}{177.0} + \frac{(47-54.2)^2}{54.2}$$
$$= \frac{(15-44.8)^2}{44.8} + \frac{(172-193.6)^2}{193.6} + \frac{(45-59.3)^2}{59.3} + \frac{(85-49.0)^2}{49.0} = 64.65 \geq \chi^2_{.01,4} = 13.277,$$
so the independence hypothesis si rejected in favor of the conclusion that political views and level of marijuana usage are related.

33.
$$\chi^2 = \Sigma\Sigma\frac{\left(N_{ij} - \hat{E}_{ij}\right)^2}{\hat{E}_{ij}} = \Sigma\Sigma\frac{N_{ij}^2 - 2\hat{E}_{ij}N_{ij} + \hat{E}_{ij}^2}{\hat{E}_{ij}} = \frac{\Sigma\Sigma N_{ij}^2}{\hat{E}_{ij}} - 2\Sigma\Sigma N_{ij} + \Sigma\Sigma\hat{E}_{ij}, \text{ but}$$
$\Sigma\Sigma\hat{E}_{ij} = \Sigma\Sigma N_{ij} = n$, so $\chi^2 = \Sigma\Sigma\frac{N_{ij}^2}{\hat{E}_{ij}} - n$. This formula is computationally efficient because there is only one subtraction to be performed, which can be done as the last step in the calculation.

34. This is a $3\times3\times3$ situation, so there are 27 cells. Only the total sample size n is fixed in advance of the experiment, so there are 26 freely determined cell counts. We must estimate $p_{..1}$, $p_{..2}$, $p_{..3}$, $p_{.1.}$, $p_{.2.}$, $p_{.3.}$, $p_{1..}$, $p_{2..}$, and $p_{3..}$, but $\Sigma p_{i..} = \Sigma p_{.j.} = \Sigma p_{..k} = 1$ so only 6 independent parameters are estimated. The rule for d.f. now gives χ^2 df = 26 – 6 = 20.

35. With p_{ij} denoting the common value of p_{ij1}, p_{ij2}, p_{ij3}, p_{ij4} (under H_o), $\hat{p}_{ij} = \frac{N_{ij}}{n}$ and $\hat{E}_{ijk} = \frac{n_k N_{ij}}{n}$. With four different tables (one for each region), there are $8 + 8 + 8 + 8 = 32$ freely determined cell counts. Under H_o, $p_{11}, \ldots, p_{33}$ must be estimated but $\Sigma\Sigma p_{ij} = 1$ so only 8 independent parameters are estimated, giving χ^2 df = 32 – 8 = 24.

36.

a

Observed

13	19	28	60
7	11	22	40
20	30	50	100

Estimated Expected

12	18	30
8	12	20

$\chi^2 = \frac{(13-12)^2}{12} + \ldots + \frac{(22-20)^2}{20} = .6806$. Because $.6806 < \chi^2_{.10,2} = 4.605$, H_o is not rejected.

b Each observation count here is 10 times what it was in **a**, and the same is true of the estimated expected counts so now $\chi^2 = 6.806 \geq 4.605$, and H_o is rejected. With the much larger sample size, the departure from what is expected under H_o, the independence hypothesis, is statistically significant – it cannot be explained just by random variation.

c The observed counts are .13n, .19n, .28n, .07n, .11n, .22n, whereas the estimated expected $\frac{(.60n)(.20n)}{n}$ = .12n, .18n, .30n, .08n, .12n, .20n, yielding $\chi^2 = .006806n$. H_o will be rejected at level .10 iff $.006806n \geq 4.605$, i.e., iff $n \geq 676.6$, so the minimum n = 677.

Supplementary Exercises

37. There are 3 categories here – firstborn, middleborn, (2nd or 3rd born), and lastborn. With p_1, p_2, and p_3 denoting the category probabilities, we wish to test H_o: $p_1 = .25$, $p_2 = .50$ (p_2 = P(2nd or 3rd born) = .25 + .25 = .50), $p_3 = .25$. H_o will be rejected at significance level .05 if $\chi^2 \geq \chi^2_{.05,2} = 5.992$. The expected counts are (31)(.25) = 7.75, (31)(.50) = 15.5, and 7.75, so $\chi^2 = \frac{(12-7.75)^2}{7.75} + \frac{(11-15.5)^2}{15.5} + \frac{(8-7.75)^2}{7.75} = 3.65$. Because 3.65 < 5.992, H_o is not rejected. The hypothesis of equiprobable birth order appears quite plausible.

38. Let p_{i1} = the proportion of fish receiving treatment i (i = 1, 2, 3) who are parasitized. We wish to test H_o: $p_{1j} = p_{2j} = p_{3j}$ for j = 1, 2. With df = (2 – 1)(3 – 1) = 2, H_o will be rejected at level .01 if $\chi^2 \geq \chi^2_{.01,2} = 9.210$.

Observed			Estimated Expected	
30	3	33	22.99	10.01
16	8	24	16.72	7.28
16	16	32	22.29	9.71
62	27	89		

This gives $\chi^2 = 13.1$. Because $13.1 \geq 9.210$, H_o should be rejected. The proportion of fish that are parasitized does appear to depend on which treatment is used.

39. H_o: gender and years of experience are independent; H_a: gender and years of experience are not independent. Df = 4, and we reject H_o if $\chi^2 \geq \chi^2_{.01,4} = 13.277$.

	Years of Experience				
Gender	1 – 3	4 – 6	7 – 9	10 – 12	13 +
Male Observed	202	369	482	361	811
Expected	285.56	409.83	475.94	347.04	706.63
$\frac{(O-E)^2}{E}$	24.451	4.068	.077	.562	15.415
Female Observed	230	251	238	164	258
Expected	146.44	210.17	244.06	177.96	362.37
$\frac{(O-E)^2}{E}$	47.680	7.932	.151	1.095	30.061

$\chi^2 = \Sigma \frac{(O-E)^2}{E} = 131.492$. Reject H_o. The two variables do not appear to be independent. In particular, women have higher than expected counts in the beginning category (1 – 3 years) and lower than expected counts in the more experienced category (13+ years).

40.

a H_o: The probability of a late-game leader winning is independent of the sport played; H_a: The two variables are not independent. With 3 df, the computed $\chi^2 = 10.518$, and the p-value < .015 is also < .05, so we would reject H_o. There appears to be a relationship between the late-game leader winning and the sport played.

b Baseball had fewer than expected late-game leader losses.

41. The null hypothesis H_o: $p_{ij} = p_{i.}\, p_{.j}$ states that level of parental use and level of student use are independent in the population of interest. The test is based on $(3-1)(3-1) = 4$ df.

Estimated Expected

119.3	57.6	58.1	235
82.8	33.9	40.3	163
23.9	11.5	11.6	47
226	109	110	445

The calculated value of $\chi^2 = 22.4$. Since $22.4 > \chi^2_{.005,4} = 14.860$, p-value < .005, so H_o should be rejected at any significance level greater than .005. Parental and student use level do not appear to be independent.

42. The estimated expected counts are displayed below, from which $\chi^2 = 197.70$. A glance at the 6 df row of Table A.7 shows that this test statistic value is highly significant – the hypothesis of independence is clearly implausible.

Estimated Expected

	Home	Acute	Chronic	
15 – 54	90.2	372.5	72.3	535
55 – 64	113.6	469.3	91.1	674
65 – 74	142.7	589.0	114.3	846
> 74	157.5	650.3	126.2	934
	504	2081	404	2989

43. This is a test of homogeneity: H_o: $p_{1j} = p_{2j} = p_{3j}$ for j = 1, 2, 3, 4, 5. The given SPSS output reports the calculated $\chi^2 = 70.64156$ and accompanying p-value (significance) of .0000. We reject H_o at any significance level. The data strongly supports that there are differences in perception of odors among the three areas.

44. The accompanying table contains both observed and estimated expected counts, the latter in parentheses.

	Age					
Want	127 (131.1)	118 (123.3)	77 (71.7)	61 (55.1)	41 (42.8)	424
Don't	23 (18.9)	23 (17.7)	5 (10.3)	2 (7.9)	8 (6.2)	61
	150	141	82	63	49	485

This gives $\chi^2 = 11.60 \ge \chi^2_{.05,4} = 9.488$. At level .05, the null hypothesis of independence is rejected, though it would not be rejected at the level .01 (.01 < p-value < .025).

45. $(n_1 - np_{10})^2 = (np_{10} - n_1)^2 = (n - n_1 - n(1 - p_{10}))^2 = (n_2 - np_{20})^2$. Therefore

$$\chi^2 = \frac{(n_1 - np_{10})^2}{np_{10}} + \frac{(n_2 - np_{20})^2}{np_{20}} = \frac{(n_1 - np_{10})^2}{n_2}\left(\frac{n}{p_{10}} + \frac{n}{p_{20}}\right)$$

$$= \left(\frac{n_1}{n} - p_{10}\right)^2 \cdot \left(\frac{n}{p_{10}p_{20}}\right) = \frac{(\hat{p}_1 - p_{10})^2}{p_{10}p_{20}/n} = z^2.$$

Chapter 15

Section 15.1

1. We test $H_0 : \mu = 100$ vs. $H_a : \mu \neq 100$. The test statistic is s_+ = sum of the ranks associated with the positive values of $(x_i - 100)$, and we reject H$_o$ at significance level .05 if $s_+ \geq 64$. (from Table A.13, n = 12, with $\alpha / 2 = .026$, which is close to the desired value of . 025), or if $s_+ \leq \frac{12(13)}{2} - 64 = 78 - 64 = 14$.

x_i	$(x_i - 100)$	ranks
105.6	5.6	7*
90.9	-9.1	12
91.2	-8.8	11
96.9	-3.1	3
96.5	-3.5	5
91.3	-8.7	10
100.1	0.1	1*
105	5	6*
99.6	-0.4	2
107.7	7.7	9*
103.3	3.3	4*
92.4	-7.6	8

S_+ = 27, and since 27 is neither ≥ 64 nor ≤ 14, we do not reject H$_o$. There is not enough evidence to suggest that the mean is something other than 100.

2. We test $H_0 : \mu = 25$ vs. $H_a : \mu > 25$. With n = 5 and $\alpha \approx .03$, reject H$_o$ if $s_+ \geq 15$. From the table below we arrive at s_+ =1+5+2+3 = 11, which is not ≥ 15, so do not reject H$_o$. It is still plausible that the mean = 25.

x_i	$(x_i - 25)$	ranks
25.8	0.8	1*
36.6	11.6	5*
26.3	1.3	2*
21.8	-3.2	4
27.2	2.2	3*

3. We test $H_0: \mu = 7.39$ vs. $H_a: \mu \neq 7.39$, so a two tailed test is appropriate. With n = 14 and $\alpha/2 = .025$, Table A.13 indicates that H$_o$ should be rejected if either $s_+ \geq 84 or \leq 21$. The $(x_i - 7.39)$'s are -.37, -.04, -.05, -.22, -.11, .38, -.30, -.17, .06, -.44, .01, -.29, -.07, and -.25, from which the ranks of the three positive differences are 1, 4, and 13. Since $s_+ = 18 \leq 21$, H$_o$ is rejected at level .05.

4. The appropriate test is $H_0: \mu = 30$ vs. $H_a: \mu < 30$. With n = 15, and $\alpha = .10$, reject H$_o$ if $s_+ \leq \frac{15(16)}{2} - 83 = 37$.

x_i	$(x_i - 30)$	ranks	x_i	$(x_i - 30)$	ranks
30.6	0.6	3*	31.9	1.9	5*
30.1	0.1	1*	53.2	23.2	15*
15.6	-14.4	12	12.5	-17.5	13
26.7	-3.3	7	23.2	-6.8	11
27.1	-2.9	6	8.8	-21.2	14
25.4	-4.6	8	24.9	-5.1	10
35	5	9*	30.2	0.2	2*
30.8	0.8	4*			

S$_+$ = 39, which is not ≤ 37, so H$_o$ cannot be rejected. There is not enough evidence to prove that diagnostic time is less than 30 minutes at the 10% significance level.

5. The data is paired, and we wish to test $H_0: \mu_D = 0$ vs. $H_a: \mu_D \neq 0$. With n = 12 and $\alpha = .05$, H$_o$ should be rejected if either $s_+ \geq 64$ or if $s_+ \leq 14$.

d_i	-.3	2.8	3.9	.6	1.2	-1.1	2.9	1.8	.5	2.3	.9	2.5
rank	1	10*	12*	3*	6*	5	11*	7*	2*	8*	4*	9*

$s_+ = 72$, and $72 \geq 64$, so H$_o$ is rejected at level .05. In fact for $\alpha = .01$, the critical value is c = 71, so even at level .01 H$_o$ would be rejected.

6. We wish to test $H_0 : \mu_D = 5$ vs. $H_a : \mu_D > 5$, where $\mu_D = \mu_{black} - \mu_{white}$. With n = 9 and $\alpha \sim .05$, H_o will be rejected if $s_+ \geq 37$. As given in the table below, $s_+ = 37$, which is ≥ 37, so we can (barely) reject H_o at level approximately .05, and we conclude that the greater illumination does decrease task completion time by more than 5 seconds.

d_i	$d_i - 5$	rank	d_i	$d_i - 5$	rank
7.62	2.62	3*	16.07	11.07	9*
8	3	4*	8.4	3.4	5*
9.09	4.09	8*	8.89	3.89	7*
6.06	1.06	1*	2.88	-2.12	2
1.39	-3.61	6			

7. $H_0 : \mu_D = .20$ vs. $H_a : \mu_D > .20$, where $\mu_D = \mu_{outdoor} - \mu_{indoor}$. $\alpha = .05$, and because n = 33, we can use the large sample test. The test statistic is $Z = \dfrac{s_+ - \frac{n(n+1)}{4}}{\sqrt{\frac{n(n+1)(2n+1)}{24}}}$, and we reject H_o if $z \geq 1.96$.

d_i	$d_i - .2$	rank	d_i	$d_i - .2$	rank	d_i	$d_i - .2$	rank
0.22	0.02	2	0.15	-0.05	5.5	0.63	0.43	23
0.01	-0.19	17	1.37	1.17	32	0.23	0.03	4
0.38	0.18	16	0.48	0.28	21	0.96	0.76	31
0.42	0.22	19	0.11	-0.09	8	0.2	0	1
0.85	0.65	29	0.03	-0.17	15	-0.02	-0.22	18
0.23	0.03	3	0.83	0.63	28	0.03	-0.17	14
0.36	0.16	13	1.39	1.19	33	0.87	0.67	30
0.7	0.5	26	0.68	0.48	25	0.3	0.1	9.5
0.71	0.51	27	0.3	0.1	9.5	0.31	0.11	11
0.13	-0.07	7	-0.11	-0.31	22	0.45	0.25	20
0.15	-0.05	5.5	0.31	0.11	12	-0.26	-0.46	24

$s_+ = 434$, so $z = \dfrac{424 - 280.5}{\sqrt{3132.25}} = \dfrac{143.5}{55.9665} = 2.56$. Since $2.56 \geq 1.96$, we reject H_o at significance level .05.

8. We wish to test $H_0 : \mu = 75$ vs. $H_a : \mu > 75$. Since n = 25 the large sample approximation is used, so H_o will be rejected at level .05 if $z \geq 1.645$. The $(x_i - 75)'s$ are –5.5, -3.1, -2.4, -1.9, -1.7, 1.5, -.9, -.8, .3, .5, .7, .8, 1.1, 1.2, 1.2, 1.9, 2.0, 2.9, 3.1, 4.6, 4.7, 5.1, 7.2, 8.7, and 18.7. The ranks of the positive differences are 1, 2, 3, 4.5, 7, 8.5, 8.5, 12.5, 14, 16, 17.5, 19, 20, 21, 23, 24, and 25, so $s_+ = 226.5$ and $\dfrac{n(n+1)}{4} = 162.5$.

Expression (15.2) for σ^2 should be used (because of the ties): $\tau_1 = \tau_2 = \tau_3 = \tau_4 = 2$, so $\sigma^2_{s_+} = \frac{25(26)(51)}{24} - \frac{4(1)(2)(3)}{48} = 1381.25 - .50 = 1380.75$ and $\sigma = 37.16$. Thus $z = \frac{226.5 - 162.5}{37.16} = 1.72$. Since $1.72 \geq 1.645$, H$_o$ is rejected. $p - value \approx 1 - \Phi(1.72) = .0427$. The data indicates that true average toughness of the steel does exceed 75.

9.

r_1	1	1	1	1	1	1	2	2	2	2	2	2
r_2	2	2	3	3	4	4	1	1	3	3	4	4
r_3	3	4	2	4	2	3	3	4	1	4	1	3
r_4	4	3	4	2	3	2	4	3	4	1	3	1
D	0	2	2	6	6	8	2	4	6	12	10	14

r_1	3	3	3	3	3	3	4	4	4	4	4	4
r_2	1	1	2	2	4	4	1	1	2	2	3	3
r_3	2	4	1	4	1	2	2	3	1	3	1	2
r_4	4	2	4	1	2	1	3	2	3	1	2	1
D	6	10	8	14	16	18	12	14	14	18	18	20

When H$_o$ is true, each of the above 24 rank sequences is equally likely, which yields the distribution of D when H$_o$ is true as described in the answer section (e.g., P(D = 2) = P(1243 or 1324 or 2134) = 3/24). Then c = 0 yields $\alpha = \frac{1}{24} = .042$ while c = 2 implies $\alpha = \frac{4}{24} = .167$.

Section 15.2

10. The ordered combined sample is 163(y), 179(y), 213(y), 225(y), 229(x), 245(x), 247(y), 250(x), 286(x), and 299(x), so w = 5 + 6 + 8 + 9 + 10 = 38. With m = n = 5, Table A.14 gives the upper tail critical value for a level .05 test as 36 (reject H$_o$ if W ≥ 36). Since $38 \geq 36$, H$_o$ is rejected in favor of H$_a$.

11. With X identified with pine (corresponding to the smaller sample size) and Y with oak, we wish to test $H_0 : \mu_1 - \mu_2 = 0$ vs. $H_a : \mu_1 - \mu_2 \neq 0$. From Table A.14 with m = 6 and n = 8, H$_o$ is rejected in favor of H$_a$ at level .05 if either $w \geq 61$ or if $w \leq 90 - 61 = 29$ (the actual α is 2(.021) = .042). The X ranks are 3 (for .73), 4 (for .98), 5 (for 1.20), 7 (for 1.33), 8 (for 1.40), and 10 (for 1.52), so w = 37. Since 37 is neither ≥ 61 nor ≤ 29, H$_o$ cannot be rejected.

12. The hypotheses of interest are $H_0 : \mu_1 - \mu_2 = 1$ vs. $H_a : \mu_1 - \mu_2 > 1$, where 1(X) refers to the original process and 2 (Y) to the new process. Thus 1 must be subtracted from each x_i before pooling and ranking. At level .05, H_o should be rejected in favor of H_a if $w \geq 84$.

x – 1	3.5	4.1	4.4	4.7	5.3	5.6	7.5	7.6
rank	1	4	5	6	8	10	15	16
y	3.8	4.0	4.9	5.5	5.7	5.8	6.0	7.0
rank	2	3	7	9	11	12	13	14

Since w = 65, H_o is not rejected.

13. Here m = n = 10 > 8, so we use the large-sample test statistic from p. 663. $H_0 : \mu_1 - \mu_2 = 0$ will be rejected at level .01 in favor of $H_a : \mu_1 - \mu_2 \neq 0$ if either $z \geq 2.58$ or $z \leq -2.58$. Identifying X with orange juice, the X ranks are 7, 8, 9, 10, 11, 16, 17, 18, 19, and 20, so w = 135. With $\frac{m(m+n+1)}{2} = 105$ and $\sqrt{\frac{mn(m+n+1)}{12}} = \sqrt{175} = 13.22$, $z = \frac{135-105}{13.22} = 2.27$. Because 2.27 is neither ≥ 2.58 nor ≤ -2.58, H_o is not rejected. $p-value \approx 2(1-\Phi(2.27)) = .0232.$

14.

x	8.2	9.5	9.5	9.7	10.0	14.5	15.2	16.1	17.6	21.5
rank	7	9	9	11	12.5	16	17	18	19	20
y	4.2	5.2	5.8	6.4	7.0	7.3	9.5	10.0	11.5	11.5
rank	1	2	3	4	5	6	9	12.5	14.5	14.5

The denominator of z must now be computed according to (15.6). With $\tau_1 = 3$, $\tau_2 = 2$, $\tau_3 = 2$, $\sigma^2 = 175 - .0219[2(3)(4) + 1(2)(3) + 1(2)(3)] = 174.21$, so $z = \frac{138.5-105}{\sqrt{174.21}} = 2.54$. Because 2.54 is neither ≥ 2.58 nor ≤ -2.58, H_o is not rejected.

15. Let μ_1 and μ_2 denote true average cotanine levels in unexposed and exposed infants, respectively. The hypotheses of interest are $H_0 : \mu_1 - \mu_2 = -25$ vs. $H_a : \mu_1 - \mu_2 < -25$. With m = 7, n = 8, H_o will be rejected at level .05 if $w \le 7(7+8+1) - 71 = 41$. Before ranking, -25 is subtracted from each x_I (i.e. 25 is added to each), giving 33, 36, 37, 39, 45, 68, and 136. The corresponding ranks in the combined set of 15 observations are 1, 3, 4, 5, 6, 8, and 12, from which w = 1 + 3 + ... + 12 = 39. Because $39 \le 41$, H_o is rejected. The true average level for exposed infants appears to exceed that for unexposed infants by more than 25 (note that H_o would not be rejected using level .01).

16.

a

X	rank	Y	rank
0.43	2	1.47	9
1.17	8	0.8	7
0.37	1	1.58	11
0.47	3	1.53	10
0.68	6	4.33	16
0.58	5	4.23	15
0.5	4	3.25	14
2.75	12	3.22	13

We verify that w = sum of the ranks of the x's = 41.

b We are testing $H_0 : \mu_1 - \mu_2 = 0$ vs. $H_a : \mu_1 - \mu_2 < 0$. The reported p-value (significance) is .0027, which is < .01 so we reject H_o. There is evidence that the distribution of good visibility response time is to the left (or lower than) that response time with poor visibility.

Section 15.3

17. n = 8, so from Table A.15, a 95% C.I. (actually 94.5%) has the form $\left(\overline{x}_{(36-32+1)}, \overline{x}_{(32)}\right) = \left(\overline{x}_{(5)}, \overline{x}_{(32)}\right)$. It is easily verified that the 5 smallest pairwise averages are $\frac{5.0+5.0}{2} = 5.00$, $\frac{5.0+11.8}{2} = 8.40$, $\frac{5.0+12.2}{2} = 8.60$, $\frac{5.0+17.0}{2} = 11.00$, and $\frac{5.0+17.3}{2} = 11.15$ (the smallest average not involving 5.0 is $\overline{x}_{(6)} = \frac{11.8+11.8}{2} = 11.8$), and the 5 largest averages are 30.6, 26.0, 24.7, 23.95, and 23.80, so the confidence interval is (11.15, 23.80).

18. With n = 14 and $\frac{n(n+1)}{2} = 105$, from Table A.15 we se that c = 93 and the 99% interval is $(\bar{x}_{(13)}, \bar{x}_{(93)})$. Subtracting 7 from each x_i and multiplying by 100 (to simplify the arithmetic) yields the ordered values –5, 2, 9, 10, 14, 17, 22, 28, 32, 34, 35, 40, 45, and 77. The 13 smallest *sums* are –10, -3, 4, 4, 5, 9, 11, 12, 12, 16, 17, 18, and 19 (so $\bar{x}_{(13)} = \frac{14.19}{2} = 7.095$) while the 13 largest sums are 154, 122, 117, 112, 111, 109, 99, 91, 87, and 86 (so $\bar{x}_{(93)} = \frac{14.86}{2} = 7.430$). The desired C.I. is thus (7.095, 7.430).

19. The ordered d_i's are –13, -12, -11, -7, -6; with n = 5 and $\frac{n(n+1)}{2} = 15$, Table A.15 shows the 94% C.I. as (since c = 1) $(\bar{d}_{(1)}, \bar{d}_{(15)})$. The smallest average is clearly $\frac{-13-13}{2} = -13$ while the largest is $\frac{-6-6}{2} = -6$, so the C.I. is (-13, -6).

20. For n = 4 Table A.13 shows that a two tailed test can be carried out at level .124 or at level .250 (or, of course even higher levels), so we can obtain either an 87.6% C.I. or a 75% C.I. With $\frac{n(n+1)}{2} = 10$, the 87.6% interval is $(\bar{x}_{(1)}, \bar{x}_{(10)}) = (.045, .177)$.

21. m – n = 5 and from Table A.16, c = 21 and the 90% (actually 90.5%) interval is $(d_{ij(5)}, d_{ij(21)})$. The five smallest $x_i - y_j$ differences are –18, -2, 3, 4, 16 while the five largest differences are 136, 123, 120, 107, 86 (construct a table like Table 15.5), so the desired interval is $(16, 86)$.

22. m = 6, n = 8, mn = 48, and from Table A.16 a 99% interval (actually 99.2%) requires c = 44 and the interval is $(d_{ij(5)}, d_{ij(44)})$. The five largest $x_i - y_j$'s are 1.52 - .48 = 1.04, 1.40 - .48 = .92, 1.52 - .67 = .85, 1.33 - .48 = .85, and 1.40 - .67 = .73, while the five smallest are –1.04, -.99, -.83, -.82, and -.79, so the confidence interval for $\mu_1 - \mu_2$ (where μ_1 refers to pine and μ_2 refers to oak) is (-.79, .73).

Section 15.4

23. Below we record in parentheses beside each observation the rank of that observation nin the combined sample.

1:	5.8(3)	6.1(5)	6.4(6)	6.5(7)	7.7(10)	$r_{1.} = 31$
2:	7.1(9)	8.8(12)	9.9(14)	10.5(16)	11.2(17)	$r_{2.} = 68$
3:	5.191)	5.7(2)	5.9(4)	6.6(8)	8.2(11)	$r_{3.} = 26$
4:	9.5(13)	1.0.3(15)	11.7(18)	12.1(19)	12.4(20)	$r_{4.} = 85$

H_o will be rejected at level .10 if $k \ge \chi^2_{.10,3} = 6.251$. The computed value of k is $k = \frac{12}{20(21)}\left[\frac{31^2 + 68^2 + 26^2 + 85^2}{5}\right] - 3(21) = 14.06$. Since $14.06 \ge 6.251$, reject H_o.

24. After ordering the 9 observation within each sample, the ranks in the combined sample are

1:	1	2	3	7	8	16	18	22	27	$r_{1.} = 104$
2:	4	5	6	11	12	21	31	34	36	$r_{2.} = 160$
3:	9	10	13	14	15	19	28	33	35	$r_{3.} = 176$
4:	17	20	23	24	25	26	29	30	32	$r_{4.} = 226$

At level .05, $H_0 : \mu_1 = \mu_2 = \mu_3 = \mu_4$ will be rejected if $k \ge \chi^2_{.05,3} = 7.815$. The computed k is $k = \frac{12}{36(37)}\left[\frac{104^2 + 160^2 + 176^2 + 226^2}{5}\right] - 3(37) = 7.587$. Since 7.587 is not ≥ 7.815, H_o cannot be rejected.

25. $H_0 : \mu_1 = \mu_2 = \mu_3$ will be rejected at level .05 if $k \ge \chi^2_{.05,2} = 5.992$. The ranks are 1, 3, 4, 5, 6, 7, 8, 9, 12, 14 for the first sample; 11, 13, 15, 16, 17, 18 for the second; 2, 10, 19, 20, 21, 22 for the third; so the rank totals are 69, 90, and 94.

$$k = \frac{12}{22(23)}\left[\frac{69^2}{10} + \frac{90^2}{6} + \frac{94^2}{5}\right] - 3(23) = 9.23$$

Since $9.23 \ge 5.992$, we reject H_o.

26.

	1	2	3	4	5	6	7	8	9	10	r_i	r_i^2
A	2	2	2	2	2	2	2	2	2	2	20	400
B	1	1	1	1	1	1	1	1	1	1	10	100
C	4	4	4	4	3	4	4	4	4	4	39	1521
D	3	3	3	3	4	3	3	3	3	3	31	961
												2982

The computed value of F_r is $\frac{12}{4(10)(5)}(2982)-3(10)(5)=28.92$, which is $\geq \chi^2_{.01,3}=11.344$, so H_o is rejected.

27.

	1	2	3	4	5	6	7	8	9	10	r_i	r_i^2
I	1	2	3	3	2	1	1	3	1	2	19	361
H	2	1	1	2	1	2	2	1	2	3	17	289
C	3	3	2	1	3	3	3	2	3	1	24	576
												1226

The computed value of F_r is $\frac{12}{10(3)(4)}(1226)-3(10)(4)=2.60$, which is not $\geq \chi^2_{.05,2}=5.992$, so don't reject H_o.

Supplementary Exercises

28. The Wilcoxon signed-rank test will be used to test $H_0:\mu_D=0$ vs. $H_0:\mu_D\neq 0$, where μ_D = the difference between expected rate for a potato diet and a rice diet. From Table A.11 with n = 8, H_o will be rejected if either $s_+\geq 32$ or $s_+\leq \frac{8(9)}{2}-32=4$. The d_i's are (in order of magnitude) .16, .18, .25, -.56, .60, .96, 1.01, and –1.24, so $s_+=1+2+3+5+6+7=24$. Because 24 is not in the rejection region, H_o is not rejected.

29. Friedman's test is appropriate here. At level .05, H_o will be rejected if $f_r\geq \chi^2_{.05,3}=7.815$. It is easily verified that $r_{1.}=28$, $r_{2.}=29$, $r_{3.}=16$, $r_{4.}=17$, from which the defining formula gives $f_r=9.62$ and the computing formula gives $f_r=9.67$. Because $f_r\geq 7.815$, $H_0:\alpha_1=\alpha_2=\alpha_3=\alpha_4=0$ is rejected, and we conclude that there are effects due to different years.

30. The Kruskal-Wallis test is appropriate for testing $H_0 : \mu_1 = \mu_2 = \mu_3 = \mu_4$. H$_o$ will be rejected at significance level .01 if $k \geq \chi^2_{.01,3} = 11.344$

Treatment	ranks					r_i
I	4	1	2	3	5	15
II	8	7	10	6	9	40
III	11	15	14	12	13	65
IV	16	20	19	17	18	90

$k = \frac{12}{420}\left[\frac{225+1600+4225+8100}{5}\right] - 63 = 17.86$. Because $17.86 \geq 11.344$, reject H$_o$.

31. From Table A.16, m = n = 5 implies that c = 22 for a confidence level of 95%, so $mn - c + 1 = 25 - 22 = 1 = 4$. Thus the confidence interval extends from the 4th smallest difference to the 4th largest difference. The 4 smallest differences are –7.1, -6.5, -6.1, -5.9, and the 4 largest are –3.8, -3.7, -3.4, -3.2, so the C.I. is (-5.9, -3.8).

32.

a $H_0 : \mu_1 - \mu_2 = 0$ will be rejected in favor of $H_a : \mu_1 - \mu_2 \neq 0$ if either $w \geq 56$ or $w \leq 6(6+7+1) - 56 = 28$.

Gait	D	L	L	D	D	L	L
Obs	.85	.86	1.09	1.24	1.27	1.31	1.39

Gait	D	L	L	L	D	D
obs	1.45	1.51	1.53	1.64	1.66	1.82

$w = 1+4+5+8+12+13 = 43$. Because 43 is neither ≥ 56 nor ≤ 28, we don't reject H$_o$. There appears to be no difference between μ_1 and μ_2.

b

Differences

		Lateral Gait						
		.86	1.09	1.31	1.39	1.51	1.53	1.64
	.85	.01	.24	.46	.54	.66	.68	.79
Diagonal	1.24	-.38	-.15	.07	.15	.27	.29	.40
gait	1.27	-.41	-.18	.04	.12	.24	.26	.37
	1.45	-.59	-.36	-.14	-.06	.06	.08	.19
	1.66	-.80	-.57	-.35	-.27	-.15	-.13	-.02
	1.82	-.96	-.73	-.51	-.43	-.31	-.29	-.18

From Table A.16, c = 35 and $mn - c + 1 = 8$, giving (-.41, .29) as the C.I.

33. (Notice the error in the text: The test statistic is Y = the number of observations that exceed 25, not 20.)

a With "success" as defined, then Y is a binomial with n = 20. To determine the binomial proportion "p" we realize that since 25 is the hypothesized median, 50% of the distribution should be above 25, thus p = .50. From the Binomial Tables (Table A.1) with n = 20 and p = .50, we see that
$\alpha = P(Y \ge 15) = 1 - P(Y \le 14) = 1 - .979 = .021$.

b From the same binomial table as in **a**, we find that
$P(Y \ge 14) = 1 - P(Y \le 13) = 1 - .942 = .058$ (a close as we can get to .05), so c = 14. For this data, we would reject H_o at level .058 if $Y \ge 14$. Y = (the number of observations in the sample that exceed 25) = 12, and since 12 is not ≥ 14, we fail to reject H_o. (note: the answer section is incorrect).

34.

a Using the same logic as in Exercise 33, $P(Y \le 5) = .021$, and $P(Y \ge 15) = .021$, so the significance level is $\alpha = .042$.

b The null hypothesis will not be rejected if the median is between the 6th smallest observation in the data set and the 6th largest, exclusive. (If the median is less than or equal to 14.4, then there are at least 15 observations above, and we reject H_o. Similarly, if any value at least 41.5 is chosen, we have 5 or less observations above.) Thus with a confidence level of 95.8% the median will fall between 14.4 and 41.5.

35.

Sample:	y	x	y	y	x	x	x	y	y
Observations:	3.7	4.0	4.1	4.3	4.4	4.8	4.9	5.1	5.6
Rank:	1	3	5	7	9	8	6	4	2

The value of W' for this data is $w' = 3 + 6 + 8 + 9 = 26$. At level .05, the critical value for the upper-tailed test is (Table A.14, m = 4, n = 5) c = 27 ($\alpha = .056$). Since 26 is not ≥ 27, H_o cannot be rejected at level .05.

36. The only possible ranks now are 1, 2, 3, and 4. Each rank triple is obtained from the corresponding X ordering by the "code" 1 = 1, 2 = 2, 3 = 3, 4 = 4, 5 = 3, 6 = 2, 7 = 1 (so e.g. the X ordering 256 corresponds to ranks 2, 3, 2).

X ordering	ranks	w'	X ordering	ranks	w'	X ordering	ranks	w'
123	123	6	156	132	66	267	221	5
124	124	7	157	131	5	345	343	10
125	123	6	167	121	4	346	342	9
126	122	5	234	234	9	347	341	8
127	121	4	235	233	8	356	332	8
134	134	8	236	232	7	357	331	7
135	133	7	237	231	6	367	321	6
136	132	6	245	243	9	456	432	9
137	131	5	246	242	8	457	431	8
145	143	8	247	241	7	467	421	7
146	142	7	256	232	7	567	321	6
147	141	6	257	231	6			

Since when H_0 is true the probability of any particular ordering is 1/35, we easily obtain the null distribution and critical values given in the answer section.

Chapter 16

Section 16.1

1. All ten values of the quality statistic are between the two control limits, so no out-of-control signal is generated.

2. All ten values are between the two control limits. However, it is readily verified that all but one plotted point fall below the center line (at height .04975). Thus even though no single point generates an out-of-control signal, taken together, the observed values do suggest that there may be a decrease in the average value of the quality statistic. Such a "small" change is more easily detected by a CUSUM procedure (see section 16.5) than by an ordinary chart.

3. P(10 successive points inside the limits) = P(1st inside) x P(2nd inside) x...x P(10th inside) = $(.998)^{10}$ = .9802. P(25 successive points inside the limits) = $(.998)^{25}$ = .9512. $(.998)^{52}$ = .9011, but $(.998)^{53}$ = .8993, so for 53 successive points the probability that at least one will fall outside the control limits when the process is in control is 1 - .8993 = .1007 > .10.

Section 16.2

4. For Z, a standard normal random variable, $P(-c \le Z \le c) = .995$ implies that $\Phi(c) = P(Z \le c) = .995 + \frac{.005}{2} = .9975$. Table A.3 then gives c = 2.81. The appropriate control limits are therefore $\mu \pm 2.81\sigma$.

5.

a P(point falls outside the limits when $\mu = \mu_0 + .5\sigma$)

$$= 1 - P\left(\mu_0 - \frac{3\sigma}{\sqrt{n}} < \overline{X} < \mu_0 + \frac{3\sigma}{\sqrt{n}}\, when \mu = \mu_0 + .5\sigma\right)$$
$$= 1 - P\left(-3 - .5\sqrt{n} < Z < 3 - .5\sqrt{n}\right)$$
$$= 1 - P(-4.12 < Z < 1.882) = 1 - .9699 = .0301.$$

b $$1 - P\left(\mu_0 - \frac{3\sigma}{\sqrt{n}} < \overline{X} < \mu_0 + \frac{3\sigma}{\sqrt{n}}\, when \mu = \mu_0 - \sigma\right)$$
$$= 1 - P\left(-3 + \sqrt{n} < Z < 3 + \sqrt{n}\right) = 1 - P(-.76 < Z < 5.24) = .2236$$

c $$1 - P\left(-3 - 2\sqrt{n} < Z < 3 - 2\sqrt{n}\right) = 1 - P(-7.47 < Z < -1.47) = .6808$$

6. The limits are $13.00 \pm \frac{(3)(.6)}{\sqrt{5}} = 13.00 \pm .80$, from which LCL = 12.20 and UCL = 13.80. Every one of the 22 $\overline{x}$ values is well within these limits, so the process appears to be in control with respect to location.

7. $\bar{\bar{x}} = 12.95$ and $\bar{s} = .526$, so with $a_5 = .940$, the control limits are

$12.95 \pm 3\frac{.526}{.940\sqrt{5}} = 12.95 \pm .75 = 12.20, 13.70$. Again, every point $(\bar{x})$ is between these limits, so there is no evidence of an out-of-control process.

8. $\bar{r} = 1.336$ and $b_5 = 2.325$, yielding the control limits

$12.95 \pm 3\frac{1.336}{2.325\sqrt{5}} = 12.95 \pm .77 = 12.18, 13.72$. All points are between these limits, so the process again appears to be in control with respect to location.

9. $\bar{\bar{x}} = \frac{2317.07}{24} = 96.54$, $\bar{s} = 1.264$, and $a_6 = .952$, giving the control limits

$96.54 \pm 3\frac{1.264}{.952\sqrt{6}} = 96.54 \pm 1.63 = 94.91, 98.17$. The value of $\bar{x}$ on the 22nd day lies above the UCL, so the process appears to be out of control at that time.

10. Now $\bar{\bar{x}} = \frac{2317.07 - 98.34}{23} = 96.47$ and $\bar{s} = \frac{30.34 - 1.60}{23} = 1.250$, giving the limits

$96.47 \pm 3\frac{1.250}{.952\sqrt{6}} = 96.47 \pm 1.61 = 94.86, 98.08$. All 23 remaining $\bar{x}$ values are between these limits, so no further out-of-control signals are generated.

11.

a $P\left(\mu_0 - \frac{2.81\sigma}{\sqrt{n}} < \bar{X} < \mu_0 + \frac{2.81\sigma}{\sqrt{n}}\ when\mu = \mu_0\right)$

$= P(-2.81 < Z < 2.81) = .995$, so the probability that a point falls outside the limits is .005 and $ARL = \frac{1}{.005} = 200$.

b P = P(a point is outside the limits)

$= 1 - P\left(\mu_0 - \frac{2.81\sigma}{\sqrt{n}} < \bar{X} < \mu_0 + \frac{2.81\sigma}{\sqrt{n}}\ when\mu = \mu_0 + \sigma\right)$

$= 1 - P\left(-2.81 - \sqrt{n} < Z < 2.81 - \sqrt{n}\right)$

$= 1 - P(-4.81 < Z < .81) = 1 - .791 = .209$. Thus $ARL = \frac{1}{.209} = 4.78$

c 1 - .9974 = .0026 so $ARL = \frac{1}{.0026} = 385$ for an in-control process, and when $\mu = \mu_0 + \sigma$, the probability of an out-of-control point is $1 - P(-3-2 < Z < 1)$

$= 1 - P(Z < 1) = .1587$, so $ARL = \frac{1}{.1587} = 6.30$.

12.

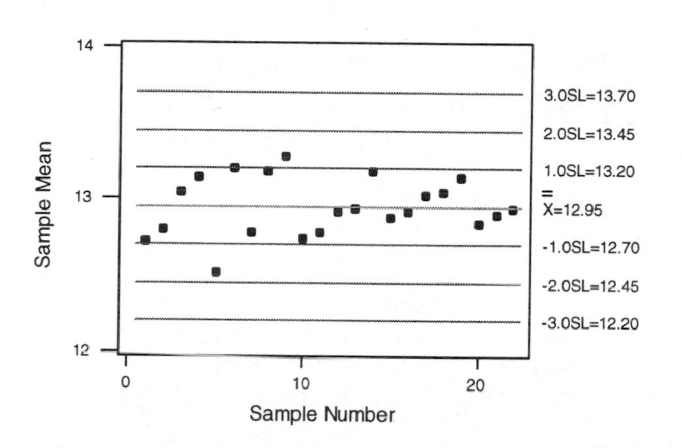

The 3-sigma control limits are from problem 7. The 2-sigma limits are $12.95 \pm .50 = 12.45, 13.45$, and the 1-sigma limits are $12.95 \pm .25 = 12.70, 13.20$. No points fall outside the 2-sigma limits, and only two points fall outside the 1-sigma limits. There are also no runs of eight on the same side of the center line – the longest run on the same side of the center line is four (the points at times 10, 11, 12, 13). No out-of-control signals result from application of the supplemental rules.

13. $\overline{\overline{x}} = 12.95$, IQR = .4273, $k_5 = .990$. The control limits are

$$12.95 \pm 3\frac{.4273}{.990\sqrt{5}} = 12.45, 13.45 = 12.37, 13.53.$$

Section 16.3

14. $\Sigma s_i = 4.895$ and $\overline{s} = \frac{4.895}{24} = .2040$. With $a_5 = .940$, the lower control limit is zero and the upper limit is $.2040 + \frac{3(.2040)\sqrt{1-(.940)^2}}{.940} = .2040 + .2221 = .4261$. Every s_i is between these limits, so the process appears to be in control with respect to variability.

15.

a $\overline{r} = \frac{85.2}{30} = 2.84$, $b_4 = 2.058$, and $c_4 = .880$. Since n = 4, LCL = 0 and UCL $= 2.84 + \frac{3(.880)(2.84)}{2.058} = 2.84 + 3.64 = 6.48$.

b $\bar{r} = 3.54$, $b_8 = 2.844$, and $c_8 = .820$, and the control limits are

$$= 3.54 \pm \frac{3(.820)(3.54)}{2.844} = 3.54 \pm 3.06 = .48, 6.60.$$

16. $\bar{s} = .5172$, $a_5 = .940$, LCL = 0 (since n = 5) and UCL =

$$.5172 + \frac{3(.5172)\sqrt{1-(.940)^2}}{.940} = .5172 + .5632 = 1.0804.$$ The largest s_i is $s_9 = .963$, so all points fall between the control limits.

17. $\bar{s} = 1.2642$, $a_6 = .952$, and the control limits are

$$1.2642 \pm \frac{3(1.2642)\sqrt{1-(.952)^2}}{.952} = 1.2642 \pm 1.2194 = .045, 2.484.$$ The smallest s_i is $s_{20} = .75$, and the largest is $s_{12} = 1.65$, so every value is between .045 and 2.434. The process appears to be in control with respect to variability.

18. $\Sigma s_i^2 = 39.9944$ and $\bar{s}^2 = \frac{39.9944}{24} = 1.6664$, so LCL = $\frac{(1.6664)(.210)}{5} = .070$, and UCL = $\frac{(1.6664)(20.515)}{5} = 6.837$. The smallest s^2 value is $s_{20}^2 = (.75)^2 = .5625$ and the largest is $s_{12}^2 = (1.65)^2 = 2.723$, so all s_i^2's are between the control limits.

Section 16.4

19. $\bar{p} = \Sigma \frac{\hat{p}_i}{k}$ where $\Sigma \hat{p}_i = \frac{x_1}{n} + \ldots + \frac{x_k}{n} = \frac{x_1 + \ldots + x_k}{n} = \frac{578}{100} = 5.78$. Thus

$$\bar{p} = \frac{5.78}{25} = .231.$$

a The control limits are $.231 \pm 3\sqrt{\frac{(.231)(.769)}{100}} = .231 \pm .126 = .105, .357$.

b $\frac{13}{100} = .130$, which is between the limits, but $\frac{39}{100} = .390$, which exceeds the upper control limit and therefore generates an out-of-control signal.

20. $\Sigma x_i = 567$, from which $\overline{p} = \frac{\Sigma x_i}{nk} = \frac{567}{(200)(30)} = .0945$. The control limits are $.0945 \pm 3\sqrt{\frac{(.0945)(.9055)}{200}} = .0945 \pm .0621 = .0324,.1566$. The smallest x_i is $x_7 = 7$, with $\hat{p}_7 = \frac{7}{200} = .0350$. This (barely) exceeds the LCL. The largest x_i is $x_5 = 37$, with $\hat{p}_5 = \frac{37}{200} = .185$. Thus $\hat{p}_5 > UCL = .1566$, so an out-of-control signal is generated. This is the only such signal, since the next largest x_i is $x_{25} = 30$, with $\hat{p}_{25} = \frac{30}{200} = .1500 < UCL$.

21. LCL > 0 when $\overline{p} > 3\sqrt{\frac{\overline{p}(1-\overline{p})}{n}}$, i.e. (after squaring both sides) $50\overline{p}^2 > 3\overline{p}(1-p)$, i.e. $50\overline{p} > 3(1-\overline{p})$, i.e. $53\overline{p} > 3 \Rightarrow \overline{p} = \frac{3}{53} = .0566$.

22. The suggested transformation is $Y = h(X) = \sin^{-1}\left(\sqrt{X/n}\right)$, with approximate mean value $\sin^{-1}\left(\sqrt{p}\right)$ and approximate variance $\frac{1}{4n}$. $\sin^{-1}\left(\sqrt{x_1/n}\right) = \sin^{-1}\left(\sqrt{.050}\right) = .2255$ (in radians), and the values of $y_i = \sin^{-1}\left(\sqrt{x_i/n}\right)$ for i = 1, 2, 3, …, 30 are

0.2255	0.2367	0.2774	0.3977
0.3047	0.3537	0.3381	0.2868
0.3537	0.3906	0.2475	0.2367
0.2958	0.2774	0.3218	0.3218
0.4446	0.2868	0.2958	0.2678
0.3133	0.3300	0.3047	0.3835
0.1882	0.3047	0.2475	
0.3614	0.2958	0.3537	

These give $\Sigma y_i = 9.2437$ and $\overline{y} = .3081$. The control limits are $\overline{y} \pm 3\sqrt{\frac{1}{4n}} = .3081 \pm 3\sqrt{\frac{1}{800}} = .3081 \pm .1091 = .2020,.4142$. In contrast ot the result of exercise 20, there I snow one point below the LCL (.1882 < .2020) as well as one point above the UCL.

23. $\Sigma x_i = 102$, $\overline{x} = 4.08$, and $\overline{x} \pm 3\sqrt{\overline{x}} = 4.08 \pm 6.06 \approx (-2.0,10.1)$. Thus LCL = 0 and UCL = 10.1. Because no x_i exceeds 10.1, the process is judged to be in control.

24. $\bar{x} - 3\sqrt{\bar{x}} < 0$ is equivalent to $\sqrt{\bar{x}} < 3$, i.e. $\bar{x} < 9$.

25. With $u_i = \frac{x_i}{g_i}$, the u_i's are 3.75, 3.33, 3.75, 2.50, 5.00, 5.00, 12.50, 12.00, 6.67, 3.33, 1.67, 3.75, 6.25, 4.00, 6.00, 12.00, 3.75, 5.00, 8.33, and 1.67 for I = 1, ..., 20, giving $\bar{u} = 5.5125$. For $g_i = .6$, $\bar{u} \pm 3\sqrt{\frac{\bar{u}}{g_i}} = 5.5125 \pm 9.0933$, LCL = 0, UCL = 14.6. For $g_i = .8$, $\bar{u} \pm 3\sqrt{\frac{\bar{u}}{g_i}} = 5.5125 \pm 7.857$, LCL = 0, UCL = 13.4. For $g_i = 1.0$, $\bar{u} \pm 3\sqrt{\frac{\bar{u}}{g_i}} = 5.5125 \pm 7.0436$, LCL = 0, UCL = 12.6. Several u_i's are close to the corresponding UCL's but none exceed them, so the process is judged to be in control.

26. $y_i = 2\sqrt{x_i}$ and the y_i's are 3/46, 5.29, 4.47, 4.00, 2.83, 5.66, 4.00, 3.46, 3.46, 4.90, 5.29, 2.83, 3.46, 2.83, 4.00, 5.29, 3.46, 2.83, 4.00, 4.00, 2.00, 4.47, 4.00, and 4.90 for I = 1, ..., 25, from which $\Sigma y_i = 98.35$ and $\bar{y} = 3.934$. Thus $\bar{y} \pm 3 = 3.934 \pm 3 = .934, 6.934$. Since every y_i is well within these limits it appears that the process is in control.

Section 16.5

27. $\mu_0 = 16$, $k = \frac{\Delta}{2} = 0.05$, $h = .20$, $d_i = \max(0, d_{i-1} + (\bar{x}_i - 16.05))$, $e_i = \max(0, e_{i-1} + (\bar{x}_i - 15.95))$.

i	$\bar{x}_i - 16.05$	d_i	$\bar{x}_i - 15.95$	e_i
1	-0.058	0	0.024	0
2	0.001	0.001	0.101	0
3	0.016	0.017	0.116	0
4	-0.138	0	-0.038	0.038
5	-0.020	0	0.080	0
6	0.010	0.010	0.110	0
7	-0.068	0	0.032	0
8	-0.151	0	-0.054	0.054
9	-0.012	0	0.088	0
10	0.024	0.024	0.124	0
11	-0.021	0.003	0.079	0
12	-0.115	0	-0.015	0.015
13	-0.018	0	0.082	0
14	-0.090	0	0.010	0
15	0.005	0.005	0.105	0

For no time r is it the case that $d_r > .20$ or that $e_r > .20$, so no out-of-control signals are generated.

28. $\mu_0 = .75$, $k = \frac{\Delta}{2} = 0.001$, $h = .003$, $d_i = \max(0, d_{i-1} + (\bar{x}_i - .751))$,
$e_i = \max(0, e_{i-1} + (\bar{x}_i - .749))$.

i	$\bar{x}_i - .751$	d_i	$\bar{x}_i - .749$	e_i
1	-.0003	0	.0017	0
2	-.0006	0	.0014	0
3	-.0018	0	.0002	0
4	-.0009	0	.0011	0
5	-.0007	0	.0013	0
6	.0000	0	.0020	0
7	-.0020	0	.0000	0
8	-.0013	0	.0007	0
9	-.0022	0	-.0002	.0002
10	-.0006	0	.0014	0
11	.0006	.0006	.0026	0
12	-.0038	0	-.0018	.0018
13	-.0021	0	-.0001	.0019
14	-.0027	0	-.0007	.0026
15	-.0039	0	-.0019	.0045*
16	-.0012	0	.0008	.0037
17	-.0050	0	-.0030	.0067
18	-.0028	0	-.0008	.0075
19	-.0040	0	-.0020	.0095
20	-.0017	0	.0003	.0092
21	-.0048	0	-.0028	.0120
22	-.0029	0	-.0009	.0129

Clearly $e_{15} = .0045 > .003 = h$, suggesting that the process mean has shifted to a value smaller than the target of .75.

29. Connecting 600 on the in-control ARL scale to 4 on the out-of-control scale and extending to the k' scale gives k' = .87. Thus $k' = \frac{\Delta/2}{\sigma/\sqrt{n}} = \frac{.002}{.005/\sqrt{n}}$ from which $\sqrt{n} = 2.175 \Rightarrow n = 4.73 = s$. Then connecting .87 on the k' scale to 600 on the out-of-control ARL scale and extending to h' gives h' = 2.8, so

$$h = \left(\frac{\sigma}{\sqrt{n}}\right)(2.8) = \left(\frac{.005}{\sqrt{5}}\right)(2.8) = .00626.$$

30. In control ARL = 250, out-of-control ARL = 4.8, from which
$k' = .7 = \frac{\Delta/2}{\sigma/\sqrt{n}} = \frac{\sigma/2}{\sigma/\sqrt{n}} = \frac{\sqrt{n}}{2}$. So $\sqrt{n} = 1.4 \Rightarrow n = 1.96 \approx 2$. Then h' = 2.85,
giving $h = \left(\frac{\sigma}{\sqrt{n}}\right)(2.85) = 2.0153\sigma$.

Section 16.6

31. For the binomial calculation, n = 50 and we wish
$$P(X \le 2) = \binom{50}{0} p^0(1-p)^{50} + \binom{50}{1} p^1(1-p)^{49} + \binom{50}{2} p^2(1-p)^{48}$$
$= (1-p)^{50} + 50p(1-p)^{49} + 1225p^2(1-p)^{48}$ when p = .01, .02, …, .10. For the hypergeometric calculation
$$P(X \le 2) = \frac{\binom{M}{0}\binom{500-M}{50}}{\binom{500}{50}} + \frac{\binom{M}{1}\binom{500-M}{49}}{\binom{500}{50}} + \frac{\binom{M}{2}\binom{500-M}{48}}{\binom{500}{50}}$$, to be calculated for M = 5, 10, 15, …, 50. The resulting probabilities appear in the answer section in the text.

32. $$P(X \le 1) = \binom{50}{0} p^0(1-p)^{50} + \binom{50}{1} p^1(1-p)^{49} = (1-p)^{50} + 50p(1-p)^{49}$$

p	.01	.02	.03	.04	.05	.06	.07	.08	.09	.10
$P(X \le 1)$	.9106	.7358	.5553	.4005	.2794	.1900	.1265	.0827	.0532	.0338

33. $$P(X \le 2) = \binom{100}{0} p^0(1-p)^{100} + \binom{100}{1} p^1(1-p)^{99} + \binom{100}{2} p^2(1-p)^{98}$$

p	.01	.02	.03	.04	.05	.06	.07	.08	.09	.10
$P(X \le 2)$	.9206	.6767	.4198	.2321	.1183	.0566	.0258	.0113	.0048	.0019

For values of p quite close to 0, the probability of lot acceptance using this plan is larger than that for the previous plan, whereas for larger p this plan is less likely to result in an "accept the lot" decision (the dividing point between "close to zero" and "larger p" is someplace between .01 and .02). In this sense, the current plan is better.

34. $\dfrac{LTPD}{AQL} = \dfrac{.07}{.02} = 3.5 \approx 3.55$, which appears in the $\dfrac{p_1}{p_2}$ column in the c = 5 row. Then $n = \dfrac{np_1}{p_1} = \dfrac{2.613}{.02} = 130.65 \approx 131.$

$$P(X > 5 \text{ when } p = .02) = 1 - \sum_{x=0}^{5} \binom{131}{x} (.02)^x (.98)^{131-x} = .0487 \approx .05$$

$$P(X \le 5 \text{ when } p = .07) = \sum_{x=0}^{5} \binom{131}{x} (.07)^x (.93)^{131-x} = .0974 \approx .10$$

35. P(accepting the lot) = $P(X_1 = 0 \text{ or } 1) + P(X_1 = 2, X_2 = 0, 1, 2, \text{ or } 3) + P(X_1 = 3, X_2 = 0, 1, \text{ or } 2) = P(X_1 = 0 \text{ or } 1) + P(X_1 = 2)P(X_2 = 0, 1, 2, \text{ or } 3) + P(X_1 = 3)P(X_2 = 0, 1, \text{ or } 2)$.

p = .01: $= .9106 + (.0756)(.9984) + (.0122)(.9862) = .9981$

p = .05: $= .2794 + (.2611)(.7604) + (.2199)(.5405) = .5968$

p = .10: $= .0338 + (.0779)(.2503) + (.1386)(.1117) = .0688$

36. P(accepting the lot) = $P(X_1 = 0 \text{ or } 1) + P(X_1 = 2, X_2 = 0 \text{ or } 1) + P(X_1 = 3, X_2 = 0)$ [since $c_2 = r_1 - 1 = 3$] $= P(X_1 = 0 \text{ or } 1) + P(X_1 = 2)P(X_2 = 0 \text{ or } 1) + P(X_1 = 3)P(X_2 = 0)$

$$= \sum_{x=0}^{1} \binom{50}{x} p^x (1-p)^{50-x} + \binom{50}{2} p^2 (1-p)^{48} \cdot \sum_{x=0}^{1} \binom{100}{x} p^x (1-p)^{100-x}$$

$$= \binom{50}{3} p^3 (1-p)^{47} \cdot \binom{100}{0} p^0 (1-p)^{100}.$$

p = .02: $= .7358 + (.1858)(.4033) + (.0607)(.1326) = .8188$

p = .05: $= .2794 + (.2611)(.0371) + (.2199)(.0059) = .2904$

p = .10: $= .0338 + (.0779)(.0003) + (.1386)(.0000) = .0038$

37.

a $AOQ = pP(A) = p[(1-p)^{50} + 50p(1-p)^{49} + 1225p^2(1-p)^{48}]$

p	.01	.02	.03	.04	.05	.06	.07	.08	.09	.10
AOQ	.010	.018	.024	.027	.027	.025	.022	.018	.014	.011

b p = .0447, AOQL = .0447P(A) = .0274

c ATI = 50P(A) + 2000(1 – P(A))

p	.01	.02	.03	.04	.05	.06	.07	.08	.09	.10
ATI	77.3	202.1	418.6	679.9	945.1	1188.8	1393.6	1559.3	1686.1	1781.6

38. $AOQ = pP(A) = p[(1-p)^{50} + 50p(1-p)^{49}]$. Exercise 32 gives P(A), so multiplying each entry in the second row by the corresponding entry in the first row gives AOQ:

p	.01	.02	.03	.04	.05	.06	.07	.08	.09	.10
AOQ	.0091	.0147	.0167	.0160	.0140	.0114	.0089	.0066	.0048	.0034

ATI = 50P(A) + 2000(1 – P(A))

p	.01	.02	.03	.04	.05	.06	.07	.08	.09	.10
ATI	224.3	565.2	917.2	1219.0	1455.2	1629.5	1753.3	1838.7	1896.3	1934.1

$\frac{d}{dp}AOQ = \frac{d}{dp}\left[pP(A) = p[(1-p)^{50} + 50p(1-p)^{49}]\right] = 0$ gives the quadratic equation

$2499p^2 - 48p - 1 = 0$, from which $p = \frac{48 + 110.91}{4998} = .0318$, and

$AOQL = .0318P(A) \approx .0167$.

Supplementary Exercises

39. n = 6, k = 26, $\Sigma\bar{x}_i = 10{,}980$, $\bar{\bar{x}} = 422.31$, $\Sigma s_i = 402$, $\bar{s} = 15.4615$, $\Sigma r_i = 1074$, $\bar{r} = 41.3077$

S chart: $15.4615 \pm \frac{3(15.4615)\sqrt{1-(.952)^2}}{.952} = 15.4615 \pm 14.9141 \approx .55, 30.37$

R chart: $41.31 \pm \frac{3(.848)(41.31)}{2.536} = 41.31 \pm 41.44$, so LCL = 0, UCL = 82.75

$\bar{X}$ chart based on $\bar{s}$: $422.31 \pm \frac{3(15.4615)}{.952\sqrt{6}} = 402.42, 442.20$

$\bar{X}$ chart based on $\bar{r}$: $422.31 \pm \frac{3(41.3077)}{2.536\sqrt{6}} = 402.36, 442.26$

40. A c chart is appropriate here. $\Sigma\bar{x}_i = 92$ so $\bar{x} = \frac{92}{24} = 3.833$, and

$\bar{x} \pm 3\sqrt{\bar{x}} = 3.833 \pm 5.874$, giving LCL = 0 and UCL = 9.7. Because $x_{22} = 10 >$ UCL, the process appears to have been out of control at the time that the 22nd plate was obtained.

41.

i	$\bar{x}_i$	s_i	r_i
1	50.83	1.172	2.2
2	50.10	.854	1.7
3	50.30	1.136	2.1
4	50.23	1.097	2.1
5	50.33	.666	1.3
6	51.20	.854	1.7
7	50.17	.416	.8
8	50.70	.964	1.8
9	49.93	1.159	2.1
10	49.97	.473	.9
11	50.13	.698	.9
12	49.33	.833	1.6
13	50.23	.839	1.5
14	50.33	.404	.8
15	49.30	.265	.5
16	49.90	.854	1.7
17	50.40	.781	1.4
18	49.37	.902	1.8
19	49.87	.643	1.2
20	50.00	.794	1.5
21	50.80	2.931	5.6
22	50.43	.971	1.9

$\Sigma s_i = 19.706$, $\bar{s} = .8957$, $\Sigma \bar{x}_i = 1103.85$, $\bar{\bar{x}} = 50.175$, $a_3 = .886$, from which an s chart has LCL = 0 and UCL = $.8957 + \dfrac{3(.8957)\sqrt{1-(.886)^2}}{.886} = 2.3020$, and $s_{21} = 2.931 > UCL$. Since an assignable cause is assumed to have been identified we eliminate the 21st group. Then $\Sigma s_i = 16.775$, $\bar{s} = .7998$, $\bar{\bar{x}} = 50.145$. The resulting UCL for an s chart is 2.0529, and $s_i < 2.0529$ for every remaining i. The $\bar{x}$ chart based on $\bar{s}$ has limits $50.145 \pm \dfrac{3(.7988)}{.886\sqrt{3}} = 48.58, 51.71$. All $\bar{x}_i$ values are between these limits.

42. $\bar{p} = .0608$, n = 100, so $UCL = n\bar{p} + 3\sqrt{n\bar{p}(1-\bar{p})} = 6.08 + 3\sqrt{6.08(.9392)}$ $= 6.08 + 7.17 = 13.25$ and LCL = 0. All points are between these limits, as was the case for the p-chart. The p-chart and np-chart will always give identical results since

$$\bar{p} - 3\sqrt{\frac{\bar{p}(1-\bar{p})}{n}} < \hat{p}_i < \bar{p} + 3\sqrt{\frac{\bar{p}(1-\bar{p})}{n}} \quad \textbf{iff}$$

$$np - 3\sqrt{n\bar{p}(1-\bar{p})} < n\hat{p}_i = x_i < n\bar{p} + 3\sqrt{n\bar{p}(1-\bar{p})}$$

43. $\Sigma n_i = 4(16)+(3)(4)=76$, $\Sigma n_i \bar{x}_i = 32{,}729.4$, $\bar{\bar{x}} = 430.65$,

$s^2 = \dfrac{\Sigma(n_i - 1)s_i^2}{\Sigma(n_i - 1)} = \dfrac{27{,}380.16 - 5661.4}{76 - 20} = 590.0279$, so s = 24.2905. For variation:

when n = 3, $UCL = 24.2905 + \dfrac{3(24.2905)\sqrt{1-(.886)^2}}{.886} = 24.29 + 38.14 = 62.43$,

when n = 4, $UCL = 24.2905 + \dfrac{3(24.2905)\sqrt{1-(.921)^2}}{.921} = 24.29 + 30.82 = 55.11$. For location: when n = 3, $430.65 \pm 47.49 = 383.16, 478.14$, and when n = 4, $430.65 \pm 39.56 = 391.09, 470.21$.

44.

a Provided the $E(\bar{X}_i) = \mu$ for each i,

$$E(W_t) = \alpha E(\bar{X}_t) + \alpha(1-\alpha)E(\bar{X}_{t-1}) + \ldots + \alpha(1-\alpha)^{t-1}E(\bar{X}_1) + (1-\alpha)^t \mu$$
$$= \mu\left[\alpha + \alpha(1-\alpha) + \ldots + \alpha(1-\alpha)^{t-1} + (1-\alpha)^t\right]$$
$$= \mu\left[\alpha\left(1 + (1-\alpha) + \ldots + (1-\alpha)^{t-1}\right) + (1-\alpha)^t\right]$$
$$= \mu\left[\alpha\sum_{i=0}^{\infty}(1-\alpha)^i - \alpha\sum_{i=t}^{\infty}(1-\alpha)^i + (1-\alpha)^t\right]$$
$$= \mu\left[\frac{\alpha}{1-(1-\alpha)} - \alpha(1-\alpha)^t \cdot \frac{1}{1-(1-\alpha)} + (1-\alpha)^t\right] = \mu$$

b $$V(W_t) = \alpha^2 V(\bar{X}_t) + \alpha^2(1-\alpha)^2 V(\bar{X}_{t-1}) + \ldots + \alpha^2(1-\alpha)^{2(t-1)}V(\bar{X}_1)$$
$$= \alpha^2\left[1 + (1-\alpha)^2 + \ldots + (1-\alpha)^{2(t-1)}\right] \cdot V(\bar{X}_1)$$
$$= \alpha^2\left[1 + C + \ldots + C^{t-1}\right] \cdot \frac{\sigma^2}{n}$$ (where $C = (1-\alpha)^2$.)
$$= \alpha^2 \frac{1-C^t}{1-C} \cdot \frac{\sigma^2}{n}$$, which gives the desired expression.

c From Example 16.8, $\sigma = .5$ (or $\bar{s}$ can be used instead). Suppose that we use $\alpha = .6$ (not specified in the problem). Then

$$w_0 = \mu_0 = 40$$
$$w_1 = .6\bar{x}_1 + .4\mu_0 = .6(40.20) + .4(40) = 40.12$$
$$w_2 = .6\bar{x}_2 + .4w_1 = .6(39.72) + .4(40.12) = 39.88$$
$$w_3 = .6\bar{x}_3 + .4w_2 = .6(40.42) + .4(39.88) = 40.20$$
$$w_4 = 40.07,\ w_5 = 40.06,\ w_6 = 39.88,\ w_7 = 39.74,\ w_8 = 40.14,\ w_9 = 40.25,$$
$$w_{10} = 40.00,\ w_{11} = 40.29,\ w_{12} = 40.36,\ w_{13} = 40.51,\ w_{14} = 40.19,$$
$$w_{15} = 40.21,\ w_{16} = 40.29$$

$$\sigma_1^2 = \frac{.6\left[1-(1-.6)^2\right]}{2-.6}\cdot\frac{.25}{4} = .0225,\ \sigma_1 = .1500,$$
$$\sigma_2^2 = \frac{.6\left[1-(1-.6)^4\right]}{2-.6}\cdot\frac{.25}{4} = .0261,\ \sigma_2 = .1616,$$
$$\sigma_3 = .1633,\ \sigma_4 = .1636,\ \sigma_5 = .1637 = \sigma_6 \ldots \sigma_{16}$$

Control limits are:

For t = 1, $40 \pm 3(.1500) = 39.55, 40.45$
For t = 2, $40 \pm 3(.1616) = 39.52, 40.48$
For t = 3, $40 \pm 3(.1633) = 39.51, 40.49$.
These last limits are also the limits for t = 4, ..., 16.

Because $w_{13} = 40.51 > 40.49 =$ UCL, an out-of-control signal is generated.